Principal Functions

THE UNIVERSITY SERIES IN HIGHER MATHEMATICS

Editorial Board

M. H. Stone, *Chairman*

L. Nirenberg S. S. Chern

HALMOS, PAUL R.—Measure Theory

JACOBSON, NATHAN—Lectures in Abstract Algebra
Vol. I—Basic Concepts *Vol. II*—Linear Algebra
Vol. III—Theory of Fields and Galois Theory

KLEENE, S. C.—Introduction to Metamathematics

LOOMIS, LYNN H.—An Introduction to Abstract Harmonic Analysis

LOÈVE, MICHEL—Probability Theory, 3rd Edition

KELLEY, JOHN L.—General Topology

ZARISKI, OSCAR, and SAMUEL, PIERRE—Commutative Algebra, Vols. I and II

GILLMAN, LEONARD, and JERISON, MEYER—Rings of Continuous Functions

RICKART, CHARLES E.—General Theory of Banach Algebras

J. L. KELLEY, ISAAC NAMIOKA, and CO-AUTHORS—Linear Topological Spaces

SPITZER, FRANK—Principles of Random Walk

SCHENKMAN, EUGENE—Group Theory

NACHBIN, LEOPOLDO—The Haar Integral

KEMENY, JOHN G., SNELL, J. LAURIE, and KNAPP, ANTHONY W.—Denumerable Markov Chains

SARIO, LEO, and NOSHIRO, KIYOSHI—Value Distribution Theory

RODIN, BURTON, and SARIO, LEO—Principal Functions

A series of advanced text and reference books in pure and applied mathematics. Additional titles will be listed and announced as published.

Principal Functions

BURTON RODIN

Associate Professor of Mathematics
University of California
San Diego, California

LEO SARIO

Professor of Mathematics
University of California
Los Angeles, California

in collaboration with

MITSURU NAKAI

Associate Professor of Mathematics
Nagoya University
Nagoya, Japan

D. VAN NOSTRAND COMPANY, INC.

PRINCETON, NEW JERSEY

TORONTO LONDON MELBOURNE

Van Nostrand Regional Offices: *New York, Chicago, San Francisco*

D. Van Nostrand Company, Ltd., *London*

D. Van Nostrand Company (Canada), Ltd., *Toronto*

D. Van Nostrand Australia Pty. Ltd., *Melbourne*

ISBN 978-1-4684-8040-5 ISBN 978-1-4684-8038-2 (eBook)
DOI 10.1007/978-1-4684-8038-2

Copyright © 1968, by D. VAN NOSTRAND COMPANY, INC.
Softcover reprint of the hardcover 1st edition 1968

Published simultaneously in Canada by
D. Van Nostrand Company (Canada), Ltd.

No reproduction in any form of the book, in whole or in part (except for brief quotation in critical articles or reviews), may be made without written authorization from the publisher.

DEDICATED TO

THE U. S. ARMY RESEARCH OFFICE—DURHAM
which has made possible research on principal functions
through the 15 years of their existence

PREFACE

During the decade and a half that has elapsed since the introduction of principal functions (Sario [8]), they have become important tools in an increasing number of branches of modern mathematics. The purpose of the present research monograph is to systematically develop the theory of these functions and their applications on Riemann surfaces and Riemannian spaces. Apart from brief background information (see below), nothing contained in this monograph has previously appeared in any other book.

The basic idea of principal functions is simple: Given a Riemann surface or a Riemannian space R, a neighborhood A of its ideal boundary, and a harmonic function s on A, the principal function problem consists in constructing a harmonic function p on all of R which imitates the behavior of s in A.

Here A need not be connected, but may include neighborhoods of isolated points deleted from R. Thus we are dealing with the general problem of constructing harmonic functions with given singularities and a prescribed behavior near the ideal boundary. The function p is called the principal function corresponding to the given A, s, and the mode of imitation of s by p.

The significance of principal functions is in their versatility. Not only can A and s be prescribed largely at will, but the same is true of the mode of imitation. The latter can even be different in the various components of A. As a result, from two central theorems, the Main Existence Theorem and the Main Extremal Theorem, we obtain simultaneously solutions to a great variety of existence, extremal, mapping, and classification problems that have each previously required separate, often quite intricate, treatment. Several new results are also gained, not accessible by earlier methods.

Another advantage of our approach is that it is purely constructive. Historically it ties in with the evolution that started with Weierstrass' criticism of Riemann's use of the then unproved Dirichlet's principle. To salvage the existence of harmonic functions with given singularities on closed Riemann surfaces, Schwarz and Neumann developed constructive methods bearing their names. However, after Hilbert succeeded in putting Dirichlet's principle on a firm foundation, constructive methods were largely set aside. The principal function method furnishes the construction not only in the classical case of closed surfaces, but also—and this is the very essence of our theory—on open surfaces and spaces.

The main advantages over Dirichlet's principle are explicit results, a unified approach, and accessibility of new problems. Moreover, especially with applications to physics now in sight, the constructivity of solutions is likely to be of practical value in the present computer age. In our book we shall not, however, consider such numerical aspects.

To many readers the concept of principal function is novel. For this reason we have given, in the Introduction, examples of principal functions and their typical applications.

In Chapter 0 we have compiled those rudiments of the theory of Riemann surfaces that will be needed in later chapters. The terminology we have used follows closely that in Ahlfors-Sario, "Riemann surfaces," to which reference is here made for a more detailed axiomatic treatment. Chapter 0 together with a standard graduate curriculum in analysis, algebra, topology, and differential geometry will be sufficient "prerequisites." In a few instances we have also made use of some more sophisticated theorem if its proof is readily available in a standard monograph; an exact reference is then given.

Of central importance in our method is the required mode of imitation of s by p. We specify this by choosing a linear operator L from the space of harmonic functions on the border α of A to the space of harmonic functions on A. The imitation is to mean that $p \mid A - s = L((p-s) \mid \alpha)$. Such an L is called a normal operator. In Chapter I we prove the Main Existence Theorem for principal functions. The fundamental operators L_0 and L_1 and the corresponding principal functions p_0 and p_1 are constructed and Ahlfors' conjecture on extreme operators discussed. Except for the statement of the main theorem itself, the results in this chapter are recent and provide more information than previous treatments (e.g. Ahlfors-Sario) concerning bounds and convergence properties for operators and functions.

By treating a variety of problems in analysis, Chapter II illustrates the wide use to which the theory of Chapter I may be put. Reproducing differentials, harmonic and holomorphic interpolation, and generalizations to open surfaces of the theorems of Abel and Riemann-Roch are some of the topics taken up here. The chapter concludes with a number of results relating principal functions and extremal length. The introductory sections 1, 2 may already be familiar to readers of Ahlfors-Sario, but we have included them here as the first applications of Chapter I, since they illustrate existence and extremal methods in purest form.

Further applications are the capacity functions which are generalized and drawn under systematic study in Chapter III. A central theorem in the theory is established and, as its consequences, proofs are obtained to a score of old and also of hitherto unsolved problems on univalent functions. The intriguing problem of the stability of boundary components is discussed.

Extremal properties of principal functions lead to inclusion relations among null classes of Riemann surfaces characterized by the nonexistence of harmonic and analytic functions with various boundedness properties. These topics are covered in Chapter IV, a brief survey of applications to the classification theory. Extremal length plays an important role in the treatment. The class O_{KD}, for example, is characterized by the property that removal of the ideal boundary of such surfaces does not change extremal distances on its compactification. In order to make this chapter self-contained, some standard material on the subject has been included, in spite of its already having received excellent treatment at the hands of various authors.

Applications of principal functions to value distribution theory are given in Chapter V. They permit the construction of proximity functions on arbitrary Riemann surfaces and thus open the way to the general theory of complex analytic mappings between abstract Riemann surfaces. Classical theorems on meromorphic functions on the plane or the disk follow as special cases. There are two reasons why we chose to include this brief chapter, despite some overlapping with the Sario-Noshiro monograph "Value distribution theory." First, the use of principal functions is the only method known today to construct the above proximity function. Thus the general value distribution theory is among the prime achievements of the theory of principal functions, and also a striking illustration of the Main Existence Theorem with estimates of $p-s$. Second, the construction of the general proximity function has also led to important potential-theoretic work

(Nakai [7], [8], [10]), to which we here specifically call the attention of the reader.

In Chapter VI we first generalize the theory of principal functions to arbitrary Riemannian spaces. After a brief discussion of locally flat spaces we then derive necessary and sufficient conditions for the existence of principal harmonic forms, fields, semifields, and tensor potentials on arbitrary Riemannian spaces. In this direction (cf. VI.3.3D) may well lie the most fertile field of further analytic and differential-geometric research in the theory.

Also of promise appears to be the largely unknown field of applications of principal functions to physics, in particular hydrodynamics, electrostatics, and thermodynamics. Such possibilities are briefly pointed out in Chapter VI, §1.

Ultimately generality of the theory of principal functions is reached in Chapter VII, devoted to such functions in abstract harmonic spaces. These are the most general spaces on which harmonicity can be considered. The principal function problem is solved by making use of the Riesz-Schauder theory for the abstract Fredholm equation of completely continuous operators.

In existing literature potential theory has been restricted to hyperbolic and parabolic surfaces, corresponding to Green's and Evans potentials respectively. In the Appendix, written by M. Nakai, a general potential theory is systematically developed, with no restrictions imposed on the surface. The potentials here have for their kernel the proximity function of Chapter V constructed on arbitrary Riemann surfaces by means of principal functions.

Before starting a systematic study of the book the reader may wish to read, for further orientation, the introductory remarks at the beginning of the chapters and sections.

We have indicated above the points of contact of our monograph with existing books. Beyond these brief topics, all material here appears for the first time in any book. To summarize (cf. Contents), in the theory proper such new topics include the generalized convergence problem, estimation of the distortion constant, characterization of extreme operators, construction of principal functions using integral equations or orthogonal projection, construction of principal functions, forms, fields, and tensor potentials on Riemannian spaces, and the solution of the principal function problem on harmonic spaces. Among the new applications are those to reproducing differentials, harmonic and holomorphic interpolation, generalized theorems of Abel and Riemann-Roch, extremal length, generalized capacity functions, exponential mappings, stability problems, some classification problems, general potential theory, and, very briefly, hydrodynamics, electrostatics, and thermodynamics.

Among writers who have made significant contributions to the theory of principal functions we refer to Ahlfors, Browder, Maeda, Nakai, and Schiffer. Several former U.C.L.A. students also helped to bring the theory into being by their doctoral dissertations and later publications: Bruckner, Emig, Fuller, Goldstein, Harmon, Nickel, Oikawa, Rao, Savage, Seewerker, and Weill have worked on Riemann surfaces; Breazeal, Glasner, Larsen, Meehan, Ow, and Smith on Riemannian spaces. A comprehensive list of writers in the field is contained in the Author Index.

Our deepest indebtedness is due to our collaborator M. Nakai who made important contributions to several parts of the theory and, in particular, wrote the Appendix. P. Loeb, A. Marden, K. V. R. Rao, R. Redheffer, M. Schiffer, and B. Walsh also read portions of the manuscript, K. Larsen, M. Glasner, and several graduate students assisted in checking it, and K. Matsumoto gave valuable help in proofreading. To all these coworkers we wish to express our sincere thanks.

We are grateful to the U. S. Army Research Office—Durham for their continued support, in particular during the period 1962–67 of the preparation of this book. Drs. J. Dawson, A. S. Galbraith, and G. Parrish of AROD cooperated in every way from the inception of the project to its completion.

Our indebtedness also goes to Professor Marshall H. Stone, the Chairman of the Editorial Board, who read the entire manuscript, to Professor S. S. Chern, who initiated the inclusion of the book in this series, and to Professors L. Paige and S. Warschawski, who made it possible for us to devote a large portion of our time to the writing of the book.

Mrs. Elaine Barth and Mrs. Bari Saccoman and their teams at U.C.L.A. and U.C.S.D. worked with skill and devotion in typing and mimeographing countless versions of the manuscript.

La Jolla and Los Angeles, September, 1967

Burton Rodin   Leo Sario

CONTENTS

Chapter II

PRINCIPAL FUNCTIONS

Chapter IV

CLASSIFICATION THEORY

CHAPTER V

ANALYTIC MAPPINGS

CHAPTER VI

PRINCIPAL FORMS AND FIELDS ON RIEMANNIAN SPACES

CHAPTER VII

PRINCIPAL FUNCTIONS ON HARMONIC SPACES

APPENDIX

SARIO POTENTIALS ON RIEMANN SURFACES

By *Mitsuru Nakai*

INTRODUCTION: WHAT ARE PRINCIPAL FUNCTIONS?

The purpose of this Introduction is to give a sampling of principal functions and their various applications. It is not needed for a systematic reading of the ensuing chapters, and some readers may prefer to return to it after acquainting themselves with the concepts and proofs in the book proper. However, we hope that this bird's-eye view, while leaving many details dim, will help the reader discern the simple common idea of principal functions in its various apparently different realizations.

The existence of certain harmonic functions with prescribed singularities and prescribed boundary behavior is central to a variety of areas of complex function theory. It provides the unifying factor for all the topics of this book. Such functions will be called principal functions. For further orientation let us consider how they are constructed and some of their uses.

Our starting point is the classical problem of constructing harmonic functions with prescribed singularities on a closed Riemann surface R. Suppose there is given a singularity function s at a point $\zeta \in R$. That is, s is harmonic in a punctured closed disk A centered at ζ. We seek a function p harmonic on $R-\{\zeta\}$ such that $p-s$ has a removable singularity at ζ.

The problem can be restated in terms which are intrinsic to the surface $S = R-\{\zeta\}$. Once this has been accomplished we shall have an existence problem with wide and useful generalizations.

To obtain the intrinsic formulation we consider the values of $p-s$ along the border α of A. Obviously they determine $p-s$ in A. Specifically, if $f \in H(\alpha)$, the space of harmonic functions on α, let Lf be the restriction to A of the solution of the Dirichlet problem in $A \cup \{\zeta\}$ for the boundary values f. Then L is a linear transformation of $H(\alpha)$ into the space $H(A)$ of functions harmonic on A. The existence problem may now be stated as follows: Given a Riemann surface S, a linear operator $L: H(\alpha) \to H(A)$, and an $s \in H(A)$, find a $p \in H(S)$ such that

$$p|A - s = L((p-s)|\alpha). \tag{1}$$

Concerning the notation, the restrictions $(p-s)|\alpha$, $p|A$ are so obvious that we may rewrite (1) as

$$p - s = L(p-s)$$

without risking confusion.

In order to obtain an existence theorem for the general problem it is necessary that L possess certain simple properties (I.1.2A) shared by the above Dirichlet operator. Then L will be called a *normal operator*. Furthermore the singularity s cannot be arbitrary. For example even the original problem clearly cannot be solved if R is the extended plane, $\zeta = \infty$, $A = \{|z| \geq 1\}$, and $s = \log |z|$. Therefore we shall require that the flux of s across α vanish. This property

$$\int_\alpha {*ds} = 0 \tag{2}$$

turns out to be necessary and sufficient and will be referred to as the *flux condition.*

In the general situation we let A be a bordered boundary neighborhood of S. By this we mean that A is the closure of an open set G such that no component of G is relatively compact, the complement of G is compact, and the boundary of G consists of analytic Jordan curves. We are now able to state the

Main Existence Theorem. *Let A, with border α, be a bordered boundary neighborhood of an open Riemann surface S. Let $L:H(\alpha) \to H(A)$ be a normal operator and let $s \in H(A)$. A necessary and sufficient condition for the existence of $p \in H(S)$ such that*

$$p - s = L(p-s)$$

in A is that s satisfy the flux condition (2).

The function p is called the principal function corresponding to s and L. It is unique up to an additive constant.

EXAMPLES

1. The classical singularity problem. Our existence theorem encompasses the solution of the singularity problem for closed Riemann surfaces. For example, let us use it to construct the fundamental potential on a closed surface R.

Choose distinct points $\zeta_0, \zeta_1 \in R$. Let Δ_0, Δ_1 be disjoint punctured closed disks centered about these points and let α_0, α_1 be the borders of these disks. Let $A = \Delta_0 \cup \Delta_1$, $\alpha = \alpha_0 \cup \alpha_1$. Define the singularity function s in A by

$$s(z) = \begin{cases} \log |z - \zeta_0| & z \in \Delta_0, \\ -\log |z - \zeta_1| & z \in \Delta_1. \end{cases}$$

The flux of s across α_0 is 2π and across α_1 it is -2π. Hence the flux condition is satisfied. Let L be the normal operator corresponding to the Dirichlet problem for A. We apply the existence theorem to the Riemann surface $S = R - \{\zeta_0, \zeta_1\}$ and obtain a harmonic function p on S which satisfies $p - s = L(p - s)$ on A. Therefore p has simple logarithmic poles at ζ_0, ζ_1 of opposite signs. It may be normalized by an additive constant so that it vanishes at a prescribed point ζ_2. The resulting function of z may be denoted by $p(z; \zeta_0, \zeta_1, \zeta_2)$. It is the *fundamental potential* on R.

2. Green's function. It is time to consider normal operators L other than the Dirichlet operators discussed so far. Let B be a compact bordered Riemann surface whose border contours are partitioned into two nonempty classes α, β. Let $A = B - \beta$. Given $f \in H(\alpha)$ define $L_1 f \in H(A)$ to be the harmonic function with boundary values f on α and a constant value k on β. This constant is to be chosen so that $L_1 f$ has vanishing flux across α.

Now let R be the interior of a compact bordered Riemann surface with border β. Let Δ_0 be a punctured disk, with border, centered at a point $\zeta_0 \in R$. Let A' be a bordered boundary neighborhood of R, disjoint from Δ_0. Denote the borders of Δ_0, A' by α_0, α_1. Set $A = \Delta_0 \cup A'$ and $\alpha = \alpha_0 \cup \alpha_1$. We may define a normal operator $L: H(\alpha) \to H(A)$ using L_1 in A' and the Dirichlet operator in Δ_0. Specifically, given $f \in H(\alpha)$ define $Lf \in H(A)$ to be the function $L_1(f|\alpha_1)$ in A', and let the function Lf solve the Dirichlet problem on Δ_0 with boundary values $f|\alpha_0$ on α_0.

Next we define the singularity function $s \in H(A)$. In Δ_0 set $s(z) = -\log|z-\zeta_0|$. In A' let s be the "harmonic measure" which is 0 on α_1 and a negative constant on β, the constant so chosen that the flux across α_1 is 2π. Then the total flux of s across α is zero. The Main Existence Theorem applied to $S = R-\{\zeta_0\}$ yields a function p which satisfies $p-s = L(p-s)$ in A. This means that p has a simple positive logarithmic pole at ζ_0. Along β the functions s and $L_1(p-s)$ are constant. Hence p is also constant on β. By subtracting this constant from p we obtain the Green's function for R with pole at ζ_0.

In defining the normal operator $L_1 : H(\alpha) \to H(A)$ we assumed for simplicity that the ideal boundary β of A could be realized as analytic Jordan curves. This restriction is actually not needed (see I.2.2). Thus there is a normal operator L_1 defined on any bordered boundary neighborhood of any Riemann surface. Even in this general case $L_1 f$ is constant on the ideal boundary in a sense that can be made precise (see II.6.2G). Using this general L_1 we may repeat the construction above for the Green's function for an arbitrary Riemann surface R. However, in defining s we would now have to assume that the harmonic measure of β is not identically zero.

3. Other fundamental functions. We return for a moment to the case of a bordered compact region B with border $\alpha \cup \beta$ and with $A = B-\beta$. Instead of requiring that $L_1 f$ be constant on the entire β we may require only that it be constant on each component of β. In that case we choose the constants so that $L_1 f$ has vanishing flux across every component of β. In this way we obtain another normal operator L_1.

More generally, we consider a partition P of β into mutually disjoint sets of components and require that $L_1 f$ be constant on each set with vanishing flux across it. The two cases of L_1 discussed above correspond to the identity partition $P = I$, where all of β is in the same set, and to the canonical partition $P = Q$, where each component of β is a separate set.

We also consider the operator $L_0 : H(\alpha) \to H(A)$ defined by the vanishing of the normal derivative $\partial L_0 f/\partial n = 0$ on β. The operator L_1 for an arbitrary P and the operator L_0 can again be extended to an arbitrary bordered boundary neighborhood A.

On an arbitrary Riemann surface R there are six principal functions which will be useful for our examples. To define them choose points ζ, ζ_0, ζ_1 on R. Let A' be a bordered boundary neighborhood of R which does not contain these points. About these three points place some bordered punctured disks Δ, Δ_0, Δ_1, respectively. We assume that Δ_0, Δ_1 are disjoint and that all of the disks are disjoint from A'.

First consider the singularity function s given by Re $(z-\zeta)^{-1}$ in Δ and $\equiv 0$ in A'. Let $A = \Delta \cup A'$ and form the normal operator L determined by L_0 in A' and the Dirichlet operator in Δ. The Main Existence Theorem yields a function p_0 harmonic on $S = R - \{\zeta\}$ and satisfying $p_0 - s = L(p_0 - s)$ in A. This means that $p_0 = L_0 p_0$ in A' and that $p_0(z) -$ Re $(z-\zeta)^{-1}$ is regular at ζ. If we choose L_1 (for a given partition P) instead of L_0 in this construction then we obtain a principal function which we denote by p_1. Here we are only interested in the cases $P = I, Q$.

Next let the singularity function s be given by $\log |z-\zeta_0|$ in Δ_0, $-\log |z-\zeta_1|$ in Δ_1, and $\equiv 0$ in A'. The above three choices for L lead to principal functions which will be denoted by g_0, g_1, the latter with $P = I, Q$. To summarize:

On any Riemann surface R there are functions p_0, p_1 which are harmonic except for a singularity Re $(z-\zeta)^{-1}$ *at a given point ζ. In a bordered boundary neighborhood A' of R they satisfy*

$$p_0 = L_0 p_0, \qquad p_1 = L_1 p_1.$$

There are functions g_0, g_1 which are harmonic on R except for a simple negative logarithmic pole at ζ_0 and a simple positive logarithmic pole at ζ_1. In A' they satisfy

$$g_0 = L_0 g_0, \qquad g_1 = L_1 g_1.$$

Both p_1 and g_1 may correspond either to the identity partition $P = I$ or the canonical partition $P = Q$ of the ideal boundary of R. The six functions so obtained are uniquely determined up to an additive constant.

4. Conformal mapping. Now assume that R is a plane region. From $p_1 = L_1 p_1$ for $P = Q$ we see that p_1 has vanishing flux across every cycle on R. Therefore it has a harmonic conjugate function p_1^*. Form the meromorphic function $P_1 = p_1 + ip_1^*$. It gives a univalent conformal mapping of R into the extended plane. Its real part is constant on each boundary component and P_1 is the *vertical slit mapping* with pole at ζ. Similarly, $P_0 = p_0 + ip_0^*$ is the *horizontal slit mapping.*

The principal functions g_0 and g_1 do not have single-valued conjugates due to their conjugate periods of $\pm 2\pi$ about ζ_0, ζ_1. However, for $P = Q$ we do obtain meromorphic functions G_0, G_1 by exponentiating the multiple-valued functions $g_0 + ig_0^*$, $g_1 + ig_1^*$. The function G_1 is the *circular slit mapping* with zero at ζ_0 and pole at ζ_1. Similarly G_0 is the *radial slit mapping.*

The function $P_0 + P_1$ also turns out to be univalent. It maps R onto a region, every complementary component of which is convex (III.4.2).

5. Extremal properties. Let R now be an arbitrary Riemann surface, planar or not. The principal functions enjoy several interesting extremal properties (Chapter II). For any function u let $B(u) = \int u * du$ where the integral is taken along the ideal boundary β of R.

Among all harmonic functions u on R with singularity $\text{Re}\,(z-\zeta)^{-1}$ at ζ the functional $B(u)$ is minimized by $u = \frac{1}{2}(p_0 + p_1)$ for $P = I$. It is minimized by $\frac{1}{2}(g_0 + g_1)$, $P = I$, among all harmonic functions with the logarithmic singularities at ζ_0, ζ_1. The functions $\frac{1}{2}(p_0 + p_1)$ and $\frac{1}{2}(g_0 + g_1)$ for $P = Q$ minimize this functional among all harmonic functions with the corresponding singularities and whose conjugate periods vanish along all dividing cycles (not separating ζ_0 and ζ_1).

For plane regions this functional takes on added interest. This is so because if u is the real part of a univalent conformal mapping, then $-B(u)$ is the area of the complement of the image region. Hence we see that $\frac{1}{2}(P_0 + P_1)$ maximizes the complementary area of the image among all univalent meromorphic functions on R with a simple pole of residue 1 at ζ.

Consider again an arbitrary Riemann surface R. The function $g = g_0 - g_1$ can be extended harmonically over ζ_0, ζ_1 and the resulting function has finite Dirichlet integral $D(g)$ over R. For $P = I$ this function minimizes the expression

$$D(u) + 4\pi(u(\zeta_0) - u(\zeta_1)) = D(u-g) - D(g) \qquad (3)$$

among all regular harmonic functions u on R, while $p = p_0 - p_1$ minimizes

$$(4) \qquad D(u) - 4\pi\left(\frac{\partial u}{\partial x}\right)_{z=\zeta} = D(u-p) - D(p)$$

among such functions. For $P = Q$ the functional (4) is minimized by $p_0 - p_1$ among all harmonic functions whose conjugate periods vanish along all dividing cycles (II.1.2).

The functionals (3), (4) may be used to generalize the classical notion of span. For example, on a plane region $P_0 - P_1$ minimizes $D(f) - 4\pi \operatorname{Re} f'(\zeta)$ among all holomorphic functions f on R. The value of this minimum is $-2\pi(P_0 - P_1)'(\zeta)$. The quantity $(P_0 - P_1)'(\zeta)$ is known as the span at ζ, and it is always real and nonnegative. In fact, the span at ζ is $(2\pi)^{-1}$ times the complementary area under the mapping $P_0 + P_1$. Thus one concludes for instance that the span vanishes if and only if every conformal mapping of R leaves a complement of zero area.

6. Classification theory. Principal functions and normal operators also play a role in the classification of Riemann surfaces. For illustrative purposes we consider two examples.

Recall the extremal property (3) of g_0, g_1. It follows that $g_0 - g_1 =$ constant for all ζ_0, ζ_1 if and only if there are no nonconstant harmonic functions on R with a finite Dirichlet integral. Furthermore, it is easily seen that g_0, g_1 are bounded in A' (I.2.1), and a fortiori $g_0 - g_1$ is bounded on R. Consequently if there is one nonconstant harmonic function on R with a finite Dirichlet integral then there is such a function which is bounded.

The Main Existence Theorem is useful for relating properties of the ideal boundary of R to global properties of R. For example, let E be a closed subset of the extended plane with connected complement R. Suppose there are no nonconstant holomorphic functions on R with finite Dirichlet integrals. Now let f be a function holomorphic on $A = G - E$ where G is a closed simply connected neighborhood of E. We apply the existence theorem to $s = \operatorname{Re} f$ and the Dirichlet operator L for G. We obtain a harmonic function p on R satisfying $p - s = L(p - s)$ in A.

It is easily seen that p has a single-valued harmonic conjugate

p^* on R (IV.2.1D). Since $D(p+ip^*) < \infty$, p must be constant. Therefore $s = Ls$ which means that $\operatorname{Re} f$, and hence f itself, is continuable to E. Thus E is a removable singularity for every holomorphic function with finite Dirichlet integral.

7. Reproducing differentials. We return to arbitrary Riemann surfaces R. The principal function p_1 has the singularity $\operatorname{Re}(z-\zeta)^{-1}$. Let q_1 denote the principal function with singularity $\operatorname{Im}(z-\zeta)^{-1}$ and with the behavior $q_1 = L_1 q_1$, $P = I$, in a bordered boundary neighborhood A'. Then the differential $\psi = dp_1 + *dq_1$ is singularity free and square integrable on R. It possesses the reproducing property

$$a(\zeta) = -\frac{1}{2\pi}\iint_R \omega \wedge *\psi, \tag{5}$$

which is valid for every square integrable harmonic differential $\omega = a\,dx + b\,dy$. Thus ψ is a generalized Bergman kernel [1].

There are many other reproducing differentials which can be expressed in terms of principal functions (see II.3.3). For example $\psi = dp_0 - dp_1$, $P = I$, enjoys property (5) with respect to all exact square integrable harmonic differentials ω.

Let c be an arc on R. Using principal functions we shall construct a harmonic differential ψ_c such that

$$\int_c \omega = 2\pi \iint_R \omega \wedge *\psi_c \tag{6}$$

for all square integrable harmonic differentials ω. First suppose that c is contained in a disk $\Delta \subset R$ and let $\partial c = \zeta_2 - \zeta_1$. We define a singularity function $\sigma(z) = \log|(z-\zeta_2)/(z-\zeta_1)|$ in Δ. Let p_σ be the principal function with singularity σ and with the behavior $p_\sigma = L_1 p_\sigma$, $P = I$, in a bordered boundary neighborhood A' of R. Next consider the singularity $\tau(z) = \arg\{(z-\zeta_2)/(z-\zeta_1)\}$ in $\Delta - c$, and $\tau(z) \equiv 0$ in A'. On applying the existence theorem to the surface $R - c$ and the normal operator consisting of L_1 in A' and the Dirichlet operator for Δ we obtain a principal function p_τ. The differential dp_τ can be extended harmonically to all of

$R-\{\zeta_1,\zeta_2\}$. We continue to denote the extension by dp_τ even though it is not exact. The differential

$$(7) \qquad \psi_c = dp_\sigma + *dp_\tau$$

has the reproducing property (6). If c is any 1-chain on R it is homologous to a finite sum $\sum n_i c_i$ where each c_i is an arc contained in a parametric disk and n_i is an integer. If we set $\psi_c = \sum n_i \psi_{c_i}$ then (6) remains valid. In particular, if R is nonplanar we may take c to be a nondividing cycle. The corresponding ψ_c demonstrates that any nonplanar surface carries a nonzero square integrable harmonic differential.

8. Interpolation. Consider the problem of constructing harmonic differentials with prescribed periods and derivatives. The differential of minimum norm which solves the problem is a combination of reproducing differentials (II.4). Hence it can often be expressed in terms of principal functions.

For example, suppose we seek a harmonic function u with prescribed values of $(\partial^n u/\partial x^n)_{z=\zeta_j}$ for a finite set of points ζ_j and integers $n \geq 0$. If R is a plane region whose complement has positive area then there exists such a u with finite Dirichlet integral. The solution which minimizes $D(u)$ is a linear combination of regular harmonic functions of the form $t_0 - t_1$ where t_0 and t_1 are principal functions with suitable singularities at ζ_j and with the properties $t_0 = L_0 t_0$, $t_1 = L_1 t_1$ in a boundary neighborhood of R. Since each $t_0 - t_1$ is bounded, so is the solution of our interpolation problem.

9. Capacities. Let $\bar{\Omega}$ be a compact bordered subregion of an open Riemann surface R and let $\zeta \in \Omega$. We construct the principal function $t_\Omega(z,\zeta)$ in $\bar{\Omega}$ with singularity $\log|z-\zeta|$ and with constant value k_Ω on the border β_Ω of Ω such that $t_\Omega(z,\zeta) - \log|z-\zeta| \to 0$ as $z \to \zeta$. As Ω exhausts R the constants k_Ω increase to a limit k_β. We call $c_\beta = e^{-k_\beta}$ the *capacity of the ideal boundary* β *of* R.

There exists a sequence Ω_n such that the corresponding functions t_{Ω_n} converge to a function t_β on R, the *capacity function of* R. In contrast with Green's function the capacity function exists on every Riemann surface, although in the case $c_\beta = 0$ the uniqueness is not assured.

Among all harmonic functions $t(z,\zeta)$ on R with $t(z,\zeta) - \log|z-\zeta| \to 0$ as $z \to \zeta$ the capacity function minimizes $\sup_R t$, the minimum being k_β. It also gives to the boundary integral $B(t) = \int_\beta t * dt$ the minimum $2\pi k_\beta$.

Let $\partial\Omega$ be so partitioned that each part is the border of exactly one component of $R - \Omega$. Let γ be a boundary component of R and let $\beta_{\Omega\gamma}$ be the part of β_Ω that separates γ from Ω. We form the principal function $t_{\Omega\gamma}$ on $\bar{\Omega}$ with the same singularity and normalization as t_Ω but with a constant value $k_{\Omega\gamma}$ on $\beta_{\Omega\gamma}$ and constant values on all other parts of β_Ω such that the flux is 2π across $\beta_{\Omega\gamma}$ and 0 across every other part. Again the $k_{\Omega\gamma}$ converge to a constant k_γ as $\Omega \to R$, and we can define the *capacity* $c_\gamma = e^{-k_\gamma}$ *of a boundary component* γ.

A boundary component γ of a planar Riemann surface R is a point under all univalent mappings of R if and only if $c_\gamma = 0$. Such a component is called *weak*. A component is called *strong* if it is a continuum under all univalent mappings of R. The remaining alternative is an *unstable* boundary component. An interesting phenomenon is that, under univalent mappings of R onto a parallel slit region of varying direction of the slits and with a fixed residue 1 at ζ, an unstable component *reduces to a point for one choice of the direction only* (III.4.1B).

An unsolved problem is to extend the concepts of strong and unstable boundary components to nonplanar Riemann surfaces.

10. Extremal length. Let $\bar{R}$ be a compact bordered Riemann surface. Let α^0, α^1 be two contours of $\bar{R}$ and suppose that the set of remaining contours is partitioned into two classes γ^0, γ^1. Using the Main Existence Theorem one can construct a harmonic function u on $\bar{R}$ such that $u = 0$ along α^0, $u = 1$ along α^1, $\partial u/\partial n = 0$ along each contour of γ^0, and u is a constant with vanishing conjugate period on each contour of γ^1. In brief, u has L_0-behavior on γ^0 and L_1-behavior with $P = Q$ on γ^1.

The extremal length of all curves in $\bar{R} - \gamma^0$ which, together with γ^1, connect α^0 with α^1 is given by $D(u)^{-1}$. The extremal length of all curves in $\bar{R} - \gamma^1$ which, together with γ^0, separate α^0 and α^1 is given by $D(u)$ (II.6.1D). Furthermore, if $\bar{R}$ is planar then

$\exp\{2\pi(u+iu^*)/D(u)\}$ is a univalent conformal mapping of $\bar{R}$ onto an annulus with radial and circular slits. In Chapter III these facts are generalized to arbitrary surfaces R.

Another connection between principal functions and extremal length concerns the reproducing differential (7). If c is a cycle then the norm of ψ_c is the square root of the extremal length of all curves homologous to c.

11. Value distribution. In the treatment of general value distribution (Chapter V) the following application of principal functions is made: Let $p(z) = -2g_0(z)$ where g_0 was defined in number 3 above. Thus p has a positive logarithmic pole at ζ_0 and a negative logarithmic pole at ζ_1; each pole is of order 2, and we normalize by $p(z)+2\log(z-\zeta_0)\to 0$ as $z\to\zeta_0$. To emphasize the dependence on ζ_1 we also write $p(z) = p(z,\zeta_1)$.

The *proximity function* is defined by

$$s(z,a) = \log\{(1+e^{p(z,\zeta_1)})(1+e^{p(a,\zeta_1)})\} - p(z,a).$$

It is a C^∞ function on $R-\{a\}$. Furthermore, it is bounded from below on R and tends to $+\infty$ at a. This function thus provides a measure of the proximity of z and a. The metric

$$\sqrt{\Delta s}|dz|$$

has Gaussian curvature 1 and gives R an area of 4π. It constitutes a generalization of the spherical metric and allows us to extend the classical theory of value distribution to Riemann surfaces.

Typically one asks how many points can a complex analytic mapping f of an open Riemann surface R into an arbitrary Riemann surface S omit. To answer this choose a parametric disk $R_0\subset R$ with border β_0 and let Ω be an adjacent regular subregion of $R-R_0$ with border $\beta_0\cup\beta_\Omega$. Take the harmonic function u on $\bar{\Omega}$ with $u|\beta_0 = 0$, $u|\beta_\Omega = k$, where k is a constant such that $\int_{\beta_0}{*du} = 1$. For $h\in[0,k]$ the level line $\beta_h: u = h$ determines a subregion Ω_h of Ω with Euler characteristic $e(h)$, say. Let $a(h)$ be the area of the multisheeted image above S of $R_h = R_0\cup\Omega_h$ in the above metric $\sqrt{\Delta s}|dz|$ on S. We denote by $E_2(h)$ and $C_2(h)$ the quantities $e(h)$ and $a(h)$ integrated thrice from 0 to h and set

$$\eta = \liminf_{R_k \to R} \frac{E_2(k)}{C_2(k)}.$$

Then the number of points of S a nondegenerate mapping f can omit is at most $2+\eta$. This is an extension to Riemann surfaces of Picard's celebrated theorem.

12. Riemannian spaces. Let V be an arbitrary noncompact Riemannian space and V_1 its bordered boundary neighborhood with border ∂V_1. Given on $\bar{V}_1$ a harmonic differential form σ characterized by $\Delta\sigma = 0$ we seek on V a harmonic form ρ that imitates the behavior of σ in V_1. More precisely, we require that the square norm $||\sigma-\rho||_{V_1}^2 = \int_{V_1}(\sigma-\rho)\wedge *(\sigma-\rho)$ be finite and that $\sigma-\rho$ have L_1-behavior on the ideal boundary β of V; by this we now mean that $\sigma-\rho$ possesses, in a sense, vanishing Dirichlet data on β. The flux condition is replaced by the following necessary and sufficient condition: For every C^∞ form φ with compact support in V

$$(8) \quad \int_{\partial V_1} \sigma\wedge *d\varphi+\delta\sigma\wedge *\varphi-\varphi\wedge *d\sigma-\delta\varphi\wedge *\sigma = O(||\Delta\varphi||).$$

The condition is satisfied by forms with strikingly simple properties. For example, if σ coincides on ∂V_1 with an exact form and with a coexact form defined in some vicinity of ∂V_1, then (8) holds.

The corresponding problem for harmonic forms with some analogue of an L_0-behavior on β remains open.

Concluding remarks

By means of these few examples we have attempted to indicate how principal functions form a convenient point of departure for a variety of investigations. In Chapter I we prove the Main

Existence Theorem and other fundamental facts about principal functions. Chapter II contains a number of problems in complex analysis which may be treated by means of these functions. These two chapters may be considered as a single unit—the second chapter providing a collection of illustrative examples to the basic theory in the first chapter.

The remaining chapters are essentially independent of each other. In Chapter VI we tried to make a start toward generalizing the preceding Riemann surface techniques to Riemannian spaces of higher dimensions. We have also included a Chapter 0 on prerequisite Riemann surface theory. It is written for the reader whose mathematical background consists only of the usual graduate courses in algebra, analysis, and topology.

CHAPTER 0
PREREQUISITE RIEMANN SURFACE THEORY

In this chapter we list the basic properties of Riemann surfaces which are needed in the remainder of the book. For the most part we have omitted detailed proofs here since the reader will often be able to supply his own or else refer to standard textbooks for them.

In §1 we define Riemann surfaces, bordered Riemann surfaces, and give some examples and standard constructions. The remainder of the section surveys the basic algebraic topology of surfaces. In §2 we discuss the Dirichlet problem, differential forms, and integration on Riemann surfaces.

For a comprehensive discussion of the axiomatics of the theory of Riemann surfaces we refer to Ahlfors-Sario [1, Chapters I and II].

§1. TOPOLOGY OF RIEMANN SURFACES

1. Definition of a Riemann surface

1A. Conformal structure. A *Riemann surface* is a pair (R,Φ) where R is a connected Hausdorff space and $\Phi = \{\varphi_\alpha\}_{\alpha\in A}$ is an indexed family of functions with the following properties: Each $\varphi_\alpha : U_\alpha \to \mathbf{C}$ is a homeomorphism of an open subset $U_\alpha \subset R$ onto an open subset $\varphi_\alpha(U_\alpha)$ of the complex plane $\mathbf{C}$. The family Φ, referred to as the *conformal structure*, must satisfy the axioms:

(A1) The domains of the φ_α form an open cover of R: $R = \bigcup_{\alpha\in A} U_\alpha$.

(A2) The functions in Φ are holomorphically related: if $\alpha,\beta \in A$ and $U_\alpha \cap U_\beta \neq \varnothing$, then $z \to \varphi_\beta \circ \varphi_\alpha^{-1}(z)$ is a holomorphic mapping of $\varphi_\alpha(U_\alpha \cap U_\beta)$ onto $\varphi_\beta(U_\alpha \cap U_\beta)$.

(A3) The family Φ is maximal: if φ is a homeomorphism of an open subset $U \subset R$ onto the open subset $\varphi(U) \subset \mathbf{C}$ such that for each $\alpha \in A$ with $U_\alpha \cap U \neq \emptyset$ the mapping $z \to \varphi_\alpha \circ \varphi^{-1}(z)$, $z \in \varphi(U_\alpha \cap U)$, is holomorphic, then $\varphi \in \Phi$.

1B. A simple example of a Riemann surface is afforded by a plane region $R \subset \mathbf{C}$. Let Φ be the set of all 1–1 holomorphic functions φ_α such that the domain of each φ_α is an open subset of R. One verifies immediately that (R, Φ) is a Riemann surface.

The above example also applies to subregions of the extended plane provided we adopt the usual definition for a function to be holomorphic at ∞. In particular the extended plane itself is a Riemann surface—more precisely, it has a natural conformal structure which makes it into a Riemann surface.

If (R, Φ) is a Riemann surface and only one conformal structure on R is under discussion then one allows the symbol R to denote this Riemann surface as well as its underlying topological space. The functions φ_α are called *local parameters* or *local coordinates*. These terms are also used to refer to a variable z whose range is the image of φ_α. If the image of φ_α is a unit disk and if U_α is relatively compact in R then U_α is called a *parametric disk*.

1C. Holomorphic functions. Let (R, Φ) be a Riemann surface. Let f be a complex-valued function defined on an open subset $G \subset R$. The function f is said to be *holomorphic* on G if to each $\zeta \in G$ there is a local parameter φ_α such that $\zeta \in U_\alpha$ and $z \to f \circ \varphi_\alpha^{-1}(z)$ is holomorphic for $z \in \varphi_\alpha(U_\alpha \cap G)$.

If f is holomorphic then $z \to f \circ \varphi_\beta^{-1}(z)$ is a holomorphic function whenever the domain of a local parameter φ_β meets the domain of f.

According to convention one usually writes $f(z)$ in place of $f \circ \varphi_\alpha^{-1}(z)$. Thus one could say that f is holomorphic on R provided $f(z)$ is a holomorphic function for every local parameter z.

1D. Holomorphic mappings. Let (R, Φ) and (S, Ψ) be Riemann surfaces. A mapping $f: R \to S$ is called *holomorphic* if for each local parameter $\psi \in \Psi$ whose domain V meets $f(R)$ the function $\psi \circ f$ is holomorphic on $f^{-1}(V)$.

Note that a function $f: R \to \mathbf{C}$ is holomorphic if and only if it is a holomorphic mapping of the Riemann surface (R, Φ) into

$(\mathbf{C},\Psi)$ where Ψ is the natural conformal structure of $\mathbf{C}$. Two Riemann surfaces are *conformally equivalent* if there is a 1–1 holomorphic mapping of one onto the other. We often identify such surfaces.

1E. Basis for conformal structure. Suppose (R,Φ_0) satisfies all the axioms for a Riemann surface except the axiom (A3) of maximality. Then Φ_0 can be enlarged to a family Φ so that (R,Φ) is a Riemann surface. Furthermore, the enlarged family Φ is uniquely determined by Φ_0.

Therefore, to construct a Riemann surface from a connected Hausdorff space R it suffices to construct a family Φ_0 of homeomorphisms which are holomorphically related and whose domains form an open cover of R. Such a family Φ_0 is called a *basis* for the conformal structure on R. For example, the natural conformal structure on a plane region R may be defined by specifying that the identity function z with domain R be a basis.

1F. Examples. We have mentioned that any subregion of the extended plane is a Riemann surface in a natural way. It is also easy to see that a subregion of any Riemann surface is again a Riemann surface in a natural way.

Another important example of a Riemann surface is the multi-sheeted image of a holomorphic function f which is often introduced in elementary complex analysis. Closely related are the surfaces encountered in the study of analytic continuation. We now discuss one general example which encompasses these and all others to be considered in later chapters.

1G. Let $f:R\to\mathbf{C}$ be a continuous function on a connected Hausdorff space R. Assume that f is a local homeomorphism on $R-\{\zeta_1,\zeta_2,\cdots\}$ where $\{\zeta_1,\zeta_2,\cdots\}$ is a discrete closed set of points on R. This means that each $\zeta\in R-\{\zeta_1,\zeta_2,\cdots\}$ has an open neighborhood in $R-\{\zeta_1,\zeta_2,\cdots\}$ which is mapped homeomorphically by f onto an open subset of $\mathbf{C}$. With an additional hypothesis on the behavior of f near each ζ_i we shall show that R possesses a unique conformal structure with respect to which f is holomorphic.

Before stating this additional hypothesis we consider the case when the sequence $\{\zeta_1,\zeta_2,\cdots\}$ is empty. Let Φ_0 consist of all restrictions $f|U_\alpha$ where U_α is open in R and $f|U_\alpha$ is a homeomorphism. Clearly Φ_0 is a basis for a conformal structure Φ on R which makes f holomorphic.

For the general case we add the hypothesis that each ζ_i has an open neighborhood G_i with the following property: There is a homeomorphism f_i of G_i onto an open set $f_i(G_i) \subset \mathbf{C}$ and an integer $n_i \geq 2$ such that

$$f(\zeta) - f(\zeta_i) = [f_i(\zeta)]^{n_i}$$

for all $\zeta \in G_i$.

Now let Φ_0 consist of all $f|U_\alpha$ where U_α is open in R and f is a homeomorphism on U_α, and of all f_i. Then Φ_0 is a basis for a conformal structure Φ on R and f is holomorphic. The reader may check that this structure is unique if f is to be holomorphic.

In the situation above one says that the Riemann surface R is a *ramified covering* of $f(R) \subset \mathbf{C}$. The function f is called the *projection map*. The points ζ_i are the *branch points* and n_i is the *multiplicity* of the branch point ζ_i. We may also consider ramified coverings of a Riemann surface rather than a plane region. We leave it to the reader to note the necessary modifications.

1H. Bordered Riemann surfaces. Let R be a plane region whose boundary consists of analytic Jordan curves. Then R is a Riemann surface. The closure of R will be an example of a bordered Riemann surface.

Let $\bar{R}$ be a connected Hausdorff space and $\bar{\Phi}$ a collection of homeomorphisms $\{\varphi_\alpha\}$ such that the domain U_α of φ_α is open in $\bar{R}$ and the image $\varphi_\alpha(U_\alpha)$ is a relatively open subset of the closed upper half-plane.

We require that $\bar{\Phi}$ satisfy the axioms (A1), (A2), (A3) for a Riemann surface with the following modifications. In axiom (A2) the sets $\varphi_\alpha(U_\alpha \cup U_\beta)$ need not be open in $\mathbf{C}$. But if we understand that a function is holomorphic on a subset of $\mathbf{C}$ if it can be extended to a holomorphic function on an open neighborhood of that set then that axiom shall still apply. In axiom (A3) the maximality must be understood to refer to the class of all homeomorphisms φ whose image is a relatively open subset of the closed upper half-plane.

Given such a pair $(\bar{R}, \bar{\Phi})$ we define the *border* $\partial\bar{R}$ of $\bar{R}$ to consist of the points $\zeta \in \bar{R}$ whose image $\varphi_\alpha(\zeta)$ lies on the real axis for some local parameter φ_α defined at ζ. Note that if $\zeta \in \partial \bar{R}$ then by the open mapping property $\varphi_\beta(\zeta)$ lies on the real axis for every local parameter φ_β defined at ζ (see e.g. Alhfors-Sario [1, p. 117]).

If the border $\partial \bar{R}$ is not empty we say that $(\bar{R}, \bar{\Phi})$, or simply $\bar{R}$, is a *bordered Riemann surface*. The notion of holomorphic function is defined as in 1C. The space $R = \bar{R} - \partial \bar{R}$ is called the *interior* of $\bar{R}$. It is a Riemann surface with the conformal structure naturally induced by $\bar{\Phi}$.

1I. Arcs. An arc σ on a Riemann surface R is a continuous mapping $\sigma: [0,1] \rightarrow R$. We often use the symbol σ for the image of this mapping as well as for the mapping itself. In context, the precise intention should be clear.

In terms of local parameters on R it makes sense to say that σ is differentiable, of class C^n, or analytic. It is said to be *smooth* if it is of class C^∞ and $\sigma' \neq 0$. An arc σ is *closed* if $\sigma(0) = \sigma(1)$. It is *simple* if it is 1–1. An *open arc* is a continuous mapping of the open unit interval into R. Note that all these definitions apply also if R is replaced by a bordered Riemann surface $\bar{R}$.

It is easy to show that the border $\partial \bar{R}$ of a bordered Riemann surface $\bar{R}$ is a union of analytic arcs.

1J. Orientation. A Riemann surface is oriented. Indeed, if φ_α, φ_β are local parameters with $U_\alpha \cap U_\beta \neq \emptyset$ then $\varphi_\alpha \circ \varphi_\beta^{-1}$ is holomorphic and hence has positive Jacobian determinant.

For example, if σ is a simple arc on the border of $\bar{R}$ then $\bar{R}$ lies either to the left or else to the right of σ. Indeed, let $\zeta \in \sigma$ and let φ_α be a local parameter defined in a neighborhood U_α of ζ. Then $\varphi_\alpha(\zeta)$ lies on the real axis and $\varphi_\alpha(U_\alpha)$ is a relatively open neighborhood of $\varphi_\alpha(\zeta)$ with respect to the closed upper half-plane. Hence the image of $t \rightarrow \varphi_\alpha \circ \sigma(t)$, defined for t in a neighborhood of $\sigma^{-1}(\zeta)$, is an interval of the real axis about the point $\varphi_\alpha(\zeta)$. If this interval is positively oriented, i.e. if $\varphi_\alpha \circ \sigma(t)$ is increasing, then we say $\bar{R}$ lies to the left of σ (at ζ). Otherwise we say it lies to the right.

Suppose $\bar{R}$ lies to the left. Let us show that this will be the case independently of the local parameter φ_α and the particular point $\zeta \in \sigma$.

For fixed ζ, if φ_β is another local parameter at ζ then $\varphi_\beta \circ \varphi_\alpha^{-1}$ extends to a holomorphic mapping of a full neighborhood of $\varphi_\alpha(\zeta)$

in $\mathbf{C}$ onto a full neighborhood of $\varphi_\beta(\zeta)$. It is easily seen (*loc. cit.*) that this mapping is increasing along the part of the real axis within its domain. Hence $\varphi_\beta \circ \sigma$ is also increasing.

Now let ζ vary on $\sigma \subset \partial\bar{R}$. It is clear that the above criterion is invariant under small displacements of ζ. By connectedness of σ we conclude that $\bar{R}$ lies to the left of σ everywhere.

1K. The double. Let $(\bar{R},\bar{\Phi})$ be a bordered Riemann surface. Let $\bar{R}^*$ be another copy of $\bar{R}$. We shall regard the spaces $\bar{R}$, $\bar{R}^*$ as disjoint. To each $\zeta \in \bar{R}$ there corresponds naturally a point in $\bar{R}^*$ which we denote by $j(\zeta)$. Consider the space $\hat{R}$ obtained from $\bar{R} \cup \bar{R}^*$ by identifying two points if and only if they both lie on a border and correspond to each other under j. Then $\hat{R}$ is a connected Hausdorff space and there is an obvious way to consider $\bar{R}$, $\bar{R}^*$ as subsets of $\hat{R}$. Thus we have $\bar{R} \cap \bar{R}^* = \partial\bar{R}$.

We shall define a conformal structure Φ on $\hat{R}$ so that $(\hat{R},\Phi)$ is a Riemann surface. To do this it suffices to specify a local parameter at each $\zeta \in \hat{R}$ and verify that they are holomorphically related. We consider separately the three cases $\zeta \in \bar{R} - \partial\bar{R}$, $\zeta \in \bar{R}^* - \partial\bar{R}$, and $\zeta \in \partial\bar{R}$.

If $\zeta \in \bar{R} - \partial\bar{R}$ then there is a neighborhood $U_\zeta \subset \bar{R} - \partial\bar{R}$ of ζ and a local parameter φ_ζ in $\bar{\Phi}$ with domain U_ζ. Let Φ contain all such φ_ζ.

If $\zeta \in \bar{R}^* - \partial\bar{R}$ then $j^{-1}(\zeta) = \zeta_0$ belongs to $\bar{R} - \partial\bar{R}$. Thus φ_{ζ_0} has been defined and we set $\varphi_\zeta = \bar{\varphi}_{\zeta_0}$ where the bar stands for complex conjugation. Let Φ contain all such φ_ζ.

Now suppose $\zeta \in \partial\bar{R}$. Let $\varphi \in \bar{\Phi}$ have as domain a neighborhood U of ζ. Then $U \cup j(U)$ is a neighborhood of ζ on $\hat{R}$. Let φ_ζ be defined on $U \cup j(U)$ by

$$\varphi_\zeta(p) = \begin{cases} \varphi(p) & \text{if} \quad p \in U, \\ \bar{\varphi}(j^{-1}(p)) & \text{if} \quad p \in j(U). \end{cases}$$

The functions $\{\varphi_\zeta\}_{\zeta \in \hat{R}}$ form a basis for a conformal structure Φ on $\hat{R}$. The Riemann surface $(\hat{R},\Phi)$ is called the *double* of $\bar{R}$.

If $\bar{R}$ is a closed disk then $\hat{R}$ is the extended complex plane. If $\bar{R}$ is a closed annulus then $\hat{R}$ is topologically a torus.

1L. Open and closed surfaces. A Riemann surface R is

called *open* or *closed* according as the underlying topological space is noncompact or compact. One may equally well speak of noncompact or compact Riemann surfaces, but the open-closed terminology is more traditional. This dichotomy applies to bordered Riemann surfaces as well. However, the compact-noncompact terminology is standard in that case.

The study of closed Riemann surfaces, in many of its aspects, can be carried out within the realm of algebra, e.g. as the theory of algebraic curves over the complex numbers (see Walker [1]) or algebraic functions of one variable (see Chevalley [1]). Open Riemann surfaces, however, seem to require techniques of analysis for their study. In this book the emphasis is always placed on open surfaces, although most theorems apply to closed surfaces as well. In fact, one aspect of Riemann surface theory is to discover generalizations to arbitrary Riemann surfaces of classical theorems on closed surfaces (see Accola [5]).

2. Compactifications

2A. We shall consider two compactifications of a Riemann surface. The first is the one-point compactification, familiar from general topology. The second is the Kerékjártó-Stoïlow compactification.

It should be noted that these compactifications are purely topological. More delicate ones, which depend on the conformal structure, form an interesting and important topic in the study of Riemann surfaces. A discussion of these topics may be found in Constantinescu-Cornea [1] and Nakai-Sario [11].

2B. One-point compactification. We recall Alexandroff's theorem. Let R be a locally compact Hausdorff space and let β be any object such that $\beta \notin R$. Let $R^* = R \cup \{\beta\}$. Define a topology on R^* which consists of all open subsets of R and all subsets $V \subset R^*$ such that the complement of V is a compact subset of R. Then R^* is a compact Hausdorff space.

If R is a Riemann surface then it is locally compact. Hence Alexandroff's theorem applies. In this case β is called the Alexandroff ideal boundary of R. Of course in general R^* cannot be made into a compact Riemann surface.

2C. Topological representatives. It is possible to give a canonical topological representative for any Riemann surface. Let P denote the extended complex plane. Let β be a closed subset of the Cantor ternary set contained in the interval $[0,1]$ of the real axis of P. Remove from P the subset β and any finite or countable collection of open disks Δ_n, with disjoint closures, in the upper half-plane. It is required that these disks not accumulate at any point of P except possibly some points of β. Remove also the open disks $\bar{\Delta}_n$ which are reflections of the disks Δ_n about the real axis.

Form a compact topological space R^* from $\mathsf{P} - \cup \Delta_n - \cup \bar{\Delta}_n$ by identifying the points z and $\bar{z}$ whenever z lies on the boundary of some Δ_n.

It is known that any Riemann surface R is homeomorphic to $R^* - \beta$ for some space R^* of the type constructed above (Richards [1]). The space R^* is called the Kerékjártó-Stoïlow compactification of R and β is the Kerékjártó-Stoïlow ideal boundary.

The number of disks Δ_n is called the *genus* of R. The *connectivity* of R is the number of points in β.

If R is closed then β is empty. Hence the genus g is finite, and we have R represented as a sphere with g handles placed symmetrically about the equator.

2D. Boundary components. It is possible to introduce the Kerékjártó-Stoïlow ideal boundary without appealing to the above canonical topological representatives.

Let R be an open Riemann surface. Let η be the collection of all regions Q on R such that Q is not relatively compact and yet the boundary of Q is compact.

A subset $\gamma \subset \eta$ is called an *ideal boundary component* if it satisfies the conditions:

(D1) If $Q_0 \in \gamma$ and $Q \supset Q_0$ then $Q \in \gamma$.

(D2) If $Q_1, Q_2 \in \gamma$ then there exists a $Q_0 \in \gamma$ such that $Q_0 \subset Q_1 \cup Q_2$.

(D3) $\bigcap_{Q \in \gamma} \operatorname{Cl} Q = \emptyset$.

Let β be the set of all ideal boundary components of R. On $R \cup \beta$ introduce a topology by means of the basis consisting of all open subsets of R and all sets of the form $Q \cup \{\gamma \in \beta | Q \in \gamma\}$. Then $R \cup \beta$ is the Kerékjártó-Stoïlow compactification and β is the Kerékjártó-Stoïlow ideal boundary.

3. Homology

3A. Chains. Let R be a Riemann surface. A 0-simplex is merely a point of R. A 1-simplex is an arc on R. A 2-simplex is a continuous map of the triangle $\Delta^2 = \{(x,y) | 0 \leq y \leq x \leq 1\}$ in the (x,y)-plane into R.

An n-chain is a finite formal sum of n-simplexes with integral coefficients. The set of n-chains forms an abelian group C_n.

For brevity we use in this book the above terms "simplex" and "chain" instead of the formal expressions "singular simplex" and "singular chain."

3B. Boundary operator. The boundary operator ∂ is a homomorphism of C_n into C_{n-1} $(n = 1, 2)$. For a 1-simplex σ^1 it is defined by $\partial\sigma^1 = \sigma^1(1) - \sigma^1(0)$. This defines ∂ on all of C_1 by linearity.

To define $\partial\sigma^2$ for a 2-simplex σ^2 we first consider some auxiliary mappings. Let l_1, l_2, l_3 be the functions from $[0,1]$ into the (x,y)-plane given by $l_1(t) = (t,0)$, $l_2(t) = (1,1-t)$, $l_3(t) = (1-t,1-t)$. Then set

$$\partial\sigma^2 = \sigma^2 \circ l_1 - \sigma^2 \circ l_2 + \sigma^2 \circ l_3.$$

By linearity this determines ∂ on C_2.

The group of cycles Z_1 is the kernel of $\partial | C_1$. The image of $\partial | C_2$ is the group of boundaries B_1. It is easily verified that $\partial\partial\sigma^2 = 0$. Hence $B_1 \subset Z_1$ and the quotient group $H_1 = Z_1/B_1$ is the *first homology group* of R.

A cycle σ^1 is *homologous to* 0 if $\sigma^1 \in B_1$. Two cycles σ^1, τ^1 are *homologous* if $\sigma^1 - \tau^1$ is homologous to 0. The set of cycles homologous to σ^1 is the *homology class* of σ^1.

3C. Intersection numbers. Let σ, τ be closed curves on R and assume they intersect each other "nicely." Then by using local parameters defined at the intersection points we can define the *intersection number* $\sigma \times \tau$ as the number of times τ crosses σ from left to right.

If σ, τ are arbitrary closed curves we may deform them slightly to obtain closed curves which intersect nicely. The intersection number $\sigma \times \tau$ is then defined in terms of the deformed curves. Of course one must show that the result is independent of the deformation.

This notion can be extended to cycles by requiring $\sigma \times \tau$ to be bilinear. The resulting function is skew symmetric, $\sigma \times \tau = -\tau \times \sigma$, and depends only on the homology class of σ, τ.

A standard way to introduce intersection numbers rigorously is to use simplicial approximation. It is known that every Riemann surface is triangulable and therefore there is a simplicial homology theory. The intersection number can be defined easily for simplicial cycles by using dual complexes. It is then shown that this number depends only on the simplicial homology classes of the cycles. Finally, the definition can be extended to singular cycles by using their simplicial approximations. For further details see Ahlfors-Sario [1, pp. 67–72].

3D. Canonical bases. A cycle σ on R is said to be a *dividing cycle* if for each compact subset $K \subset R$ there is a cycle in $R-K$ which is homologous to σ. If R is closed then no cycle is dividing.

A set of cycles $\sigma_1, \sigma_2, \cdots$ on R is said to be a *homology basis* if the cycles are homologously independent and if every cycle on R is homologous to a finite integral linear combination of them. The independence means that there is no nontrivial integral linear combination of them which is homologous to 0.

This set is a *homology basis modulo dividing cycles* if the σ_i are independent and if to each cycle τ on R there is a dividing cycle τ_0 such that $\tau - \tau_0$ is homologous to a finite integral linear combination of $\sigma_1, \sigma_2, \cdots$.

If R is a closed surface then there is a finite homology basis $A_1, B_1, A_2, B_2, \cdots, A_g, B_g$ such that $A_i \times A_j = 0 = B_i \times B_j, A_i \times B_i = 1$, and if $i \neq j$ then $A_i \times B_j = 0$. Such a basis is called *canonical.* By referring to the canonical topological representation of R (see 2C) the reader will be able to construct such a basis and recognize that g is the genus of R. If R is an open surface the corresponding fact is that there exists a finite or countable homology basis modulo dividing cycles, say $A_1, B_1, A_2, B_2, \cdots$, with the same intersection properties as above.

§2. ANALYSIS ON RIEMANN SURFACES

1. Harmonic functions

1A. Harmonic functions. Let R be a Riemann surface. A real-valued function u defined on a subregion of R is *harmonic* if it is locally the real part of a holomorphic function. Equivalently, for each local parameter z defined within the domain of u the function $u(z)$ is harmonic in the usual sense.

All the local properties of harmonic functions in the plane apply to such functions on Riemann surfaces. For example, a harmonic function on a Riemann surface cannot achieve its maximum unless it is a constant.

1B. Subharmonic functions. A real-valued upper semicontinuous function v is *subharmonic* in a region $G \subset R$ if for any harmonic function u in a region $G' \subset G$ the function $v-u$ fails to achieve a maximum in G' or else is a constant.

1C. The Dirichlet problem. Perron's method for solving the Dirichlet problem is often discussed in basic complex analysis (Ahlfors [7, p. 240]). It works equally well for Riemann surfaces as we shall see.

Let G denote a relatively compact subregion of R. Let f be a continuous real-valued function defined on ∂G. Form the family $\mathcal{V}$ of all subharmonic functions in G which satisfy

$$\limsup_{z \to \zeta} v(z) \leq f(\zeta) \tag{1}$$

for each $\zeta \in \partial G$. Let u be the upper envelope of $\mathcal{V}$. That is,

$$u(z) = \sup_{v \in \mathcal{V}} v(z).$$

Then u is harmonic in G (e.g. Ahlfors-Sario [1, p. 139]).

If there exists a barrier (*loc. cit.*) at a point $\zeta_0 \in \partial G$ then

$$\lim_{z \to \zeta_0} u(z) = f(\zeta_0).$$

Since the existence of a barrier at ζ_0 is a local property the criteria for plane regions will apply. In particular, if ∂G is a union of analytic curves then u extends to a continuous function on Cl G and agrees with f on ∂G. The function u with this property is obviously unique.

1D. Regular subregions and partitions. Let G be a subregion of R. If ∂G is a nonempty union of analytic curves we say Cl G is a *bordered subregion* of R. If in addition G is relatively compact and each component of $R-G$ is noncompact we say G is a *regular subregion* and Cl G is a *regular bordered subregion.* We note that there are no regular subregions if R is closed.

Let G be a regular subregion of an open Riemann surface R. If the boundary of each component of $R-G$ is connected then G is said to be a *canonical subregion.* It can be shown that this condition is equivalent to the requirement that each contour of ∂G be a dividing cycle.

Suppose that for each regular subregion Ω of R we have a partition $P(\Omega)$ of the contours of Ω. Such a system of partitions will be quite useful for later work if it satisfies the following conditions:

(D1) All contours of a component of $R-\Omega$ belong to the same part in $P(\Omega)$.

(D2) If $\Omega\subset\Omega'$ then all contours in a given part of $P(\Omega')$ belong to components of $\Omega'-\Omega$ whose contours on $\partial\Omega$ are all contained in the same part of $P(\Omega)$.

A family of partitions with these properties will be called a *consistent system.* An important example is the *canonical partition* $Q(\Omega)$ wherein two contours of $\partial\Omega$ belong to the same part if and only if they lie on the boundary of a single component of $R-\Omega$. A second illustration is provided by the identity partition I which collects all contours of $\partial\Omega$ into a single part.

Consider now the compactification $R\cup\beta$ of R (1.2C, D). Let P be a partition of β into closed subsets of β. Then P is called a *regular partition of the ideal boundary of R.* It induces a consistent system of partitions for the regular subregions Ω of R in the following way. Let us call two components Q_1, Q_2 of $R-\Omega$ related if the components Q_1^*, Q_2^* of $(R\cup\beta)-\Omega$ which contain them are such that there exist ideal boundary points γ_1, γ_2 which belong to the same part of P and satisfy $\gamma_1\in Q_1^*$, $\gamma_2\in Q_2^*$. Then $P(\Omega)$ is defined by the condition that two contours of $\partial\Omega$ belong to the same part of $P(\Omega)$ if and only if they lie on the boundaries of related components of $R-\Omega$. For the proof that $P(\Omega)$ is consistent see Ahlfors-Sario [1, pp. 87–90]. It is also proved there that every consistent system of partitions is induced in this way by a regular partition of the ideal boundary.

1E. Directed limits. It can be shown that for each compact subset $K \subset R$ there is a regular subregion of R which contains K (Ahlfors-Sario [1, p. 145]). Consequently the regular subregions form a directed set under inclusion (Kelley [1, p. 65]). Therefore directed limits can be defined as follows:

Suppose $\{f_\Omega\}$ is a family of functions indexed by all regular subregions Ω which contain some fixed compact set $K \subset R$. Assume that each f_Ω is a complex-valued function defined on Ω. Then

$$f(\zeta) = \lim_{\Omega \to R} f_\Omega(\zeta)$$

means that for each $\varepsilon > 0$ there is a regular subregion Ω_0 containing ζ such that $|f(\zeta) - f_\Omega(\zeta)| < \varepsilon$ whenever $\Omega \supset \Omega_0$.

The notion that $f_\Omega \to f$ uniformly on compacta of R is defined analogously. If each f_Ω is harmonic or holomorphic, then such convergence implies that f has the corresponding property.

A subfamily of the regular subregions is called *cofinal* if every compact subset of R is contained in some member of the subfamily. Clearly, we can take directed limits with respect to any cofinal family of regular subregions. In particular, it can be shown that the canonical subregions are cofinal among the regular subregions (cf. Ahlfors-Sario [1, p. 295]).

1F. Countability. A theorem of Radó [1] asserts that the underlying topological space of a Riemann surface R satisfies the second axiom of countability. That is, there is a countable base for the topology of R. Thus R is separable and also has the Lindelöf property: every open cover of R has a countable subcover.

An immediate consequence is that the normal family criteria from elementary complex analysis carry over to Riemann surfaces (e.g. see 3H). Thus a family of harmonic functions on R is normal if it is uniformly bounded on compacta.

Using Radó's theorem one can show that there is a *countable exhaustion* $\{\Omega_n\}_{n=1}^{\infty}$ of R by canonical subregions Ω_n. By this we mean that $\mathrm{Cl}\,\Omega_n \subset \Omega_{n+1}$ and $\cup \Omega_n = R$.

A simple proof of Radó's theorem can be obtained by using harmonic functions. Let V be a parametric disk on R, let $G = R - \bar{V}$, and let f be any nonconstant continuous function on ∂V. By modifying Perron's method we can obtain a nonconstant harmonic function u on G with boundary values f on ∂V, even though G is not relatively compact. Of course such a u will not be unique. For example, u can be defined as the upper envelope of the family $\mathcal{V}$ of subharmonic functions which satisfy (1) and also

$$\limsup_{v \to \beta} v(z) \leq 0$$

where β is the Alexandroff ideal boundary.

As a surface, G possesses a universal covering surface $\tilde{G}$ (see e.g. Ahlfors-Sario [1, p. 37]). By 1.1G we may consider $\tilde{G}$ as a Riemann surface and the projection π of $\tilde{G}$ onto G will be holomorphic. The function u lifts to a harmonic function $\tilde{u} = u \circ \pi$ on $\tilde{G}$. Since $\tilde{G}$ is simply connected there is a harmonic conjugate function $\tilde{u}^*$ such that $f = \tilde{u} + i\tilde{u}^*$ is holomorphic on $\tilde{G}$.

Now f makes $\tilde{G}$ a ramified covering of the plane region $f(\tilde{G})$. It follows that $\tilde{G}$ satisfies the second axiom of countability since $f(\tilde{G})$ does. As a consequence, G also satisfies this countability axiom since it is the image of $\tilde{G}$ under a local homeomorphism. Finally $R = G \cup \bar{V}$ satisfies the countability axiom since G and V do.

2. Differential forms

2A. Notational conventions. Let φ_α be a local parameter on an open set $U_\alpha \subset R$. For $\zeta \in U_\alpha$ write $z_\alpha = \varphi_\alpha(\zeta)$. As remarked in 1.1C, we often write $f(z_\alpha)$ in place of $f \circ \varphi_\alpha^{-1}(z_\alpha)$ when f is a function on U_α. We may also omit the subscript α and write $f(z)$ for a generic notation.

This identification of z with $\varphi^{-1}(z)$ is often carried a step further. Let $\zeta_0 \in R$. One speaks of the function $f(z) = z - \zeta_0$ defined in a neighborhood U of ζ_0. The precise meaning is of course $f(\zeta) =$

$\varphi(\zeta) - \varphi(\zeta_0)$ where φ is a local parameter in U. The function f actually depends on the choice of the parameter.

As another example consider a complex-valued function f defined on a subregion $G \subset R$. Then $f(\zeta)$ is a well-defined number for $\zeta \in R$. However, $f'(z)$, which by convention means the derivative of $f \circ \varphi^{-1}(z)$, depends on the choice of local parameter z and is not a function on R. But if f is holomorphic then $f'(z)dz$ has invariant meaning as we shall see.

2B. Differentials. A *zero order differential* on R is merely a complex-valued function. A *first order differential* ω on R is a rule which prescribes to each local parameter $z_\alpha = x_\alpha + iy_\alpha$ a formal expression

$$a_\alpha(z_\alpha)dx_\alpha + b_\alpha(z_\alpha)dy_\alpha. \tag{2}$$

The functions a_α, b_α are called *coefficients* of ω. In general they may be complex-valued. If $z_\beta = x_\beta + iy_\beta$ is another local parameter then z_β is a function of z_α, namely $z_\beta = \varphi_\beta \circ \varphi_\alpha^{-1}(z_\alpha)$ for $z_\alpha \in \varphi_\alpha(U_\alpha \cap U_\beta)$, and we require that the coefficients of ω satisfy the transformation conditions

$$\begin{aligned} a_\alpha &= a_\beta \frac{\partial x_\beta}{\partial x_\alpha} + b_\beta \frac{\partial y_\beta}{\partial x_\alpha}, \\ b_\alpha &= a_\beta \frac{\partial x_\beta}{\partial y_\alpha} + b_\beta \frac{\partial y_\beta}{\partial y_\alpha}, \end{aligned} \tag{3}$$

throughout $\varphi_\alpha(U_\alpha \cap U_\beta)$.

The relations (3) follow formally by writing

$$a_\alpha dx_\alpha + b_\alpha dy_\alpha = a_\beta dx_\beta + b_\beta dy_\beta$$

and then expressing dx_β and dy_β in terms of dx_α and dy_α. Thus we write in generic notation

$$\omega = a\,dx + b\,dy. \tag{4}$$

If all the coefficients of ω are infinitely differentiable we say ω is *smooth*.

For example, if f is a function on R of class C^∞ then

$$df = \frac{\partial f}{\partial x} dx + \frac{\partial f}{\partial y} dy$$

is a smooth first order differential on R.

Two differentials may be added by adding corresponding coefficients. Obviously the sum will again be a differential. For a function f on R we may define $f\omega = (fa)dx + (fb)dy$ if ω is given by (4). With these definitions the set of smooth first order differentials forms a module over the ring of C^∞ functions on R. We also define the complex conjugate of ω by

$$\bar{\omega} = \bar{a}\, dx + \bar{b}\, dy$$

and the star conjugate by

$$*\omega = -b\, dx + a\, dy.$$

These operations also lead to differential forms.

Let u be a harmonic function on R and suppose that u has a single-valued harmonic conjugate u^* on R. The reader should check that $*du = d(u^*)$.

2C. A second order differential Ω associates a form

$$c_\alpha(z_\alpha)\, dx_\alpha dy_\alpha$$

to each local parameter $z_\alpha = x_\alpha + iy_\alpha$ on R. On $\varphi_\alpha(U_\alpha \cap U_\beta)$ the coefficients c_α must satisfy

$$c_\alpha = c_\beta \frac{\partial(x_\beta, y_\beta)}{\partial(x_\alpha, y_\alpha)}. \tag{5}$$

In generic notation we write simply $\Omega = c\, dxdy$. If each c_α is continuous, we say Ω is continuous. Such Ω's form a module over the ring of continuous functions on R with the obvious definitions for sum and product.

2D. Exterior algebra. Let $\omega_1 = a_1 dx + b_1 dy$, $\omega_2 = a_2 dx + b_2 dy$

be first order differentials. Their *exterior product* is

$$\omega_1 \wedge \omega_2 = (a_1 b_2 - a_2 b_1)\, dx dy.$$

One easily checks that $\omega_1 \wedge \omega_2$ is then a second order differential. Furthermore $\omega_1 \wedge \omega_2 = -\omega_2 \wedge \omega_1$.

If ω of (4) is smooth, then we define the *exterior derivative* of ω to be

$$d\omega = \left(\frac{\partial b}{\partial x} - \frac{\partial a}{\partial y}\right) dx dy.$$

It is easily seen to be a continuous second order differential. If f is of class C^∞ then one has

$$\begin{aligned} d(f\omega) &= df \wedge \omega + f\, d\omega, \\ d(df) &= 0. \end{aligned} \tag{6}$$

2E. A smooth first order differential ω is *closed* if $d\omega = 0$. It is *exact* if $\omega = df$ for a C^∞ function f on R. If $*\omega$ is closed then we say ω is *coclosed*. From (6) we see that an exact differential is closed. For a second order differential Ω we set $d\Omega = 0$.

If ω is of the form $a\, dx + ia\, dy$, then we write $\omega = a\, dz$ and say ω is *pure*. If a pure differential $a\, dz$ is closed then one sees immediately that a is holomorphic. In this case we say ω is *holomorphic* or *analytic*. In particular, if f is a holomorphic function on R then $df = f'(z) dz$ and df is holomorphic.

3. Integration

3A. Line integrals. Let σ be a smooth 1-simplex on R. Let $\omega = a\, dx + b\, dy$ be a smooth differential. We define

$$\int_\sigma \omega = \int_0^1 [a(z(t))x'(t) + b(z(t))y'(t)]dt \tag{7}$$

if the image of σ lies entirely within the domain of the parameter z. Here $z(t) = \varphi(\sigma(t))$. In general we may subdivide [0,1] into a finite number of subintervals so that σ maps each subinterval into the domain of some local parameter. Then we may reparametrize each subinterval by [0,1] and define $\int \omega$ over a subinterval as in (7). Then $\int_\sigma \omega$ is defined as the sum of the integrals over the subregions. The compatibility conditions (3) show that our definition is independent of the subdivision, the local parameters, and the reparametrizations.

If $\sigma = \sum n_i \sigma_i$ is a smooth 1-chain then set

$$\int_\sigma \omega = \sum n_i \int_{\sigma_i} \omega.$$

Thus $\int \omega$ is a linear functional on the module C_1 of smooth 1-chains. We say that $\int_\sigma \omega$ is the *period* of ω along σ.

3B. Area integrals. If $\Omega = c\, dxdy$ is a continuous second order differential and σ^2 is a smooth 2-simplex whose image is contained within the domain of a local parameter z we define

$$\iint_{\sigma^2} \Omega = \iint_{\Delta^2} c(x(u,v),y(u,v)) \frac{\partial(x,y)}{\partial(u,v)}\, dudv$$

where Δ^2 is the domain of σ^2 in the (u,v)-plane. By subdividing each Δ^2 we can extend the definition to any smooth 2-simplex. By linearity it is extended to smooth 2-chains. The transformation equation (5) then shows that integration is well defined.

3C. Stokes' theorem. Let σ^2 be a smooth 2-chain and let ω be a smooth first order differential. Then Stokes' theorem for the triangle Δ^2 leads immediately to the general form of Stokes' theorem

(8) $$\iint_{\sigma^2} d\omega = \int_{\partial\sigma^2} \omega.$$

By (6) we have also the formula for integration by parts

(9) $$\iint_{\sigma^2} df \wedge \omega = \int_{\partial\sigma^2} f\omega - \iint_{\sigma^2} f\, d\omega.$$

3D. Let ω be a closed differential. By Stokes' theorem $\int_\sigma \omega = 0$

if σ is homologous to zero. On a simply connected subregion G of R we may define

$$u(\zeta) = \int_{\zeta_0}^{\zeta} \omega$$

where $\zeta_0 \in G$ and the integral is taken over any path from ζ_0 to ζ. Then u is a well-defined smooth function on G and $du = \omega$ there. Thus a closed differential is locally exact. Similarly we see that if all the periods of ω are zero then ω is exact on R.

A differential ω is *harmonic* if it is closed and coclosed. The reader should check that an equivalent condition is that locally $\omega = du$ where u is a complex-valued harmonic function. That is, for each simply connected region G there are real-valued harmonic functions p, q on G such that $\omega = d(p+iq)$.

Similarly, if ω is holomorphic then locally $\omega = df$ where f is a holomorphic function.

3E. Integration over open sets. Let G be the interior of a compact bordered subregion of R. Let Ω be a continuous second order differential and ω a smooth first order differential. We wish to define

$$\iint_G \Omega \tag{10}$$

and prove Stokes' theorem

$$\iint_G d\omega = \int_{\partial G} \omega. \tag{11}$$

There are two ways to proceed. It can be shown that there is a smooth 2-chain $\sigma^2 = \sum \sigma_i^2$ such that each 2-simplex σ_i is a homeomorphism and their images form a triangulation of G. Then $\partial G = \partial \sigma^2$ if the contours of ∂G are oriented so that G lies to the left of them. Thus we may define $\iint_G \Omega = \iint_{\sigma^2} \Omega$ and it is clear from the transformation equations (5) that this definition is independent of the particular σ^2 chosen. Now (11) follows from (8).

The second way to handle the problem is as follows. We first note that the definition of (10) is obvious if Ω is zero outside of a parametric disk on R. Thus if $\{e_i\}$ is a smooth partition of unity such that each e_i has its support within a parametric disk we can define

$$\iint_G \Omega = \iint_G \sum e_i\Omega = \sum \iint_G e_i\Omega.$$

The proof of (11) by means of a partition of unity is very easy (Springer [1, pp. 158–163]).

The second method is perhaps more elegant. However, the first method yields also the fact that the oriented boundary of a relatively compact bordered subregion is homologous to zero.

Having defined (10) in one way or the other we then define

$$\iint_R \Omega = \lim_{G\to R} \iint_G \Omega \tag{12}$$

if the limit exists. It is immaterial whether this directed limit is with respect to all regular subregions G or all canonical subregions.

Since the Jacobian appearing in the transformation equation (5) is positive, $|\Omega| = |c|dxdy$ is also a second order differential. It can be seen that the limit (12) exists if and only if $\lim_{G\to R}\int\int|\Omega|$ exists.

3F. The Dirichlet integral. For a smooth ω we define its norm $||\omega||$ by

$$||\omega||^2 = \iint_R \omega \wedge *\bar{\omega}.$$

This integral exists or is $+\infty$. The space of all such norm-finite ω is unitary with hermitian inner product

$$(\omega_1,\omega_2) = \iint_R \omega_1 \wedge *\bar{\omega}_2.$$

Let u be a real harmonic function on R. Then we write $D(u) =$

$||du||^2$ and refer to $(D(u))^{1/2}$ as the *Dirichlet norm* of u. Note that it is not actually a norm on the harmonic functions since $D(u) = 0$ if u is any constant.

If G is a regular subregion of R we write

$$D_G(u) = \iint_G du \wedge *du.$$

In generic notation this may be expressed as

$$D_G(u) = \iint_G \left[\left(\frac{\partial u}{\partial x}\right)^2 + \left(\frac{\partial u}{\partial y}\right)^2\right] dxdy.$$

For harmonic u, Stokes' theorem (11) yields

(13) $$D_G(u) = \int_{\partial G} u * du$$

and

(14) $$\int_{\partial G} * du = 0.$$

If u, v are harmonic functions we define the mixed Dirichlet integrals $D(u,v) = (du,dv)$ and $D_G(u,v) = \iint_G du \wedge *\overline{dv}$. Thus $D(u) = D(u,u)$. For real harmonic u, v we have

(15) $$D_G(u,v) = \int_{\partial G} u * dv = \int_{\partial G} v * du$$

by Stokes' theorem and the symmetry of the inner product for real forms. The formulas (13), (14), (15) are referred to as Green's formulas.

The reader should be familiar with the formula

$$* du = \frac{\partial u}{\partial n} ds$$

which holds along any arc with right normal vector n and arc length element ds (Ahlfors [7, p. 162]). The formulas

$$d*du = \Delta u\, dxdy, \qquad du \wedge *du = |\operatorname{grad} u|^2 dxdy$$

are also easily derived.

If f is holomorphic on G we define $D_G(f) = D_G(u)$ where u is the real part of f. Thus in generic notation

$$D_G(f) = \iint_G |f'(z)|^2 dxdy.$$

Geometrically $D_G(f)$ is the area of the ramified covering given by $f:G \to \mathbf{C}$. Thus if G is a punctured disk and $D_G(f) < \infty$ we know that f cannot have a pole or essential singularity at the puncture. The reader may verify this analytically by computing $D_G(f)$ explicitly in terms of the Laurent expansion.

As another illustration of the use of the Dirichlet integral we mention a useful inequality connecting it with $|f'|$. Let Δ denote the unit disk and suppose f is holomorphic in Δ. Since $(f')^2$ is a complex-valued harmonic function it possesses the mean value property

$$f'(z_0)^2 = \frac{1}{2\pi} \int_0^{2\pi} f'(z_0 + re^{i\theta})^2 d\theta$$

for $|z_0| < \frac{1}{2}$, $0 < r < \frac{1}{2}$. Multiply by r and integrate over $r \in (0, \frac{1}{2})$. This gives

$$f'(z_0)^2 = \frac{4}{\pi} \iint_{|z-z_0|<1/2} f'(z)^2 dxdy,$$

and hence

$$(16) \qquad |f'(z_0)| \le (4\pi^{-1} D_\Delta(f))^{1/2}.$$

Thus if v is any harmonic function in Δ and z_1, z_2 are points of $\Delta' = \{|z| < \frac{1}{2}\}$ we may apply (16) to a holomorphic function f whose real part is v. Since $|v(z_1) - v(z_2)| \le \int_{z_1}^{z_2} |df|$ we obtain

$$(17) \qquad |v(z_1) - v(z_2)| \le (4\pi^{-1} D_\Delta(v))^{1/2} \qquad (z_1, z_2 \in \Delta').$$

Next we consider some consequences of this inequality.

3G. Convergence in Dirichlet norm. The exact harmonic differentials of finite norm form a complete metric space. In terms of harmonic functions this fact can be stated in a slightly stronger form. Let G be an arbitrary subregion of R and suppose that for each sufficiently large regular subregion Ω we are given a function u_Ω harmonic in G. Note that we may take directed limits of families such as $D_G(u_\Omega)$ as $\Omega \to R$. Indeed, the definition in 1E may be applied to the constant function $f_\Omega(\zeta) = D_G(u_\Omega)$. Assume that $\lim D_G(u_\Omega - u_{\Omega'}) = 0$ as $\Omega, \Omega' \to R$. We shall show that there is a harmonic function u on G such that as $\Omega \to R$ one has for $z_0 \in G$

(G1) $D_G(u_\Omega - u) \to 0$,

(G2) $D_G(u_\Omega) \to D_G(u)$,

(G3) $u_\Omega(z) - u_\Omega(z_0) \to u(z) - u(z_0)$ uniformly on compacta of G.

The function u is unique up to addition of a constant if G is connected.

Let K be a compact subset of G. Let V be a parametric disk on R represented by $\Delta = \{|z| < 1\}$ and let V' be the subdisk corresponding to $\Delta' = \{|z| < \frac{1}{2}\}$. The set K can be covered by a finite number of such V'. Pick a point in each V' and join it to z_0 by an arc. Each such arc can again be covered by finitely many V'. Hence there is an N such that any $z \in K$ can be connected to z_0 by $\leq N$ disks V'. On applying (17) to $v = u_\Omega - u_{\Omega'}$ we obtain

$$|u_\Omega(z) - u_{\Omega'}(z) - u_\Omega(z_0) + u_{\Omega'}(z_0)| \leq N(4\pi^{-1} D_G(u_\Omega - u_{\Omega'}))^{1/2}$$

for all $z \in K$. Therefore $u_\Omega(z) - u_\Omega(z_0)$ converges uniformly on K to a harmonic function $u(z)$. This proves (G3).

Since uniform convergence of harmonic functions implies uniform convergence of their partial derivatives we also have (G1).

Recall (3F) that $(D_G(u))^{1/2}$ is a norm on du. Hence by the triangle inequality, assuming these integrals are finite, we obtain

$$(D_G(u))^{1/2} \leq (D_G(u - u_\Omega))^{1/2} + (D_G(u_\Omega))^{1/2}.$$

It follows that $\lim_{\Omega \to R} D_G(u_\Omega) \geq D_G(u)$. By symmetry the opposite inequality also holds and (G2) is proved.

For the uniqueness suppose that $\tilde{u}$ also satisfies the condition (G1) for u. By the triangle inequality $D_G(u-\tilde{u}) = 0$. This can happen only if $u-\tilde{u}$ is a constant on each component of G.

3H. A normal family criterion. Let $\{u_n\}$ be a family of harmonic functions on R such that $D(u_n)$ and $u_n(z_0)$ are bounded for some $z_0 \in R$. Then the family is normal.

For the proof we observe by (17) and the boundedness of $u_n(z_0)$ that $\{u_n\}$ is uniformly bounded on compacta. From this the assertion follows.

CHAPTER I
THE NORMAL OPERATOR METHOD

In this chapter the basic tools are created which will be used throughout the remainder of the book. The central topic is the Main Existence Theorem for principal functions, given in §1. The hypotheses of this theorem require the existence of normal operators. That such operators always exist is a nontrivial fact; its proof is given in §2 by constructing the operators L_0 and L_1 on an arbitrary Riemann surface. The method used there, which is typical of such problems, consists of constructing operators on compact bordered subregions and passing to a limit.

The Main Existence Theorem applied to a normal operator yields a principal function. When the operator is defined by a limiting process a harmonic function may also be obtained by applying the main theorem to each approximating operator and forming a limit of the resulting functions. It will be important to know when these processes commute. A general criterion to this effect is given in §2. That this criterion applies to L_0 and L_1 is the main result of §3. The final section treats some related topics and open problems. The results of that section are not required for the succeeding chapters.

The normal operator method and principal functions were introduced by Sario [8], [11], [27]; the algebraic existence proof and the general convergence theorems by Rodin-Sario [2]. The estimates for q were obtained by Oikawa [2], [4], the study of the space of normal operators was initiated by Ahlfors [4], and the use of orthogonal projection in the existence proof was given by Nakai-Sario [1].

§1. THE MAIN EXISTENCE THEOREM

After proving a lemma on harmonic functions we proceed directly to the definition of normal operators and a proof of the main theorem. Since no examples or applications are given in this section the reader is referred to the Introduction (page 1) for preliminary orientation.

1. A lemma on harmonic functions

1A. The q-lemma. The so-called rigidity theorems of classical function theory show that analytic and harmonic functions cannot be distorted too drastically. Perhaps the simplest such statement is Harnack's inequality which expresses the fact that a positive harmonic function in the unit disk $\{|z|<1\}$ cannot grow faster than $(1-|z|)^{-1}$. Of the same nature is the following estimate (Sario [8], [11]):

Lemma. *Let α be a compact subset of a Riemann surface R. There exists a positive constant $q<1$ such that all harmonic functions u on R which change sign on α satisfy*

$$q \inf_R u \le u|\alpha \le q \sup_R u. \tag{1}$$

The important feature of this lemma is that q depends only on α and R. Some recent research (Oikawa [2], [4]) has been concerned with the problem of estimating the best possible value for q in terms of conformal-geometric invariants of α and R. Except for the case when R is simply connected the question is not completely settled. This topic will be discussed more fully in §4.

1B. First we show that to each compact set α on a Riemann surface R there corresponds a constant $c>1$ such that

$$c^{-1} \le v(\xi)/v(\eta) \le c \tag{2}$$

for all points ξ and η in α and all positive harmonic functions v. We make use of Harnack's inequality: If v is a positive harmonic function in the unit disk then

$$\frac{1-|z|}{1+|z|}\, v(0) \le v(z) \le \frac{1+|z|}{1-|z|}\, v(0).$$

Note that we may replace α by a larger connected compact subset, for example by using an exhaustion of R. Thus we may assume at the start that α is connected. By compactness α can be covered by a finite number n of parametric disks Δ_i with centers ζ_i such that the subdisks Δ_i' corresponding to $\{|z|<\frac{1}{2}\}$ also form an open cover of α. By Harnack's inequality $\frac{1}{3}<v(\xi_i)/v(\zeta_i)<3$ for $\xi_i \in \Delta_i'$, and consequently $\frac{1}{9}<v(\xi_i)/v(\eta_i)<9$ if ξ_i and η_i are in Δ_i'. For any pair of points ξ and η in α there is a sequence $\xi = \pi_0, \cdots, \pi_n = \eta$ from α such that for each j $(0 \leq j \leq n-1)$ the points π_j and π_{j+1} belong to the same Δ_i'. In fact for fixed ξ the set of such η's is seen to be relatively open and closed in α. Thus 9^n can be a value for c in (2).

On applying the left inequality in (2) to a function $1-u$ we obtain $u(\xi) \leq c^{-1}u(\eta)+(1-c^{-1})$ if $u \leq 1$ on R. Consequently, there is a positive $q<1$ such that $u(\xi) \leq (1-q)u(\eta)+q \sup_R u$ for any harmonic function u on R, as may be seen by applying the previous inequality to $u/\sup_R u$ and setting $q = 1-c^{-1}$. This last formula may be applied to $-u$ and yields

$$q \inf_R u+(1-q)\, u(\eta') \leq u(\xi) \leq (1-q)\, u(\eta)+q \sup_R u$$

for any $\xi, \eta, \eta' \in \alpha$. Since u changes sign on α we may choose η, η' so that $u(\eta) \leq 0 \leq u(\eta')$ thus obtaining (1). Actually, we have proved the stronger inequality

$$q \inf_R u+(1-q) \max_\alpha u \leq u|\alpha \leq (1-q) \min_\alpha u+q \sup_R u.$$

1C. A short proof of a weaker form of Lemma 1A can be based on normal families. We shall show that $\max_\alpha |u| \leq q \sup_R |u|$ for all harmonic u which change sign on α.

It suffices to prove $\max_\alpha |u| \leq q$ for all such u which satisfy $|u|<1$. These functions form a normal family. If the assertion were false there would exist a subsequence $\{u_n\}$ with $\max_\alpha |u_n| \to 1$. Any limit function v would satisfy $\max_\alpha |v| = 1 = \sup_R |v|$, hence $v = \pm 1$. But it is impossible that $u_n \to \pm 1$ uniformly on α since each u_n changes sign there.

2. The main theorem

2A. Normal operators. Let A be a union of disjoint bordered Riemann surfaces. Let α denote the border of A, $C(\alpha)$ the space of continuous real-valued functions on α, and $H_1(A)$ the space of real-valued functions which are continuous on A and harmonic on its interior. In a natural way we may consider $C(\alpha)$ and $H_1(A)$ as real vector spaces and introduce (Sario [8]):

Definition. *A normal operator L for A is a linear transformation of $C(\alpha)$ into $H_1(A)$ such that for all $f \in C(\alpha)$*

(A1) $(Lf)|\alpha = f$,

(A2) $\min_\alpha f \leq Lf \leq \max_\alpha f$,

(A3) $\int_\beta *dLf = 0$.

In condition (A3) β is any cycle on A homologous to α. By Green's formula the integral is then well defined; it is called the flux of Lf across the ideal boundary of A.

The family of all normal operators for A is a convex set. In the next section we show that it is nonempty. Ahlfors [4] has posed some problems concerning a topology for this set and a characterization of its extreme points. We shall return to these interesting questions in §4.

Remark. Condition (A2) has the following consequence. Suppose α_0 is a compact subset of A and the linear spaces $C(\alpha)$, $C(\alpha_0)$ are normed by the sup norms $||\cdot||_\alpha$, $||\cdot||_{\alpha_0}$. Then the composite mapping $r_{\alpha_0} \circ L$, where $r_{\alpha_0}: H_1(A) \to C(\alpha_0)$ denotes restriction to α_0, is a bounded linear transformation of $C(\alpha)$ into $C(\alpha_0)$ and hence continuous with respect to these topologies of uniform convergence.

2B. The main theorem. Let us agree to call a subset A of a Riemann surface R a *bordered boundary neighborhood* if $R - A$ is a regular subregion of R. We denote by $H(R)$ the vector space of harmonic functions on R and, as in 2A, by $H_1(A)$ the space of functions which are continuous on A and harmonic on its interior. We are ready to state the Main Existence Theorem for principal functions (Sario [8]):

Theorem. *Let L be a normal operator for a bordered boundary neighborhood A of a Riemann surface R, and let $s \in H_1(A)$. A necessary and sufficient condition for the existence of a principal function*

$p \in H(R)$ characterized by

$$(3) \qquad p - s = L(p-s) \quad \textit{in} \quad A$$

is $\int_\beta *ds = 0$. The function is uniquely determined up to an additive constant. It reduces to a constant if and only if $s \equiv Ls$.

To be precise (3) should be written $p|A - s = L(p|\alpha - s|\alpha)$ where $\alpha = \partial A$, but we shall often omit these references to restriction operators when it is safe to do so. The function p will also be referred to as the *L-principal function* corresponding to the singularity s; in symbols $p = p[s,L]$.

The theorem may be given an algebraic formulation. Recall that the cokernel of a transformation L is the quotient space of the range of L by the image of L; in symbols coker L = rg L/im L. There is a natural linear transformation of $H(R)$ into coker L by $p \to p|A +$ im L. Theorem 2B is equivalent to the statement that this transformation has as its kernel the space of real-valued constant functions and as its image the space of cosets $s +$ im L with $\int_\beta *ds = 0$. We shall give the proof in terms of this reformulation.

2C. Suppose p is harmonic on R and in the kernel of the mapping $p \to p|A +$ im L. Then $p|A \in$ im L and so achieves its maximum on $\alpha = \partial A$ according to (A2). This maximum also dominates $p|R - A$. Since p has a maximum at an interior point of R it is constant. By (A2) every constant is in im L and therefore the kernel of the mapping in question is precisely the space of real-valued constant functions.

The flux $\int_\beta *ds$ of a coset $s +$ im L is independent of the representative s according to (A3). By Green's formula $\int_\beta *dp = 0$ for any $p \in H(R)$. Hence the image of the mapping $p \to p|A +$ im L is contained in the subspace of coker L consisting of cosets with vanishing flux. To complete the proof we must show that every coset with vanishing flux has a global representative, that is, a representative which can be extended harmonically to all of R.

2D. Reduction of the problem. Given a coset $s +$ im L in coker L with $\int_\beta *ds = 0$ we must find a p harmonic on R such that $p - s$ is in the image of L. Without loss of generality we may assume s vanishes on α since $s - Ls$ does and it is a representative of the same coset. Let A_0 be the closure of a regular subregion of R which contains Cl $(R-A)$ in its interior. We shall now show that our problem can be reduced to that of finding a function $\tilde{p}$ on $\alpha_0 = \partial A_0$ with the property

(4) $\tilde{p} = LK\tilde{p}+s$

where K is the "Dirichlet operator" which associates to each continuous function f on α_0 the solution of the Dirichlet problem in A_0 with boundary values f. The notation $LK\tilde{p}$ in (4) is of course an abbreviation for $L((K\tilde{p})|\alpha)|\alpha_0$ and s stands for $s|\alpha_0$.

, Suppose $\tilde{p}$ is a continuous function on α_0 which satisfies (4). Then the functions $K\tilde{p}$ and $LK\tilde{p}+s$, with domains A_0 and A respectively, are equal to $K\tilde{p}$ on α and $\tilde{p}$ on α_0. By the maximum principle they must therefore be identical in $A_0\cap A$. Hence

(5) $p = \begin{cases} K\tilde{p} & \text{in } A_0, \\ \\ LK\tilde{p}+s & \text{in } A, \end{cases}$

is a well-defined harmonic function on R. Since p has the representation $LK\tilde{p}+s$ in A it will evidently serve as the required global representative of the given coset $s+$ im L.

2E. An invertible operator. Let T be the linear operator $f \rightarrow LKf$ of $C(\alpha_0)$ into itself. Let I be the identity operator on this space. In order to solve equation (4) we must show that s is in the image of $I-T$.

Recall that if X is a Banach space and $T: X \rightarrow X$ is a linear operator with $||T|| < 1$ then $I-T: X \rightarrow X$ has an inverse $(I-T)^{-1}$ whose norm is no greater than $(1-||T||)^{-1}$. Indeed, $(I-T)^{-1} = I+T+T^2+\cdots$. In our case $C(\alpha_0)$ becomes a Banach space under the supremum norm $||\cdot||_{\alpha_0}$. However, the inequality $||T|| < 1$ needed to prove that $I-T$ is invertible, and hence that $s \in$ im $(I-T)$, is not valid. Clearly this difficulty can be overcome if we exhibit a subspace X of $C(\alpha_0)$ with the following properties:

(a) X is a Banach subspace.
(b) When restricted to X the operator T has norm <1.
(c) $T(X) \subset X$.
(d) $s \in X$.

When this is accomplished we shall have proved Theorem 2B and we shall also have the estimate

$$||(I-T)^{-1}|| \leq (1-||T||)^{-1} \tag{6}$$

which will be used later (2G).

2F. Existence of X. Let ω be the harmonic function on $A \cap A_0$ with boundary values 0 on α and 1 on α_0. Let X be the kernel of the continuous linear functional $f \to \int_{\alpha_0} f * d\omega$ on $C(\alpha_0)$. Then (a) holds since X is a closed subspace of $C(\alpha_0)$.

If we prove that Kf changes sign on α then (b) will follow. Indeed, we could then apply the q-lemma to obtain

$$||Tf||_{\alpha_0} = ||LKf||_{\alpha_0} \leq ||Kf||_{\alpha} \leq q||f||_{\alpha_0}.$$

Before verifying that $Kf|\alpha$ changes sign we pause to derive a useful property of normal operators. Let g be continuous on α. Green's formula applied to the set $A_0 \cap A$, with oriented boundary $\alpha_0 - \alpha$, and to the functions Lg, ω yields

$$\int_{\alpha} g * d\omega = \int_{\alpha_0} (Lg) * d\omega.$$

A measure-theoretic interpretation of this formula will be discussed in 4.3. Note that it implies

$$\int_{\alpha_0} f * d\omega = \int_{\alpha} (Kf) * d\omega$$

since K is also a normal operator.

Along α the measure $*d\omega$ is positive. Thus the condition $f \in X$, which is equivalent to the vanishing of $\int_{\alpha} (Kf) * d\omega$, implies that Kf is not of constant sign on α. This proves (b).

From the two preceding formulas it is seen that $f \in X$ implies $LKf \in X$. This proves (c).

Since $s = 0$ along α we have

$$\int_{\alpha_0} s * d\omega = \int_{\alpha_0} * ds$$

by Green's formula for s and ω. Hence the hypothesis that s has vanishing flux leads to the condition $s \in X$. Thus (d) holds and the proof of the Main Existence Theorem is complete.

2G. The main theorem with estimates. From the above proof we see that the principal function p is determined up to an additive constant by the function $\tilde{p}$ on α_0. Indeed, one such p was defined by (5). Suppose we have a bound for $\tilde{p}$, say $|\tilde{p}| \leq M$. Then by (5) and the normalization $s|\alpha = 0$ we obtain bounds for p

$$||p||_{A_0} \leq M, \qquad ||p-s||_A \leq M.$$

We have seen that $\tilde{p} = (I-T)^{-1}s$ where $T = LK$ is restricted to X. From (6) we obtain $||\tilde{p}||_{\alpha_0} \leq (1-||T||)^{-1}||s||_{\alpha_0}$. In 2F we found that $||T|| \leq q$ where $q < 1$ is the constant determined by the compact subset α of the interior of A_0. We have proved (Sario [27]):

Theorem. *In addition to the hypothesis of Theorem* 2B *assume* $s|\alpha = 0$. *Then the principal function* p *may be normalized by an additive constant so that*

$$||p||_{A_0} \leq (1-q)^{-1}||s||_{\alpha_0},$$

$$||p-s||_A \leq (1-q)^{-1}||s||_{\alpha_0}.$$

The constant q *is determined by Lemma* 1A *applied to* α *and the interior of* A_0.

The normalization on p which makes this theorem apply is given by (5) or, equivalently, by the condition $\int_{\alpha_0} p * d\omega = 0$ (see 2F).

§2. NORMAL OPERATORS

This section deals with the existence of certain canonically defined normal operators. They may be constructed on a compact bordered Riemann surface using solutions to suitable Dirichlet problems. Properties of the Dirichlet norm are the main tools used in extending these operators to open surfaces.

1. Operators on compact bordered surfaces

1A. The operator L_0. Let A denote a bordered Riemann surface with compact border α. We wish to construct normal operators $C(\alpha)\to H_1(A)$. At first we consider a special case and assume that A is obtained from a compact bordered Riemann surface B by omitting a set β of contours of B. Thus $\partial B = \alpha\cup\beta$. We also assume β is not empty since that is the only case of interest here.

When a function u on the interior B^0 of B has a continuous extension to A and a C^1 extension to $B^0\cup\beta$ we shall often refer to the boundary values on α and to the normal derivative on β of these extensions simply as those of u.

Let $\hat{A}$ be the compact bordered surface obtained by welding together two copies of B along β. For $\zeta\in\hat{A}$ let ζ^* denote its image under the natural involutory self-mapping of $\hat{A}$; thus $\partial\hat{A} = \alpha\cup\alpha^*$. We consider A as a subset of $\hat{A}$.

Suppose f is a continuous real-valued function on α. Let $L_0 f$ be the restriction to A of the solution to the Dirichlet problem in $\hat{A}$ with boundary values $f(\zeta) = f(\zeta^*)$ on $\partial\hat{A}$. Obviously $L_0\colon C(\alpha)\to H_1(A)$ is a linear transformation and the other conditions for a normal operator are easily verified. The operator L_0 may be characterized by the properties that $L_0 f$ is equal to f on α and has identically vanishing normal derivative on β.

1B. The operator L_1. Consider α, β as 1-chains on B, oriented so that $\partial B = \beta-\alpha$. Let $f\in C(\alpha)$. We shall show that there exists a unique $u\in H_1(B)$ such that $u|\alpha = f$, $u|\beta = c$ where c is a constant, and $\int_\beta *du = 0$. The mapping $f\to u|A$ will be seen to define a normal operator $L_1\colon C(\alpha)\to H_1(A)$.

Fix $f\in C(\alpha)$ and consider the transformation of the real number field into itself given by $c\to\int_\beta *du_c$ where u_c is the solution of the Dirichlet problem in B with boundary values f on α, c on β. This transformation is continuous. Indeed, by the reflection principle u_c extends to $\hat{A}$ where it is characterized in $H_1(\hat{A})$ by its boundary values $f(\zeta)$ on α and $2c-f(\zeta^*)$ on α^*. Therefore $\lim c_n = c$ implies $\lim u_{c_n} = u_c$ uniformly on $\hat{A}$, which in turn gives $\lim \int_\beta *du_{c_n} = \int_\beta *du_c$.

If $c = \min_\alpha f$ then clearly the normal derivative $\partial u_c/\partial n$ is ≤ 0 along β and hence $\int_\beta *du_c \leq 0$. Similarly the flux of u_c across β is nonnegative if c is $\max_\alpha f$. Therefore the flux of u_c vanishes for some constant c between these extremes. Define $L_1 f = u_c$ for this choice of c.

To see that c is unique consider $v = u_c - u_{c'}$ for another such constant c'. Then $v \in H_1(B)$, $v|\alpha = 0$, $v|\beta =$ constant, and $\int_\beta *dv = 0$. For the Dirichlet integral of v we have $D(v) = \int_{\beta-\alpha} v * dv = 0$ and therefore $v \equiv 0$ and $u_c \equiv u_{c'}$.

Using the maximum principle one easily checks that L_1 is a normal operator. It will occasionally be convenient to write $L_1[A]$ to emphasize that L_1 is the above defined operator for A.

1C. Partitions of β. Denote the components of β by $\beta_1, \cdots, \beta_n$ and orient them so that $\beta = \beta_1 + \cdots + \beta_n$. Let these β_j be collected into mutually disjoint sets $\beta^1, \cdots, \beta^k$. In terms of homology, $\beta = \sum_1^k \beta^i$ and each β^i is a sum of cycles on the border of B. Designate this partition by P.

Lemma. *Let $f \in C(\alpha)$. There exist uniquely determined constants $c_1, \cdots, c_k$ such that the function in $H_1(B)$ with boundary values f on α and c_i on β^i for $1 \leq i \leq k$ has vanishing flux across each β^i.*

For the proof consider, in the interior B^0 of B, disjoint bordered neighborhoods $U, V_1, \cdots, V_k$ of $\alpha, \beta^1, \cdots, \beta^k$, respectively, with corresponding borders $\delta, \gamma_1, \cdots, \gamma_k$. Thus on B we have $\partial U = \delta \cup \alpha$, $\partial V_i = \gamma_i \cup \beta^i$. Let $G = U \cup V_1 \cup \cdots \cup V_k$. Then G is a bordered boundary neighborhood in B^0. We construct a normal operator L for G as follows: Given $g \in C(\delta \cup \gamma_1 \cup \cdots \cup \gamma_k)$ define Lg by

$$Lg|U = L_1[U](g|\delta),$$

$$Lg|V_i = L_1[V_i](g|\gamma_i).$$

Now apply the Main Existence Theorem (1.2B) to the normal operator L, the bordered boundary neighborhood G, the Riemann surface B^0, and the singularity function s defined as follows: In U let s be any harmonic function which has a continuous extension to α, agrees with f there, and has vanishing flux; in $V_1 \cup \cdots \cup V_k$ set $s \equiv 0$. We obtain a function p harmonic in B^0 which satisfies $p - s = L(p - s)$ in G. This implies that p has a continuous extension to B and that the extended function is constant on each β^i, has vanishing flux across each β^i, and on α differs from f by a constant.

Subtracting the latter constant produces the required function. That its boundary values $c_1,\cdots,c_k$ on $\beta^1,\cdots,\beta^k$ are uniquely determined may be seen as follows: If v is harmonic on B with $v|\alpha = 0$, $v|\beta^i = c_i$ and $\int_{\beta^i} *dv = 0$ for $1 \le i \le k$ then by Green's formula

$$D(v) = \int_{\beta-\alpha} v * dv = 0$$

where $D(v)$ denotes the Dirichlet integral of v. Thus $v \equiv 0$.

Using this lemma we obtain a transformation which takes each $f \in C(\alpha)$ to a $u \in H_1(A)$. The function u has a harmonic extension to β and satisfies $u|\alpha = f$, $u|\beta^i = c_i$ and $\int_{\beta^i} *du = 0$ for each $1 \le i \le k$. The transformation $f \to u$ depends on the partition P; we denote it by $(P)L_1$. The proof that $(P)L_1$ is a normal operator is left to the reader.

The *identity partition* I which puts all contours of β into the same class induces the operator $(I)L_1$ which is identical with L_1 as defined in 1B. The *canonical partition* Q is defined to be the one which puts each contour of β into a class by itself. When a fixed partition P is under discussion we often shorten the notation $(P)L_1$ to simply L_1.

2. Operators on open surfaces

2A. Limits of normal operators. Let A be a bordered boundary neighborhood of an open Riemann surface R. Let Ω denote a generic regular subregion containing Cl $(R-A)$, and let $L_0[\Omega \cap A]$ be the normal operator for $\Omega \cap A$ (see 1A). For brevity we shorten the notation to $L_{0\Omega}$. We shall show that for any continuous function f on the border α of A the net $\{L_{0\Omega} f\}_\Omega$ converges uniformly on compacta of A. That is, given an $\varepsilon > 0$ and a compact subset E there is an Ω_0 such that $\|L_{0\Omega} f - L_{0\Omega'} f\|_E < \varepsilon$ whenever $\Omega, \Omega' \supset \Omega_0$. We may then define a normal operator L_0 for A by setting $L_0 f = \lim_\Omega L_{0\Omega} f$.

2B. Convergence of $L_{0\Omega}$. The family $\{L_{0\Omega} f\}_\Omega$ is uniformly bounded by $\|f\|$ and hence it is normal. We wish to prove there is a unique limit function.

Let $L_{0\Omega} f = u_\Omega$. For our purposes a subnet of $\{u_\Omega\}_\Omega$ shall mean an indexed subfamily $\{u_\Psi\}_\Psi$ where Ψ varies over a cofinal subset of the Ω's (see 0.2.1E). Suppose u_Ψ, $u_{\Psi'}$ are two subnets which converge to limit functions u, u' respectively. For compact subsets E interior to $\Omega \cap A$ we have $D_E(u_\Omega) \leq D_\Omega(u_\Omega)$. Here and in the sequel D_Ω stands for $D_{\Omega \cap A}$, and D shall signify D_A. Taking the limit over the subnet Ψ' and the supremum over sets E we obtain

$$D(u') \leq \lim_{\Psi'} D_{\Psi'}(u_{\Psi'}). \tag{7}$$

Assume temporarily that f is actually harmonic on α. Then since $u'|\alpha = u_\Omega|\alpha = f$ and $*du_\Omega$ vanishes along $\partial\Omega$ we have by Green's formula $D_\Omega(u', u_\Omega) = -\int_\alpha f * du_\Omega = D_\Omega(u_\Omega)$. Consequently

$$0 \leq D_E(u' - u_\Omega) \leq D_\Omega(u' - u_\Omega) = D_\Omega(u') - D_\Omega(u_\Omega). \tag{8}$$

If we take the limit in (8) for Ω in Ψ' we see that (7) holds with equality. By the symmetry of u, u' we must also have $D(u) = \lim_\Psi D_\Psi(u_\Psi)$. Use this fact and take the limit of (8) for Ω in Ψ. The result is $D_E(u' - u) \leq D(u') - D(u)$. By symmetry it follows that $D_E(u' - u) = 0$. Thus u, u' differ by a constant. But since they agree on α they must be identical.

We can therefore define L_0 by $L_0 f = \lim_\Omega L_{0\Omega} f$. This L_0 satisfies all conditions for a normal operator except that its domain consists only of functions harmonic on α. However, L_0 can be extended to all of $C(\alpha)$ by uniform continuity since the functions harmonic on α are dense in $C(\alpha)$. The extended operator is again normal.

The representation $L_0 f = \lim_\Omega L_{0\Omega} f$ remains valid for arbitrary $f \in C(\alpha)$. To prove this choose sequences $\{g_n\}$, $\{h_n\}$ of harmonic functions on α which satisfy $g_n \leq f \leq h_n$ and $\lim g_n = \lim h_n = f$. For any Ω we have

$$L_{0\Omega} g_n \leq L_{0\Omega} f \leq L_{0\Omega} h_n.$$

For any subnet Ψ such that $L_{0\Psi} f$ converges

$$L_0 g_n \leq \lim_\Psi L_{0\Psi} f \leq L_0 h_n$$

for all n. On the other hand we also have $L_0 g_n \leq L_0 f \leq L_0 h_n$. Therefore $L_{0\Omega} f$ has the unique limit function $L_0 f$, and the proof is complete.

2C. Consistent partitions. We wish to obtain corresponding results for the family $\{(P)L_1[\Omega \cap A]\}_\Omega$. This would be possible for the partition I, but in general a condition of consistency is required among the $P(\Omega)$. In the first place we require that the border of any component of $R-\Omega$ belong to only one part of $P(\Omega)$. Suppose Ω and Ω' are two regular subregions satisfying $\Omega \subset \Omega'$. It is clear how a partition of $\partial\Omega$ and $\partial\Omega'$ determines a partition of the components of $R-\Omega$ and $R-\Omega'$. In terms of this induced partition we may state the consistency requirement: the components of $R-\Omega'$ which make up a part for $P(\Omega')$ are contained in components of $R-\Omega$ belonging to the same part for $P(\Omega)$. In 0.2.1D it was shown that every consistent system is induced by a regular partition P of the ideal boundary.

The canonical partition Q of the ideal boundary of R is the system of partitions $P(\Omega)$ which puts the border of each component of $R-\Omega$ into exactly one part. Note that Q is a consistent system.

Let $L_{1\Omega} = (P)L_1[\Omega \cap A]$ where the partitions $P = P(\Omega)$ are consistent. Let f be harmonic on α; set $u_\Omega = L_{1\Omega} f$. Note the following consequence of our consistency hypothesis: If $\beta^i \subset \partial\Omega$ is a part for $P(\Omega)$ and $\Omega' \supset \Omega$ then $\int_{\beta^i} {*du_{\Omega'}} = 0$.

2D. Convergence of $L_{1\Omega}$. We continue using the notation of 2B. Let u, u' be limiting functions of the net $\{u_\Omega\}$. They are determined as limits, uniformly on compacta, of subnets u_Ψ, $u_{\Psi'}$ respectively. If f is harmonic, one consequence of this uniformity is

$$\lim_{\Psi'} \int_\alpha f * d(u' - u_{\Psi'}) = 0.$$

Indeed, by reflection $u' - u_{\Psi'} \to 0$ uniformly on some fixed neighborhood of α. Thus

$$\int_{-\alpha} f * du' = \lim_{\Psi'} \int_{-\alpha} f * du_{\Psi'} = \lim_{\Psi'} D_{\Psi'}(u_{\Psi'}).$$

The consistency hypothesis also yields $D_\Omega(u',u_\Omega) = \int_{-\alpha} f * du'$. Hence $D_\Omega(u',u_\Omega) = \lim_{\Psi'} D_{\Psi'}(u_{\Psi'})$. As a consequence we have for any compact $E \subset \Omega \cap A$

$$(9) \quad 0 \leq D_E(u'-u_\Omega) \leq D_\Omega(u'-u_\Omega)$$

$$= D_\Omega(u') + D_\Omega(u_\Omega) - 2 \lim_{\Psi'} D_{\Psi'}(u_{\Psi'}).$$

Letting $\Omega \to R$ through the Ψ' we arrive at

$$0 \leq D(u') - \lim_{\Psi'} D_{\Psi'}(u_{\Psi'}).$$

Equation (7) also holds and therefore $D(u') = \lim_{\Psi'} D_{\Psi'}(u_{\Psi'})$. The corresponding equation for u, Ψ is valid as well. Now take the limit of (9) through Ψ. This gives $D_E(u'-u) \leq D(u) - D(u')$. By symmetry we have the opposite inequality. Hence

$$D_E(u'-u) = 0$$

and we conclude $u' = u$.

We define $L_1[A]$ by $L_1 f = \lim_\Omega L_{1\Omega} f$ if f is harmonic on α. An approximation argument as in 2B shows that the definition actually applies to any $f \in C(\alpha)$ and yields a normal operator.

2E. Direct sum operators. If the boundary neighborhood A has several components we may combine normal operators for the components to obtain a normal operator for A. More generally, suppose that A is a disjoint union $A_1 \cup \cdots \cup A_n$ where each A_i is a union of components of A. For $1 \leq i \leq n$ let the border of A_i be α_i and let $L^i : C(\alpha_i) \to H_1(A_i)$ be a normal operator for A_i. Define $L = L^1 \oplus \cdots \oplus L^n$ by

$$Lf|A_i = L^i(f|\alpha_i)$$

for $1 \leq i \leq n$, $f \in C(\alpha)$. The operator L is called the direct sum of $L^1, \cdots, L^n$. It is obviously a normal operator for A.

3. Convergence of principal functions

3A. Convergence of operators. Let A be a bordered boundary neighborhood of R and $\{\Omega\}$ an exhaustion of R by regular subregions. We have seen above how a family L_Ω of normal operators for $\Omega \cap A$ can be used to define a normal operator L for A. Suppose a singularity function s_Ω with vanishing flux is given on each $\Omega \cap A$. The Main Existence Theorem (1.2B) gives functions p_Ω on Ω such that $p_\Omega - s_\Omega \in \operatorname{im} L_\Omega$. Suppose further that s_Ω tends uniformly on compacta to a function s harmonic in the interior of A. The existence theorem yields also a function p harmonic on R such that $p - s \in \operatorname{im} L$. We wish to find conditions on $\{L_\Omega\}$ which will ensure that $p_\Omega \to p$ uniformly on compacta.

We shall write $L_\Omega \Rightarrow L$ provided to each compact set E in A and each $\varepsilon > 0$ there is an Ω_0 such that

$$||Lf - L_\Omega f||_E < \varepsilon$$

for all $f \in C(\alpha)$ satisfying $||f|| = 1$ and all $\Omega \supset \Omega_0$.

3B. Convergence of p_Ω. We are considering a family $\{L_\Omega\}$ of normal operators for $A \cap \Omega$. In $A \cap \Omega$ we are given harmonic functions s_Ω, and in Ω principal functions p_Ω with singularity s_Ω. We maintain (Rodin-Sario [2]):

Theorem. *Suppose $L_\Omega \Rightarrow L$ and s_Ω tends to a limit function s uniformly on compacta. Then for a suitably normalized family $\{p_\Omega\}$ of functions with singularity s_Ω the limit*

$$p = \lim_\Omega p_\Omega$$

exists uniformly on compacta of R where p is the principal function $p[s,L]$.

3C. For the proof we choose any compact subset E of R. Let A_0 be the closure of a regular subregion of R which contains Cl $(R - A)$ and E in its interior. As in Theorem 1.2G, we normalize p_Ω, p so that along $\alpha_0 = \partial A_0$ they belong to the space X defined in 1.2F. Thus on α_0 we have $p = (I - T)^{-1} s$ and, for sufficiently large Ω, $p_\Omega = (I - T_\Omega)^{-1} s_\Omega$ where $T = LK$, $T_\Omega = L_\Omega K$, and K is the Dirichlet operator $K: X \to H_1(A_0)$. It suffices to prove that $\lim_{\Omega \to R} ||p - p_\Omega||_{\alpha_0} = 0$.

The hypothesis $L_\Omega \Rightarrow L$ means that $||T - T_\Omega|| \to 0$ as $\Omega \to R$. Indeed, given $\varepsilon > 0$ there is an Ω_0 such that

$$||LKg - L_\Omega Kg||_{\alpha 0} \leq \varepsilon ||Kg||_\alpha$$

for all $g \in C(\alpha_0)$ and $\Omega \supset \Omega_0$. Since $||Kg||_\alpha \leq ||g||_{\alpha 0}$ we obtain

$$||Tg - T_\Omega g||_{\alpha 0} \leq \varepsilon ||g||_{\alpha 0}.$$

We have the estimate

$$||p_\Omega - p||_{\alpha 0} = ||(I - T_\Omega)^{-1} s_\Omega - (I - T)^{-1} s||_{\alpha 0}$$

$$\leq ||(I - T_\Omega)^{-1} - (I - T)^{-1}|| \cdot ||s_\Omega||_{\alpha 0} + ||(I - T)^{-1}|| \cdot ||s_\Omega - s||_{\alpha 0},$$

and this upper bound is seen to approach 0 as $\Omega \to R$. Indeed, this is obviously the case for the term $||(I - T)^{-1}|| \cdot ||s_\Omega - s||_{\alpha 0}$. The norm $||(I - T_\Omega)^{-1} - (I - T)^{-1}||$ is bounded by

$$||T - T_\Omega|| \cdot (1 - ||T||)^{-1} (1 - ||T_\Omega||)^{-1}$$

since $||T|| < 1$ and $||T_\Omega|| < 1$. The convergence to 0 now follows from the earlier remark that $||T - T_\Omega|| \to 0$. This completes the proof.

§3. THE PRINCIPAL FUNCTIONS p_0 AND p_1

We consider a bordered boundary neighborhood A of a Riemann surface R. On A we construct some harmonic functions g_0, g_1 with logarithmic poles. The construction of these auxiliary functions is accomplished by using the Main Existence Theorem. With the help of these functions it is possible to derive an integral representation of the normal operators L_0, L_1. This representation is needed to obtain certain estimates necessary for proving the convergence $L_{0\Omega} \Rightarrow L_0$, $L_{1\Omega} \Rightarrow L_1$. Finally, some consequences of this convergence are explored.

1. Integral representations

1A. Auxiliary functions. Let A be a bordered boundary neighborhood of R. Let α be the border of A, oriented so that A lies to the right. Given a regular subregion $\Omega \supset \mathrm{Cl}\,(R - A)$ we shall construct functions $g_{0\Omega}$, $g_{1\Omega}$ which are harmonic in $A \cap \Omega$ except for logarithmic poles and have the reproducing property $L_{i\Omega} f = \int_{-\alpha} f * dg_{i\Omega}$ for $f \in C(\alpha)$, $i = 0, 1$.

Lemma. *Let z be a parameter for a neighborhood of a point a in $A \cap \Omega - \alpha$. For $i = 0$, 1 there exist functions $g_{i\Omega}(\cdot, a)$ harmonic in $A \cap \Omega - \{a\}$ with L_i behavior along $\partial\Omega$, vanishing boundary values on α, and with the singularity $(2\pi)^{-1} \log |z - a|$ at a.*

More precisely, $g_{0\Omega}(\cdot, a)$ has vanishing normal derivative along $\partial\Omega$ and $g_{1\Omega}(\cdot, a)$ has a constant value on each part of $\partial\Omega$ associated with some partition which we need not specify. In addition $g_{1\Omega}(\cdot, a)$ has zero flux across each part of $\partial\Omega$. The functions $h_{i\Omega}$ defined near a by

$$h_{i\Omega}(z) = g_{i\Omega}(z, a) - \frac{1}{2\pi} \log |z - a|$$

are supposed to have harmonic extensions to $z = a$.

For the proof consider the surface $S = A \cap \Omega - \alpha - \{a\}$. On S take disjoint bordered neighborhoods U of α, V of $\partial\Omega$, and Δ of a. We may assume Δ corresponds to $\{0 < |z| \leq 1\}$. Let u be any harmonic function in $U \cup \alpha$ with constant value on α and satisfying $\int_\alpha * du = -1$. Define a harmonic function s in $U \cup V \cup \Delta$ by

$$s = \begin{cases} u & \text{in } U, \\ \dfrac{1}{2\pi} \log |z - a| & \text{in } \Delta, \\ 0 & \text{in } V. \end{cases} \tag{10}$$

Note that s has total flux 0 across the ideal boundary of S.

Let D be the Dirichlet operator for Δ. That is, for $f \in C(\partial\Delta)$ let Df be the restriction to Δ of the solution to the Dirichlet problem in $\Delta \cup \{a\}$. Let $L_1[U]$ correspond to the identity partition and consider the normal operator $L = L_1[U] \oplus D \oplus L_i[V]$ $(i = 0,1)$. By the Main Existence Theorem we find a principal function p_i which satisfies $p_i - s = L(p_i - s)$. This means that p_i has the properties required of $g_{i\Omega}$ except that its value on α, when extended there harmonically, may be some constant other than 0. We merely subtract such a constant to obtain $g_{i\Omega}(\cdot, a)$.

1B. Integral representations. Let u be harmonic in $A \cap \mathrm{Cl}\,\Omega$. For $0<r<1$ let Δ_r correspond to $\{|z|<r\}$. Green's formula may be applied to u and $g_{i\Omega} = g_{i\Omega}(\cdot,a)$ on the surface $A \cap \mathrm{Cl}\,\Omega - \Delta_r$. If we note that

$$\lim_{r\to 0} \int_{\partial\Delta_r} u *dg_{i\Omega} = u(a),$$

$$\lim_{r\to 0} \int_{\partial\Delta_r} g_{i\Omega} *du = 0,$$

we obtain the reproducing formulas

$$u(a) = \int_{-\alpha} u *dg_{0\Omega} - \int_{\partial\Omega} g_{0\Omega} *du,$$

$$u(a) = \int_{-\alpha} u *dg_{1\Omega} + \int_{\partial\Omega} u *dg_{1\Omega} - g_{1\Omega} *du.$$

In case u has L_0 or L_1 behavior on $\partial\Omega$ these formulas simplify. In particular we have

$$L_{i\Omega} f(a) = -\int_{\zeta\in\alpha} f(\zeta) *dg_{i\Omega}(\zeta,a) \tag{11}$$

for f harmonic on α, $i = 0, 1$.

1C. Convergence of auxiliary functions. For a fixed $a \in A - \alpha$ we shall prove that $\{g_{i\Omega}(\cdot,a)\}_\Omega$ converges to a function $g_i(\cdot,a)$ uniformly on compacta.

First, we use the estimates of the Main Existence Theorem to see that $\{g_{i\Omega} - g_{i\Omega'}\}_{\Omega,\Omega'}$ is a normal family. In 1A we constructed $g_{i\Omega}$ as a principal function on the surface $S = A \cap \Omega - \alpha - \{a\}$ for the singularity function s. This s depends on Ω in a nonessential manner: it is identically 0 near $\partial\Omega$. For this reason the estimates in Theorem 1.2G give a bound for $g_{i\Omega}$ which does not depend on Ω, provided $g_{i\Omega}$ is normalized with an additive constant c_Ω determined by the normalization condition of that theorem. Denote the normalized function by $\tilde{g}_{i\Omega}$. The constant c_Ω is clearly the value of $\tilde{g}_{i\Omega}$ on α. Thus the bounds on $\tilde{g}_{i\Omega}$ can be used to bound

c_Ω, and we conclude that $\{g_{i\Omega}\}$ is uniformly bounded on compacta of $A-\{a\}$.

We shall now show that $\{g_{i\Omega}\}$ has a unique limit function. Suppose g_i, g_i' are two limit functions. The limit process means uniformly on compacta. Note that the convergence is uniform on α since $g_{i\Omega}$ can be extended harmonically to the symmetric surface obtained by welding together two copies of $A\cap\Omega$ along α. In §2 we showed the convergence of $L_{i\Omega}f$. Therefore, from (11) we see that

$$\int_{\zeta\in\alpha} f(\zeta) * d(g_i(\zeta,a) - g_i'(\zeta,a)) = 0$$

for all $f\in C(\alpha)$. Thus $g_i - g_i'$ has vanishing normal derivative at each point of α. Since it is 0 on α and harmonic, it is 0 everywhere. This completes the proof.

Denote the limit of $g_{i\Omega}(\cdot,a)$ by $g_i(\cdot,a)$. As a consequence of the uniform convergence on α we have by (11)

$$L_i f(a) = -\int_{\zeta\in\alpha} f(\zeta) * dg_i(\zeta,a) \tag{12}$$

for $f\in C(\alpha)$, $a\in A-\alpha$, $i=0,1$ (Sario [27]). Therefore for fixed $a\in A-\alpha$, $L_{i\Omega}f(a)$ converges to $L_i f(a)$ as $\Omega\to R$. The convergence is uniform for all $f\in C(\alpha)$ with $|f|\le 1$.

2. Convergence of p_0, p_1

2A. Proof of $L_{i\Omega}\Rightarrow L_i$. We are now able to prove the convergence of $L_{i\Omega}$ in the sense of 2.3A. Since $\|L_{i\Omega}f\|_A\le\|f\|_\alpha$ the family $\{L_{i\Omega}f\}_{\Omega,f}$ is normal if $\|f\|_\alpha\le 1$. From this fact, together with the pointwise convergence $L_{i\Omega}f(a)\to L_i f(a)$ which is uniform in f when $\|f\|\le 1$, it follows that $L_{i\Omega}\Rightarrow L_i$.

We have shown that $L_{i\Omega}\Rightarrow L_i$. In conjunction with Theorem 2.3B we obtain (Rodin-Sario [2]):

Theorem. *Let A be a bordered boundary neighborhood of R. For each regular subregion Ω of R let s_Ω be a singularity function on*

$A \cap \Omega$ and suppose $\lim_{\Omega \to R} s_\Omega = s$ where s is a singularity function on A and the convergence is uniform on compact subsets of A. Then for $i = 0, 1$ there are principal functions $p_\Omega = p_\Omega[s_\Omega, L_{i\Omega}]$, $p = p[s, L_i]$ such that $p_\Omega \to p$ uniformly on compacta.

2B. Principal functions with singularities. We shall now derive some consequences of the preceding theorem for an important special case. On an open Riemann surface R consider a finite set of points $\zeta_1, \cdots, \zeta_n$. Let Δ_j be parametric disks about the ζ_j $(j = 1, \cdots, n)$ with disjoint closures. Let s' be a harmonic function defined in $\cup(\text{Cl } \Delta_j - \{\zeta_j\})$ and assume s' satisfies the flux condition $\int_{\Sigma \partial \Delta_j} {*ds'} = 0$. Let Ω_0 be a regular subregion of R which contains $\cup \text{ Cl } \Delta_j$ and consider an exhaustion $\{\Omega\}$ of R with $\text{Cl } \Omega_0$ contained in each Ω. Let s_Ω be the harmonic function defined by

$$s_\Omega = \begin{cases} s' & \text{in} \quad \cup (\text{Cl } \Delta_j - \{\zeta_j\}), \\ 0 & \text{in} \quad \Omega - \Omega_0. \end{cases} \tag{13}$$

We shall apply the Main Existence Theorem to the function s_Ω defined in the bordered boundary neighborhood

$$A_\Omega = (\Omega - \Omega_0) \cup (\cup (\text{Cl } \Delta_j - \{\zeta_j\}))$$

of the Riemann surface $\Omega - \{\zeta_1, \cdots, \zeta_n\}$. For the normal operator we take the direct sum of the Dirichlet operators for $\text{Cl } \Delta_j$ $(j = 1, \cdots, n)$ together with the $L_{0\Omega}$ or $L_{1\Omega}$ operator for $\Omega - \Omega_0$. Denote the resulting principal function by $p_{0\Omega}$ or $p_{1\Omega}$ respectively. Thus, for example, $p_{0\Omega}$ is harmonic on $\Omega - \{\zeta_1, \cdots, \zeta_n\}$, satisfies $p_{0\Omega} = L_0(p_{0\Omega})$ on $\Omega - \Omega_0$, and $p_{0\Omega} - s_\Omega$ has removable singularities at $\zeta_1, \cdots, \zeta_n$. A similar statement applies to $p_{1\Omega}$. In this case we assume the operators $L_{1\Omega}$ are defined with respect to a consistent system of partitions of $\{\partial\Omega\}$ (see 2.2C).

If we replace Ω by R in the above construction we obtain principal functions p_0, p_1. By Theorem 2A we see that $p_{i\Omega} \to p_i$ $(i = 0,1)$ uniformly on compact subsets of $R - \{\zeta_1, \cdots, \zeta_n\}$. As a consequence we obtain the following useful limit relations (Sario [9], Rodin-Sario [2]):

Theorem. *The principal functions constructed above satisfy the following conditions:*

(a) $\lim_\Omega D_{\Omega-\Omega_0}(p_i - p_{i\Omega}) = 0 \qquad (i = 0,1).$

(b) $D_{R-\Omega_0}(p_i) = -\int_{\partial\Omega_0} p_i * dp_i = \lim_\Omega D_{\Omega-\Omega_0}(p_{i\Omega}) \quad (i = 0,1).$

(c) *If f is a function on $R-\Omega_0$ with $D(f) < \infty$ then*

$$\int_\beta f * dp_0 = 0,$$

and if $\int_\beta * df = 0$ then

$$\int_\beta p_1 * df = 0$$

where $\int_\beta = \lim_\Omega \int_{\partial\Omega}$ and p_1 corresponds to the identity partition.

2C. For the proof we make repeated use of the uniform convergence of the $p_{i\Omega}$ and their normal derivatives along $\partial\Omega_0$. Thus in the case of p_0 we have

$$\lim_\Omega D_{\Omega-\Omega_0}(p_0, p_{0\Omega}) = -\int_{\partial\Omega_0} p_0 * dp_0.$$

From $D_E(p_{0\Omega}) \leq D_{\Omega-\Omega_0}(p_{0\Omega})$ for a compact set $E \subset \Omega - \Omega_0$ we infer $D_{\Omega-\Omega_0}(p_0) \leq \lim_\Omega D_{\Omega-\Omega_0}(p_{0\Omega})$ by the uniform convergence on E. Hence from

$$0 \leq D_{\Omega-\Omega_0}(p_0 - p_{0\Omega}) = D_{\Omega-\Omega_0}(p_0) + D_{\Omega-\Omega_0}(p_{0\Omega}) - 2D_{\Omega-\Omega_0}(p_0, p_{0\Omega})$$

and

$$D_{\Omega-\Omega_0}(p_0) \leq \lim_\Omega D_{\Omega-\Omega_0}(p_{0\Omega}) = \lim_\Omega \int_{-\partial\Omega_0} p_{0\Omega} * dp_{0\Omega}$$

we obtain

$$0 \leq D_{\Omega-\Omega_0}(p_0 - p_{0\Omega}) \leq o(1) \tag{14}$$

where $o(1) \to 0$ as $\Omega \to R$. This proves (a) for the case $i = 0$.

To prove (b) we apply Green's formula to $D_{\Omega-\Omega_0}(p_0-p_{0\Omega})$ and find that it reduces to $\int_{\partial\Omega}(p_0-p_{0\Omega})*dp_0+o(1)$. This line integral can be written

$$\int_{\partial\Omega} p_0*dp_0+\int_{\partial\Omega} p_0*dp_{0\Omega}-p_{0\Omega}*dp_0.$$

The second integral can be transferred to $\partial\Omega_0$ and therefore can be absorbed in $o(1)$. We finally obtain

$$D_{\Omega-\Omega_0}(p_0-p_{0\Omega}) = \int_{\partial\Omega} p_0*dp_0+o(1)$$

and hence by (a)

$$\lim_{\Omega}\int_{\partial\Omega} p_0*dp_0 = 0.$$

Now $D_{\Omega-\Omega_0}(p_0) = \int_{\partial\Omega-\partial\Omega_0} p_0*dp_0$ and therefore $D_{R-\Omega_0}(p_0) = -\int_{\partial\Omega_0} p_0*dp_0 = \lim_\Omega D_{\Omega-\Omega_0}(p_{0\Omega})$, which proves (b) in our case.

We now consider (c) for the function p_0. To obtain an upper bound for

$$\left|\int_{-\partial\Omega_0} f*dp_0\right| = |\lim_{\Omega} D_{\Omega-\Omega_0}(f,p_{0\Omega})|$$

we apply the Schwarz inequality. This gives

$$\left|\int_{\partial\Omega_0} f*dp_0\right| \leq \{\lim_{\Omega} D_{\Omega-\Omega_0}(f)D_{\Omega-\Omega_0}(p_{0\Omega})\}^{1/2}.$$

By (b) this upper bound is $\{D_{R-\Omega_0}(f)D_{R-\Omega_0}(p_0)\}^{1/2}$. Now (c) follows by letting $\Omega_0\to R$.

The proofs of (a), (b), (c) in the case of p_1 are analogous and will be left to the reader.

§4. SPECIAL TOPICS

In this section we discuss some problems and open questions relating to the preceding topics of this chapter. In 1 we show how the Main Existence Theorem is related to a Fredholm integral equation. In 2 we consider estimates of the constant q in the q-lemma 1.1. A general problem concerning the structure of the space of normal operators on a Riemann surface is discussed in 3. Finally, it is shown that the L_1-principal function can be constructed by the method of orthogonal projection.

The topics in this section have interest in their own right. However, they are not needed for the understanding of subsequent chapters.

1. An integral equation

1A. Consider the normal operators L_0, L_1 defined for a bordered boundary neighborhood A of an open Riemann surface R. Let s be a singularity function in A and let α be the border of A. Thus s is harmonic in A, $\int_\alpha *ds = 0$, and we shall assume $s|\alpha = 0$. If h, k are nonnegative real numbers satisfying $h+k = 1$, the transformation $L_{hk} = hL_0+kL_1$ is a normal operator. By the Main Existence Theorem there is a principal function p on R which satisfies $p = s+L_{hk}p$ in A. We now show that p may also be obtained as the solution of a Fredholm integral equation.

Let A_0 be the closure of a regular subregion which contains Cl $(R-A)$. It was shown in 1.2D that the problem of constructing p reduces to that of finding a function $\tilde{p}$ on $\alpha_0 = \partial A_0$ which satisfies

$$(15) \qquad \tilde{p} = s+L_{hk}K\tilde{p}$$

where K is the Dirichlet operator which assigns to each continuous function on α_0 the harmonic function in A_0 with these boundary values. In fact, if $\tilde{p}$ satisfies (15) then $p|A_0 = K\tilde{p}$ and $p|A = s+L_{hk}K\tilde{p}$.

1B. Assume that the contours α are parametrized by a function δ defined on a union of closed intervals. We may assume the intervals are of the form $[t_i,t_{i+1}]$ with $0 = t_1 < t_2 \cdots < t_n = 1$. Then δ is piecewise continuous on $[0,1]$. Similarly we may parametrize α_0 by a piecewise continuous function γ on $[0,1]$. We choose the orientation so that α and α_0 leave $A\cap A_0$ to the right and left respectively.

Let $g(z,\zeta)$ be Green's function for A_0 with singularity at ζ.

Let $g_0(z,\zeta)$ and $g_1(z,\zeta)$ be the auxiliary functions defined in 3.1A. Let $\partial/\partial n$ denote the normal derivative on α and α_0 interior to $A_0 \cap A$. Set $g_{hk} = hg_0 + kg_1$. Given continuous functions u and v on γ and δ, respectively, the operators K and L_{hk} have the following integral representations:

$$Ku(\delta(t)) = \int_0^1 u(\gamma(y)) \frac{\partial g(\gamma(y),\delta(t))}{\partial n} dy,$$

$$L_{hk}v(\gamma(x)) = -\int_0^1 v(\delta(t)) \frac{\partial g_{hk}(\delta(t),\gamma(x))}{\partial n} dt.$$

We introduce the kernel

$$\varphi(x,y) = -\int_0^1 \frac{\partial g(\gamma(y),\delta(t))}{\partial n} \frac{\partial g_{hk}(\delta(t),\gamma(x))}{\partial n} dt \tag{16}$$

and have

$$L_{hk}Ku(\gamma(x)) = \int_0^1 \varphi(x,y)\, u(\gamma(y))dy.$$

Then (15) becomes (Sario [27])

$$\bar{p}(\gamma(x)) = s(\gamma(x)) + \int_0^1 \varphi(x,y)\, \bar{p}(\gamma(y))dy. \tag{17}$$

Thus we are dealing with a Fredholm integral equation. It is known that the solution is $\bar{p} = \sum_0^\infty (L_{hk}K)^n s$ provided the series converges uniformly. Indeed, the operator $L_{hk}K$ may then be applied term by term and gives $L_{hk}K\bar{p} = \sum_1^\infty (L_{hk}K)^n s = \bar{p} - s$. The proof of uniform convergence follows from the q-lemma as in 1.2F and will be omitted here.

2. Estimates of q

2A. The Poincaré metric. Given a compact set α on a Riemann surface R, we are interested in finding the smallest constant q, $0<q<1$, such that $||u||_\alpha \leq q||u||_R$ for all u harmonic on R which change sign on α. That such q's exist is the content of Lemma 1.1; let q_0 denote the smallest one. We first obtain an upper bound for q_0 in terms of the Poincaré metric on R. This bound is sharp in case R is the unit disk. In 2F we use a different metric for estimating q_0.

The unit disk may be made into a complete Riemannian space by introducing the Poincaré metric

$$ds = \frac{|dz|}{1-|z|^2}.$$

If $\zeta = T(z)$ is a biholomorphic self-mapping of the disk and $t \to \delta(t)$ a smooth curve then $\int_\delta ds = \int_{T\circ\delta} ds$. In fact, if $T(a) = b$ then

$$\left|\frac{\zeta - b}{1-\bar{b}\zeta}\right| = \left|\frac{z-a}{1-\bar{a}z}\right|,$$

from which it follows immediately that

$$\frac{|d\zeta|_{\zeta=b}}{1-|b|^2} = \frac{|dz|_{z=a}}{1-|a|^2}.$$

It is easily verified that the shortest path from $z = 0$ to $z = r>0$ is the line segment $[0,r]$. Therefore, due to the invariance of the metric under biholomorphic self-mappings, the geodesic path between any two points is an arc of the circle which passes through these points and is orthogonal to the unit circle.

Consider a Riemann surface R whose universal covering surface is the unit disk. The Poincaré metric, since it is invariant under linear transformations of the unit disk, induces a unique metric on R. It is called the hyperbolic metric.

2B. Poincaré diameter. Let a, b be points of the unit disk and $d(a,b)$ the distance between them in the Poincaré metric. The *Poincaré diameter* of a subset α of the disk is

$$\sup \{d(a,b) \,|\, a\in\alpha,\ b\in\alpha\}.$$

The following result (Oikawa [4]) gives an exact value for q_0 in case R is the unit disk:

Theorem. *Let D be the Poincaré diameter of a compact subset α of the unit disk. Then*

$$q_0 = \frac{2}{\pi} \tan^{-1} \sinh 2D. \tag{18}$$

2C. For the proof consider any harmonic function u which changes sign on α and satisfies $||u||_R = 1$. Let a and b be points of α such that $u(a) < 0 < u(b)$. There is a point c on the geodesic arc C connecting a and b such that $u(c) = 0$. Note that $d(c,k) \leq D$, for all $k \in \alpha$. Indeed, consider the linear transformation T of the disk onto itself which sends k to the origin. The image $T(C)$ is a circular arc concave to the origin, or a line segment through it. Since geodesics from the origin are radial segments it is evident that

$$d(0,Tc) \leq \max \{d(0,Ta), d(0,Tb)\}.$$

Due to the invariance of the metric this implies

$$d(k,c) \leq \max \{d(k,a), d(k,b)\} \leq D.$$

Let S be a biholomorphic self-mapping of the disk which carries c to the origin. Then $S(\alpha)$ is contained in the concentric disk of radius r where r is determined by $d(0,r) = D$. One has $d(0,r) = \frac{1}{2} \log [(1+r)/(1-r)]$, and hence $r = \tanh D$. That the right side of (18) is a possible value for q is obtained immediately on applying the following lemma (Oikawa [4]) to $u \circ S^{-1}$ and noting that $2r/(1-r^2) = \sinh 2D$.

Lemma. *Let u be harmonic in the unit disk with $u(0) = 0$, $|u(z)| < 1$. Then*

$$|u(z)| \leq \frac{2}{\pi} \tan^{-1} \frac{2|z|}{1-|z|^2},$$

and this estimate is sharp.

The proof can be based on the Schwarz lemma. Choose a

harmonic conjugate v of u such that $w = u+iv$ satisfies $w(0) = 0$. Then

$$f(z) = \frac{\exp[(\pi i/2)w(z)]-1}{\exp[(\pi i/2)w(z)]+1} = i \tan \frac{\pi}{4} w(z)$$

maps the disk into itself keeping the origin fixed. Consequently $|f(z)| \leq |z|$. Solving the inequality $f(z)\bar{f}(z) \leq |z|^2$ one obtains

$$\frac{1-|z|^2}{1+|z|^2} \cosh \frac{\pi}{2} v \leq \cos \frac{\pi}{2} u.$$

The inequality in Lemma 2C now follows since $\cosh \frac{1}{2}\pi v \geq 1$.

To see that the estimate of Lemma 2C is sharp, consider the function

$$u_0 = \operatorname{Re} \frac{2i}{\pi} \log \frac{1+z}{1-z}.$$

It satisfies the hypotheses and takes the value

$$\frac{2}{\pi} \cdot 2 \tan^{-1} \rho = \frac{2}{\pi} \tan^{-1} \frac{2\rho}{1-\rho^2}$$

at $z = -i\rho$.

2D. The proof of Theorem 2B will be complete if the estimate $q = 2\pi^{-1} \tan^{-1} \sinh 2D$ is shown to be the best possible. To this end take points a and b in α with $d(a,b) = D$ and a positive number $\varepsilon < D$. Let C be the geodesic connecting a and b, and c_ε the point on C with $d(a,c_\varepsilon) = \varepsilon$. Let T be the linear transformation of the unit disk which sends c_ε to the origin, and a to the positive imaginary axis. Let u_0 be the function introduced in 2C and set $u_\varepsilon = u_0 \circ T$. Then $u_\varepsilon(a) < 0$ and an explicit computation shows that $u_\varepsilon(b) = 2\pi^{-1} \tan^{-1} \sinh 2(D-\varepsilon)$. Since ε was arbitrary the sharpness of the estimate is established.

2E. Theorem 2B may be used to obtain an upper bound for q_0 in the case of an arbitrary Riemann surface R. If R carries nonconstant bounded harmonic functions, the only case of interest, then obviously R has the disk as its universal covering surface.

This allows us to introduce the hyperbolic metric (see 2A) on R.

Suppose α is a compact subset of R with diameter D in the hyperbolic metric. Consider any harmonic function u on R which changes sign on α and satisfies $|u|<1$. Let a and b be points in α with $u(a)<0<u(b)$. There exist points $\tilde{a}$ and $\tilde{b}$ in the universal covering of R which lie over a and b respectively, and whose distance in the Poincaré metric is less than D. Apply Theorem 2B to the compact set $\{\tilde{a},\tilde{b}\}$ and the function u lifted to the universal covering surface. This gives the following bound for u (Oikawa [4]):

Theorem. *Let α be a compact subset of a Riemann surface R with hyperbolic metric. If α has diameter D in this metric then*

$$\frac{2}{\pi}\tan^{-1}\sinh 2D$$

is a possible value for q.

2F. Harmonic metric. Suppose R carries nonconstant bounded harmonic functions. Let B be the family of harmonic functions u with $\|u\|_R \leq 1$. For each $u \in B$ we consider the metric (with zeros) $d\sigma_u = |\operatorname{grad} u||dz|$.

The quotient of two metrics is a function. For $a \in R$ we may therefore define $d\sigma_u(a) \leq d\sigma_v(a)$ to mean $d\sigma_u(a)/d\sigma_v(a) \leq 1$. Let $d\sigma(a) = \sup\{d\sigma_u(a) | u \in B\}$. Since B is a normal family $d\sigma$ is everywhere finite. Evidently $d\sigma$ is an invariant metric, although possibly with zeros. It is called the *harmonic metric* for R.

In case R is simply connected we have (Oikawa [2]):

Lemma. *Let ds be the Poincaré metric on a simply connected surface R. Then $d\sigma = 4\pi^{-1}ds$.*

For the proof we may take R to be the unit disk. It suffices to prove

$$\sup_{u\in B} |\operatorname{grad} u(z_0)| = \frac{4}{\pi(1-|z_0|^2)} \tag{19}$$

for $z_0 \in R$. Form the analytic function $f = u+iv$ where v is the harmonic conjugate of u. The function

$$w = \varphi(\zeta) = \frac{2i}{\pi}\log\frac{1+\zeta}{1-\zeta}$$

maps $\{|\zeta|<1\}$ conformally onto $\{|\operatorname{Re} w|<1\}$.

Consider also the function $\zeta = F(z) = \varphi^{-1}[f(z)]$ which satisfies $|F|<1$ in R. By Pick's theorem (see Hille [1, II, p. 239]) ds is nonincreasing under analytic mappings $R\to R$. Hence

$$\frac{|d\zeta|}{1-|\zeta|^2} \le \frac{|dz|}{1-|z|^2}.$$

Thus at z_0, $\zeta_0 = F(z_0)$

$$|f'(z_0)|(1-|z_0|^2) \le |\varphi'(\zeta_0)|(1-|\zeta_0|^2) = \frac{4}{\pi}\frac{(1-|\zeta_0|^2)}{|1-\zeta_0^2|} \le \frac{4}{\pi},$$

whence $|\operatorname{grad} u(z_0)| \le 4/(\pi(1-|z_0|^2))$. There is a $u\in B$ which makes this relation an equality, namely

$$u = \operatorname{Re}\varphi\left(\frac{z-z_0}{1-\bar{z}_0 z}\right).$$

This proves (19) and hence the lemma.

2G. Harmonic diameter. Let Δ_α denote the diameter of a subset $\alpha\subset R$ in the metric $d\sigma$. It is called the *harmonic diameter* of α. We claim (Oikawa [2]):

Theorem. *Let α be a compact subset of a Riemann surface R. Then $q_0\le\Delta_\alpha$.*

For the proof we need only show that $\|u\|_\alpha \le \Delta_\alpha$ for all $u\in B$ which change sign on α. We may assume $\max_\alpha u = \|u\|_\alpha$ and then choose a and b in α such that $u(b) = \|u\|_\alpha > 0 > u(a)$. Then

$$\|u\|_\alpha < u(b) - u(a),$$

and for any path from a to b we have

$$u(b)-u(a) = \int_a^b du \le \int_a^b d\sigma_u \le \int_a^b d\sigma.$$

For any $\varepsilon > 0$ there is a path from a to b such that $\int_a^b d\sigma < \Delta_\alpha + \varepsilon$. Consequently $||u||_\alpha \leq \Delta_\alpha$ and the theorem is proved.

2H. Comparison. In the case of the unit disk or any simply connected Riemann surface Theorem 2B gives the exact value of q_0. That Theorems 2E, 2G are not sharp can be seen as follows.

Let R be the punctured unit disk and α the slit $[a,b]$ where $0 < a < b < 1$ and $\Delta_{[0,b]} < 1$. As a tends to 0 the diameter D of $[a,b]$, in the hyperbolic metric of R, tends to ∞. Hence we may find an a so small that $\Delta_{[a,b]} < 2\pi^{-1} \tan^{-1} \sinh 2D$. Theorem 2G therefore shows that Theorem 2E is not sharp.

On the other hand Theorem 2B shows that Theorem 2G is not sharp. In fact, $2\pi^{-1} \tan^{-1} \sinh \frac{1}{2}\pi\Delta_\alpha < \Delta_\alpha$, and the left-hand side is the value of q_0 for the unit disk by Lemma 2F and Theorem 2B.

3. The space of normal operators. Ahlfors' problem

3A. The space $N(A)$. Let R be an open Riemann surface with ideal boundary β and A a bordered boundary neighborhood of R with border α. Denote by $N(A)$ the space of all normal operators for A. Ahlfors [4] suggested the following approach to the problem of determining the space $N(A)$.

The set $N(A)$ is clearly convex, i.e. for two operators L_1 and L_2 in $N(A)$ the segment $\lambda L_1 + (1-\lambda) L_2$ $(0 \leq \lambda \leq 1)$ is contained in $N(A)$. An element $L \in N(A)$ is, by definition, an extreme point of the convex set $N(A)$ if

$$L = \lambda L_1 + (1-\lambda) L_2 \quad (\lambda \in [0,1]; L_1, L_2 \in N(A) \text{ with } L_1 \neq L_2)$$

implies either $\lambda = 0$ or $\lambda = 1$. We will denote by $E(K)$ the set of all extreme points in a convex set K.

Suppose there exists a locally convex linear topological space $\tau = \tau(N(A))$ containing $N(A)$ as a compact subset. By the Krein-Milman theorem (see e.g. Royden [9, p. 179]) the convex hull of $E(N(A))$ is dense in $N(A)$. Moreover, if τ is metrizable then $E(N(A))$ is a G_δ subset of τ and there exists a unit Baire measure μ_L on τ with $\mu_L(E(N(A))) = 1$ such that

$$L = \int_{E(N(A))} l \, d\mu_L(l)$$

for every $L \in N(A)$ (see Choquet [1]).

The space $N(A)$ is therefore completely determined if the following two problems are settled:

(I) *Find the shape of* $E(N(A))$.

(II) *Find an appropriate space* $\tau(N(A))$.

No definitive solutions to these important and fascinating problems have been given as yet. For related work we refer to Ahlfors [4], Phelps [1], Ryff [1], Savage [3], Nakai-Sario [2], and Goldstein [2].

3B. The space $N(\alpha,\beta)$. Hereafter in 3 we suppose that β is realized as a finite union of disjoint analytic Jordan curves, i.e. R is the interior of a compact bordered Riemann surface $\bar{R}$ with border β. Moreover we shall assume that $R_0 = R-A$ is a parametric disk.

Let ω be the harmonic function on $A\cup\beta$ such that $\omega|\alpha = 0$ and $\omega|\beta = k$ (a positive constant) with $\int_\alpha *d\omega = 1$. Along α and β, $*d\omega$ gives positive unit measures $d\mu$ and $d\nu$ respectively.

Take an $L\in N(A)$ and an $f\in C(\alpha)$. We know that Lf is bounded on R and that $d\nu$ is equivalent to the Lebesgue linear measure on β, considered locally. We conclude by the Fatou theorem (see e.g. Tsuji [5]) that Lf has angular limits $d\nu$-a.e. on β; we denote them by $T_L f$. Clearly $T_L f\in L^\infty(\beta) = L^\infty(\beta,d\nu)$, and T_L is a linear transformation of $C(\alpha)$ into $L^\infty(\beta)$. Conversely let T be a linear transformation of $C(\alpha)$ into $L^\infty(\beta)$. By a generalization of the Schwarz theorem (see e.g. Tsuji [5]) there exists a unique bounded function $L_T f\in H_1(A)$ such that $L_T f$ has the angular limits Tf $d\nu$-a.e. on β. Clearly L_T is a linear transformation of $C(\alpha)$ into $H_1(A)$. If L_T is normal for A we call T *normal* for (α,β). In this case $T_{L_T} = T$. We denote by $N(\alpha,\beta)$ the space of all normal operators for (α,β). We have the following characterization of $N(\alpha,\beta)$ (Ahlfors [4]):

Theorem. *In order that T be an element of $N(\alpha,\beta)$ it is necessary and sufficient that T be a linear transformation of $C(\alpha)$ into $L^\infty(\beta)$ such that for all $f\in C(\alpha)$*

(B1) $\min_\alpha f\leq Tf\leq \max_\alpha f$ *$d\nu$-a.e. on β,*

(B2) $\int_\alpha f\,d\mu = \int_\beta Tf\,d\nu$.

For the proof take a linear transformation T of $C(\alpha)$ into $L^\infty(\beta)$ and consider $L = L_T$, which is a linear transformation of $C(\alpha)$ into the space of bounded functions in $H_1(A)$ with (A1) of 1.2A. If L satisfies (A2) then clearly T satisfies (B1). Conversely suppose T satisfies (B1). Let $E\subset\beta$ be the exceptional set in (B1) and U an arbitrary open set in β with $U\supset E$. Let u_U be the bounded function harmonic on A with boundary values $u_U|\alpha\cup(\beta-\bar U) = 0$ and $u_U|U = 1$. By the maximum principle

$$\min_\alpha f - cu_U \le Lf \le \max_\alpha f + cu_U$$

on A where $c = \sup_A |Lf|$. Since $\nu(E) = 0$, u_U can be made arbitrarily small at each point of A by choosing $\nu(U)$ sufficiently small, and we infer the validity of (A2) of 1.2A for L.

Let β_t be the level line $\{\omega = t\}$ of ω with $t\in(0,k)$. By virtue of Green's formula

$$\int_{\alpha-\beta_t}(Lf *d\omega - \omega *dLf) = 0$$

and we have

$$\int_\alpha f\,d\mu - \int_\beta (Lf)(\pi_t(z)) *d\omega(\pi_t(z)) = t\int_\alpha *dLf \tag{20}$$

where $\pi_t(z)$ is a point on β_t which lies on the line passing through $z\in\beta$ with $*d\omega = 0$. Since $(Lf)(\pi_t(z))$ is uniformly bounded and converges to $(Tf)(z)$ $d\nu$-a.e. on β and $*d\omega(\pi_t(z))\to d\nu(z)$ vaguely (see Bourbaki [1]) on β as $t\to k$, we obtain on passing to the limit

$$\int_\alpha f\,d\mu - \int_\beta Tf\,d\nu = \int_\alpha *dLf.$$

Therefore (A3) of 1.2A for L is equivalent to (B2) for T.

The proof of Theorem 3B is herewith complete.

Remark 1. Goldstein [2] showed that $T\in N(\alpha,\beta)$ can be extended to a linear transformation of $L^p(\alpha,d\mu)$ into $L^p(\beta,d\nu)$ with

(B1) and (B2) for every $p \in [1, \infty]$. For $p = \infty$ see also Nakai-Sario [2].

Remark 2. Let β^w be the Wiener harmonic boundary of R (see e.g. Constantinescu-Cornea [1, p. 98]). It is easily seen that $L^\infty(\beta) = C(\beta^w)$ as Banach spaces with supremum norms, and that ν gives a distorted harmonic measure ν^w on β^w, i.e. a regular Borel measure on β^w which is equivalent to the harmonic measure. Observe that bounded functions in $H_1(A)$ are continuously extendable to $A \cup \beta^w$. Hence we can define the space $N(\alpha,\beta^w)$ by replacing Tf by $Tf = (Lf)|\beta^w$ and Theorem 3B can be restated in the following form:

$T \in N(\alpha,\beta^w)$ if and only if T is a linear transformation of $C(\alpha)$ into $C(\beta^w)$ such that for all $f \in C(\alpha)$

(B1′) $\min_\alpha f \leq Tf \leq \max_\alpha f$,

(B2′) $\int_\alpha f\, d\mu = \int_{\beta^w} Tf\, d\nu^w$.

In this form the assertion is also valid for an arbitrary open Riemann surface R and an arbitrary regular subregion R_0 (Nakai-Sario [3]), with ν^w some distorted harmonic measure on β^w. The border α may also be replaced by the Wiener harmonic boundary α^w (*loc. cit.*).

3C. The space $M(\alpha,\beta)$. The separability of α implies that of $C(\alpha)$ with respect to the supremum norm, and consequently that of $T(C(\alpha))$. A fortiori we can find a fixed $d\nu$-null set e such that Tf is defined uniquely as the angular limit of $L_T f$ at each point of $\beta - e$. We set $\beta_e = \beta - e$. For a fixed $z \in \beta_e$, $f \to (Tf)(z)$ defines a bounded linear functional on $C(\alpha)$. Thus by the Riesz representation theorem (see e.g. Royden [9, p. 256]), there exists a unique unit Borel measure $\xi_T(\cdot,z)$ such that

$$(Tf)(z) = \int_\alpha f\, d\xi_T(\cdot,z).$$

Conversely, given a family of unit Borel measures $\xi(\cdot,z)$, $z \in \beta_e$, on α, set

$$(T_\xi f)(z) = \int_\alpha f\, d\xi(\cdot,z).$$

If $T_\xi \in N(\alpha,\beta)$ then we call the family $\xi = \{\xi(\cdot,z)|z\in\beta_e\}$ *normal*, and denote the space of all normal families ξ by $M(\alpha,\beta)$. By Theorem 3B it is easy to see that

$$\xi = \{\xi(\cdot,z)|z\in\beta_e\} \in M(\alpha,\beta)$$

if and only if every $\xi(\cdot,z)$ *is a unit Borel measure such that for every* $f\in C(\alpha)$

(C1) $T_\xi f \in L^\infty(\beta)$,

(C2) $\int_\beta T_\xi f\,d\nu = \int_\alpha f\,d\mu$.

3D. Ahlfors' problem. The problem of determining $E(N(A))$ is clearly equivalent to that of determining either $E(N(\alpha,\beta))$ or $E(M(\alpha,\beta))$.

Suppose that g is a measure preserving map of the measure space $(\beta,d\nu)$ into the measure space $(\alpha,d\mu)$. Set

$$T_g f = f\circ g$$

for $f\in C(\alpha)$. Then evidently $T_g\in N(\alpha,\beta)$ and

$$\xi_{T_g}(\cdot,z) = \varepsilon_{g(z)}(\cdot)$$

for $z\in\beta_e$ where $\varepsilon_{g(z)}$ is the point measure at $g(z)\in\alpha$. Now assume that $T_g = \lambda S+(1-\lambda)T$ ($\lambda\in[0,1]$; $S,T\in N(\alpha,\beta)$ with $S\neq T$). Then

$$\varepsilon_{g(z)}(\cdot) = \lambda\xi_S(\cdot,z)+(1-\lambda)\xi_T(\cdot,z).$$

If $\lambda\in(0,1)$ we have $\xi_S(\cdot,z) = \xi_T(\cdot,z) = \varepsilon_{g(z)}(\cdot)$ and therefore $\lambda = 0$ or 1. This shows that $T_g\in E(N(\alpha,\beta))$.

There naturally arises the question (Ahlfors [4]) of whether the converse of this is valid:

For each $T\in E(N(\alpha,\beta))$ *does there exist a measure preserving map* g_T *of the measure space* $(\beta,d\nu)$ *into the measure space* $(\alpha,d\mu)$ *such that* $Tf = f\circ g_T$?

3E. A counterexample. Ahlfors (*loc. cit.*) conjectured that the answer may be in the affirmative. This would imply that problem (I) in 3A is solved at least for a compact bordered $\bar{R}$.

However, Savage [3] succeeded in constructing a counterexample contradicting Ahlfors' conjecture. Before proving Savage's result we make the following remark:

We can parametrize α by $*d\omega$ so as to identify it with $[0,1]$. Then $d\mu$ is the usual linear measure dx on $[0,1]$. The space $C(\alpha)$ may be replaced by $C'[0,1]$, the space of continuous functions $f(x)$ on $[0,1]$ with $f(0) = f(1)$. Similarly $L^\infty(\beta)$ can be replaced by $L^\infty[0,1]$ with respect to the usual linear measure dy. We will use the same notation e for the image of the exceptional set $e \subset \beta$. Thus $T \in N(\alpha,\beta)$ is a linear transformation of $C'[0,1]$ into $L^\infty[0,1]$ such that for all $f \in C'[0,1]$

$$\min_\alpha f \leq (Tf)|([0,1]-e) \leq \max_\alpha f, \tag{21}$$

$$\int_0^1 f(x)dx = \int_0^1 (Tf)(y)dy. \tag{22}$$

The counterexample $T \in E(N(\alpha,\beta))$ which is not of the form T_g is given for $f \in C'[0,1]$ by the operator

$$(Tf)(y) = \frac{1}{2}\left(f\left(\frac{y}{2}\right) + f\left(\frac{y+1}{2}\right)\right), \tag{23}$$

introduced by Ryff [1]. It is easy to see that $T \in N(\alpha,\beta)$.

We first show that T is extreme, i.e. $T \in E(N(\alpha,\beta))$. Suppose this were not the case. Then we could find $\lambda \in (0,1)$ and $U,V \in N(\alpha,\beta)$ with $U \neq V$ such that $T = \lambda U + (1-\lambda)V$. Let $\varepsilon > 0$ be so small that $\lambda+\varepsilon$, $\lambda-\varepsilon \in (0,1)$. Let

$$U' = (\lambda+\varepsilon)U + (1-(\lambda+\varepsilon))V$$

and $V' = (\lambda-\varepsilon)U + (1-(\lambda-\varepsilon))V$. Clearly U', $V' \in N(\alpha,\beta)$, $T = \frac{1}{2}U' + \frac{1}{2}V'$, and $U' \neq V'$. Thus we may assume from the outset that $\lambda = \frac{1}{2}$, i.e.

$$T = \tfrac{1}{2}U + \tfrac{1}{2}V.$$

By

$$\xi_T(\cdot,y) = \tfrac{1}{2}\varepsilon_{y/2}(\cdot) + \tfrac{1}{2}\varepsilon_{(y+1)/2}(\cdot)$$

we obtain

$$\varepsilon_{y/2}(\cdot) + \varepsilon_{(y+1)/2}(\cdot) = \xi_U(\cdot,y) + \xi_V(\cdot,y).$$

As a consequence

$$
(24) \qquad
\begin{aligned}
1 &= \xi_U\left(\frac{y}{2},y\right) + \xi_V\left(\frac{y}{2},y\right),\\
1 &= \xi_U\left(\frac{y+1}{2},y\right) + \xi_V\left(\frac{y+1}{2},y\right).
\end{aligned}
$$

Let

$$a(y) = \xi_U\left(\frac{y}{2},y\right) \quad \text{and} \quad b(y) = \xi_U\left(\frac{y+1}{2},y\right).$$

Since $\xi_U(\cdot,y)$ is a unit measure, $a(y)+b(y) = 1$ on $[0,1]-e$. Observe that

$$\xi_U(\cdot,y) = a(y)\,\varepsilon_{y/2}(\cdot) + b(y)\,\varepsilon_{(y+1)/2}(\cdot)$$

and thus

$$(Uf)(y) = \int_0^1 f(x)\,d\xi_U(x,y) = a(y)f\left(\frac{y}{2}\right) + b(y)f\left(\frac{y+1}{2}\right)$$

for every $f \in C'[0,1]$. By (22),

$$\int_0^1 f(x)\,dx = \int_0^1 \left[a(y)f\left(\frac{y}{2}\right) + (1-a(y))f\left(\frac{y+1}{2}\right)\right] dy.$$

Take $f \in C'[0,1]$ with $f|[\tfrac{1}{2},1] = 0$. Then

$$\int_0^1 f(x)\,dx = \int_0^1 a(y)f\left(\frac{y}{2}\right) dy.$$

Since

$$\int_0^1 f(x)\,dx = \int_0^{1/2} f(t)\,dt, \qquad \int_0^1 a(y)f\left(\frac{y}{2}\right) dy = \int_0^{1/2} 2a(2t)f(t)\,dt,$$

we obtain

$$\int_0^{1/2} (2a(2t)-1)f(t)\,dt = 0.$$

This is valid for every $f \in C[0,\frac{1}{2}]$ with $f(0) = f(\frac{1}{2}) = 0$, and it follows that $a(2t) = \frac{1}{2}$ a.e. in $[0,\frac{1}{2}]$, or equivalently $a(y) = \frac{1}{2}$ a.e. in $[0,1]$. Thus $b(y) = \frac{1}{2}$ a.e. in $[0,1]$. We conclude that (24) implies

$$\xi_U\left(\frac{y}{2},y\right) = \xi_V\left(\frac{y}{2},y\right) = \frac{1}{2},$$

$$\xi_U\left(\frac{y+1}{2},y\right) = \xi_V\left(\frac{y+1}{2},y\right) = \frac{1}{2},$$

a.e. in $[0,1]$, or everywhere in $[0,1]-e'$ with null set $e' \supset e$. This shows that $U = V$, a contradiction, and we have proved our assertion $T \in E(N(\alpha,\beta))$.

Finally we show that T is not of the form T_g with g a measure preserving map of the measure space $(\beta,d\nu) = ([0,1],dy)$ into the measure space $(\alpha,d\mu) = ([0,1],dx)$. If this were the case $\xi_T(\cdot,y) = \xi_{T_g}(\cdot,y)$ would have the form $\varepsilon_{g(y)}(\cdot)$ (cf. 3D), and we would obtain

$$\varepsilon_{g(y)}(\cdot) = \tfrac{1}{2}\varepsilon_{y/2}(\cdot) + \tfrac{1}{2}\varepsilon_{(y+1)/2}(\cdot).$$

This would imply $\varepsilon_{g(y)}(y/2) = \varepsilon_{g(y)}((y+1)/2) = \frac{1}{2}$, which is clearly a contradiction.

4. Principal functions and orthogonal projection

4A. Weyl's lemma. Let A be a bordered boundary neighborhood of an open Riemann surface R with $\alpha = \partial A$ and $R_0 = R-A$. Let $L_1 = (I)L_1$ be the normal operator for A defined in 2.1B and 2.2D. In the present No. 4 we will show that the L_1-principal function can be constructed by the method of orthogonal projection (Nakai-Sario [1]).

We remark in passing that L_1 is also characterized as the normal operator L such that Lf takes a constant boundary value on the Royden harmonic boundary of R (*loc. cit.*). This fact, however, will not be needed for the following construction.

We denote by $\Gamma = \Gamma(R)$ the Hilbert space of all real measurable first order differentials ω on R with norm

$$||\omega|| = \left(\iint_R \omega \wedge *\omega\right)^{1/2} < \infty,$$

the inner product (ω_1, ω_2) of $\omega_1, \omega_2 \in \Gamma$ being given by $\iint_R \omega_1 \wedge *\omega_2$. Let $\Gamma_e^1 = \Gamma_e^1(R)$ be the subspace of Γ of continuous exact differentials:

$$\Gamma_e^1 = \{df | f \in C^1(R), df \in \Gamma\}.$$

The closure of Γ_e^1 in Γ is denoted by Γ_e. We also consider the subspace Γ_{e0}^1 of Γ_e^1 of continuous exact differentials with compact supports in R,

$$\Gamma_{e0}^1 = \{df | f \in C_0^1(R)\},$$

where $C_0^1(R)$ is the subspace of $C^1(R)$ consisting of all functions with compact supports. We denote by Γ_{e0} the closure of Γ_{e0}^1 in Γ. Finally we consider the subspace Γ_{he} of Γ of harmonic exact differentials on R:

$$\Gamma_{he} = \{du | u \in H(R), du \in \Gamma\}.$$

We have the following orthogonal decomposition which is equivalent to Weyl's lemma (see e.g. Ahlfors-Sario [1, p. 281]):

$$\Gamma_e(R) = \Gamma_{he}(R) \oplus \Gamma_{e0}(R). \tag{25}$$

4B. Poincaré type inequality. Before embarking on the existence proof by the method of orthogonal projection we need the following inequality of Poincaré type (Nakai-Sario [1]); in our construction it plays a role of importance equal to that of the q-lemma in 1.1A.

Let ω be a fixed continuous first order differential defined on α such that

$$\int_\alpha \omega = 0. \tag{26}$$

Then there exists a constant $c = c(\omega, R_0)$ depending only on ω and R_0 such that

$$\left| \int_\alpha \varphi\omega \right| \leq c\|d\varphi\| \tag{27}$$

for every $\varphi \in C^1(R)$.

For the proof it is sufficient to show that

$$\left| \int_\alpha \varphi\omega \right| \leq c\|d\varphi\|_{R_0} \tag{28}$$

for every $\varphi \in C^1(\bar{R}_0)$. Set

$$H_0 = \{\varphi | \varphi \in H(R_0) \cap C^1(\bar{R}_0),\ \varphi(z_0) = 0\}$$

with a fixed $z_0 \in R_0$. In view of (26) and the Dirichlet principle we only have to prove (28) for $\varphi \in H_0$.

Let g be Green's function on R_0 with pole z_0. Set

$$r(z) = \exp(-g(z)), \qquad d\theta(z) = -*dg(z).$$

Fix a positive number a with $0 < 2a < 1$ and such that $U = \{z | z \in R_0, r(z) < 2a\}$ is a parametric disk in R_0. We write $\varphi_r(\theta) = \varphi(re^{i\theta})$ for $\varphi \in H_0$ and a fixed $r \in (0,1]$. It can be considered an element of $L^2(0,2\pi)$ with norm

$$|\varphi_r|_2 = \sqrt{\int_0^{2\pi} |\varphi_r(\theta)|^2 d\theta}.$$

Except for a finite number of values of θ we have

$$\varphi(e^{i\theta}) - \varphi(ae^{i\theta}) = \int_a^1 \frac{\partial}{\partial r} \varphi(re^{i\theta})\, dr,$$

and the Schwarz inequality gives

$$|\varphi_1(\theta) - \varphi_a(\theta)|^2 \leq \int_a^1 \frac{dr}{r} \int_a^1 \left| \frac{\partial}{\partial r} \varphi(re^{i\theta}) \right|^2 r\, dr.$$

Since $\left|\frac{\partial}{\partial r}\varphi(re^{i\theta})\right|^2 rdrd\theta \leq d\varphi \wedge *d\varphi$, it follows that

$$|\varphi_1 - \varphi_a|_2 \leq \sqrt{-\log a}\,||d\varphi||_{R_0}. \tag{29}$$

In the same fashion as in 0.2.3F we obtain

$$|\varphi_a(\theta)|^2 \leq \frac{1}{\pi}\,||d\varphi||^2_{R_0}.$$

Therefore

$$|\varphi_a|_2 \leq \sqrt{2}\,||d\varphi||_{R_0}.$$

By virtue of this inequality together with (29), the triangle inequality $|\varphi_1|_2 \leq |\varphi_1 - \varphi_a|_2 + |\varphi_a|_2$ implies that

$$\left(\int_0^{2\pi} |\varphi(e^{i\theta})|^2 d\theta\right)^{1/2} \leq \left(\sqrt{2} + \sqrt{-\log a}\right)||d\varphi||_{R_0}. \tag{30}$$

Take a piecewise analytic representation of α with parameter θ in $[0,2\pi)$. Then we can write $\omega = \tilde{\omega}(\theta)d\theta$ on α, with $\tilde{\omega}(\theta)$ a bounded piecewise continuous function on $[0,2\pi)$, say $|\tilde{\omega}(\theta)| \leq k < \infty$. By the Schwarz inequality we have

$$\left|\int_\alpha \varphi\omega\right|^2 = \left|\int_0^{2\pi} \varphi(e^{i\theta})\tilde{\omega}(\theta)d\theta\right|^2 \leq 2\pi k^2 \int_0^{2\pi} |\varphi(e^{i\theta})|^2 d\theta. \tag{31}$$

We set $c = (\sqrt{2} + \sqrt{-\log a})\sqrt{2\pi}k$ and conclude by (30) and (31) that the assertion (28) holds.

Remark. It is known (Nakai [6]) that (27) is valid without (26) for a hyperbolic R, and that (27) is valid for a parabolic R if and only if (26) holds.

4C. Existence proof by orthogonal projection. Let

$$s \in H_1(A) \cap C^1(A)$$

and assume that

$$\int_\alpha *ds = 0. \tag{32}$$

We are going to use the method of orthogonal projection to prove the existence of a function $p \in H(R)$ such that

$$(33) \qquad L_1(p-s) = p-s \quad \text{in} \quad A.$$

First we extend s to all of the surface R so as to have $s \in C^1(R)$. For $\sigma \in \Gamma^1_{e0}$ we set

$$(34) \qquad t(\sigma) = -\iint_R \sigma \wedge *ds = -\iint_R ds \wedge *\sigma;$$

this is well defined because σ has compact support and $\sigma \wedge *ds = ds \wedge *\sigma$ is a continuous second order differential with compact support. Clearly t gives rise to a linear functional on Γ^1_{e0}.

We shall next show that t is a bounded linear functional on Γ^1_{e0}. Take a regular region Ω_σ containing $\bar{R}_0$ and the support of $\sigma \in \Gamma^1_{e0}$. Then

$$(35) \qquad -t(\sigma) = \int_{R_0} \sigma \wedge *ds + \int_{\Omega_\sigma - R_0} \sigma \wedge *ds.$$

By the Schwarz inequality we obtain

$$(36) \qquad \left| \int_{R_0} \sigma \wedge *ds \right| \leq \left(\int_{R_0} ds \wedge *ds \right)^{1/2} ||\sigma||.$$

For $\sigma = d\varphi$ with $\varphi \in C_0^1(R)$ we have by Stokes' theorem

$$\iint_{\Omega_\sigma - \bar{R}_0} \sigma \wedge *ds = \iint_{\Omega_\sigma - \bar{R}_0} d(\varphi *ds) = -\int_\alpha \varphi *ds.$$

In view of $\int_\alpha *ds = 0$, (27) yields

$$\left| \iint_{\Omega_\sigma - R_0} \sigma \wedge *ds \right| \leq c||\sigma||.$$

This with (35) and (36) implies

$$|t(\sigma)| \leq K||\sigma||$$

for every $\sigma \in \Gamma^1_{e0}$ with the constant $K = c + \int_{R_0} ds \wedge *ds$.

We conclude that $t(\cdot)$ is a bounded linear functional on Γ^1_{e0}, and thus has a unique extension T to $\Gamma_{e0} = \overline{\Gamma^1_{e0}}$ as a bounded linear functional. Hence by the Riesz theorem (see e.g. Royden [9, p. 256]) there exists a unique element $\lambda \in \Gamma_{e0}$ such that

$$T(\cdot) = (\cdot,\lambda)$$

on Γ_{e0}. In particular $T(\sigma) = (\sigma,\lambda)$ for $\sigma \in \Gamma^1_{e0}$, i.e.

$$\int_R (\lambda+ds) \wedge *\sigma = 0. \tag{37}$$

Let $\{R_n\}_0^\infty$ be an exhaustion of R. Although $\lambda+ds$ is not an element of $\Gamma_e(R)$ in general, we can conclude that

$$\lambda+ds \in \Gamma_e(R_n),$$

because $\lambda \in \Gamma_{e0}(R) \subset \Gamma_e(R) \subset \Gamma_e(R_n)$ and $ds \in \Gamma_e(R_n)$. Since (37) is of course true for $\sigma \in \Gamma^1_{e0}(R_n)$, we have $((\lambda+ds),\sigma)_{R_n} = 0$ for every $\sigma \in \Gamma_{e0}(R_n)$. Thus by (25) there exists a $q_n \in H(R_n)$ with $dq_n = \lambda+ds$ on R_n. Clearly $dq_{n+m} = dq_n$ on R_n for $m \geq 1$. Let c_n be a constant such that $q_{n+1} = q_n+c_n$ on R_n. Set $p_1 = q_1$ in R_1, $p_n = q_n-c_{n-1}$ in R_n for $n = 2, 3,\cdots$. Then $p_n \in H(R_n)$, $dp_n = \lambda+ds$ on R_n, and $p_{n+m} = p_n$ on R_n for every $m = 1, 2,\cdots$. Hence if we put

$$p(z) = p_n(z)$$

for $z \in R_n$ then $p(z)$ does not depend on the choice of R_n to which z belongs. It follows that $p \in H(R)$ and $dp = \lambda+ds$. The function $u = p-s$ belongs to $C^1(R)$ and clearly $u \in H_1(A) \cap C^1(A)$. By (32) and $\int_\alpha *dp = 0$ we see that

$$\int_\alpha *du = 0. \tag{38}$$

We next define v_n and v on R as follows: $v_n|R_0 = v|R_0 = u$, $v_n|A \cap R_n = L_{1R_n}u$, $v|A = L_1u$, and $v_n|A-R_n = v_n|\partial R_n$ (const.).

It is easily seen that $dv_n \in \Gamma_{e0}(R)$. Since $||dv - dv_n|| \to 0$ as $n \to \infty$ (cf. 2.2D), $dv \in \Gamma_{e0}(R)$. Clearly

$$\int_\alpha * dv = 0. \tag{39}$$

Set $w = u - v$. Since $du = \lambda$ is an element of $\Gamma_{e0}(R)$, so is dw. Observe that $w|R_0 = 0$ and, by (38) and (39),

$$\int_\alpha * dw = 0. \tag{40}$$

By virtue of $dw \in \Gamma_{e0}$ we can find a sequence $\{\varphi_m\} \subset C_0^1(R)$ with $||d\varphi_m - dw|| \to 0$ as $m \to \infty$. Hence

$$||dw||^2 = \lim_{m\to\infty} (d\varphi_m, dw). \tag{41}$$

Let Ω_m be a regular subregion of R with $\Omega_m \supset \bar{R}_0$ such that the support of φ_m is contained in Ω_m. By Stokes' theorem

$$(d\varphi_m, dw) = (d\varphi_m, dw)_{\Omega_m - \bar{R}_0} = -\int_\alpha \varphi_m * dw.$$

In view of (40) we can apply (28) to conclude that

$$|(d\varphi_m, dw)| \le c||d\varphi_m||_{R_0} = c||d\varphi_m - dw||_{R_0} \le c||d\varphi_m - dw||.$$

Consequently $\lim_{m\to\infty} (d\varphi_m, dw) = 0$ and a fortiori by (41), $dw = 0$, or $w = 0$. We infer that $u = v$ on R and in particular

$$u = L_1 u$$

on A, or equivalently (33) holds on A.

Remark. A lucid account of other ways to apply the method of orthogonal projection to these topics may be found in Royden [10].

CHAPTER II
PRINCIPAL FUNCTIONS

In this chapter we discuss various problems of complex analysis which may be treated quite naturally with principal functions.

We begin in §1 with a general treatment of extremal problems which are solved by principal functions. These results are applied in §2 to the special case of plane regions and lead to the construction of canonical conformal mappings.

The remaining topics center around square-integrable differentials and extremal length. In §§3–5 we use principal functions to construct reproducing differentials, to solve interpolation problems, and to generalize the theorems of Riemann-Roch and Abel to arbitrary Riemann surfaces. In §6 we discuss extremal length, use it to characterize p_0 and p_1, and relate the extremal length of homology classes to the norms of reproducing differentials.

§1. MAIN EXTREMAL THEOREM

Among all harmonic functions with the same singularities on a Riemann surface, the principal functions can be distinguished as solutions of important extremal problems. Such extremal properties have applications to conformal mapping, interpolation problems, and classification theory.

1. An extremal functional

1A. Principal functions. Let $\{\zeta_1,\cdots,\zeta_n\}$ be a finite subset of an open Riemann surface R. For each j let Δ_j be a bordered disk about ζ_j and let Δ_0 be a bordered boundary neighborhood of R. Assume Δ_0 and Δ_j are mutually disjoint. Let s denote a harmonic function defined every in $\Delta_0\cup\cdots\cup\Delta_n$ except at $\zeta_1,\cdots,\zeta_n$ and satisfying $s\equiv0$ in Δ_0, $\Sigma\int_{\partial\Delta_j}*ds = 0$. Then the domain of s is a bordered boundary neighborhood of the punctured surface $R-\{\zeta_1,\cdots,\zeta_n\}$ and s may be considered as an admissible singularity function. As a normal operator take the direct sum of the Dirichlet operators in $\Delta_j-\{\zeta_j\}$ $(1\leq j\leq n)$ and the operator L_0 or $(P)L_1$ in Δ_0. Here P refers to any fixed partition of the ideal boundary of R.

The Main Existence Theorem I.1.2B applies and yields a principal function which is harmonic on $R-\{\zeta_1,\cdots,\zeta_n\}$. This function is uniquely determined up to an additive constant. It shall be denoted by p_0 or p_1 according to the choice of the operator L_0 or $(P)L_1$ in Δ_0.

Let h and k be real numbers with $h+k = 1$. Denote by p_{hk} the function hp_0+kp_1. Consider the mapping of functions p, harmonic on R except for the singularity s, into the Dirichlet integral $D(p-p_{hk})$. This functional obviously takes its minimum for $p = p_{hk}$. We shall see that this functional may be expressed in terms of the power series coefficients of p, p_0, p_1, and s at the points ζ_j. The exact formula will be the content of the next theorem.

1B. The extremal functional. At first suppose R is the interior of a compact bordered Riemann surface $\bar{R}$ with border β. We shall derive the extremal properties of p_0, p_1 in this case and then remove the restriction that R be contained in such an $\bar{R}$.

There is an apparent inconsistency in the use of L_0, L_1 for Δ_0 in this case. If we consider R as an open Riemann surface then L_0, L_1 for Δ_0 were defined as limits of $L_{0\Omega}$, $L_{1\Omega}$ for an exhaustion Ω of R (see I.2.2B, D). On the other hand, since Δ_0 is the interior of a compact bordered subregion of $\bar{R}$ we might use the direct definition as in I.2.1A, B. The two definitions are equivalent, however, as may be easily seen by the reasoning of I.2.2B, 2D used to prove the convergence of $u_\Omega = L_{i\Omega}f$ for any $f\in C(\partial\Delta_0)$. The direct definition shows immediately that p_0, p_1 extend harmonically to β; we continue to write p_0, p_1 for these extended functions. On β, p_0 has vanishing normal derivative and p_1 has constant values on all contours in the same part determined by the partition P. The flux of p_1 across each part is zero.

Let p be any harmonic function on $\bar{R}-\{\zeta_1,\cdots,\zeta_n\}$ which satisfies $k\int_\beta p_1 * dp = 0$ and which has the singularity s, i.e., $p-s$ has a removable singularity at each ζ_j. We call such functions *admissible.* Note that if $k = 0$ or $P = I$, the identity partition, then there is no restriction. Using the fact that $\int_\beta p*dp_0 = 0$ we obtain, for any admissible p,

$$(1)\quad D(p-p_{hk}) - \int_\beta p*dp = h\int_\beta p*dp_0 - p_0*dp$$

$$+k\int_\beta p_1*dp \quad -p*dp_1 + hk\int_\beta p_0*dp_1 - p_1*dp_0.$$

By Green's theorem each line integral on the right-hand side of (1) depends only on the homology class of β. Each may be transferred to $\bigcup_n^1 \partial\Delta_j$ and evaluated as a residue. Introduce the notation $R(p,q) = \int p*dq - q*dp$, where the integration is taken over $\bigcup_n^1 \partial\Delta_j$. If q is harmonic in $\bigcup(\Delta_j - \{\zeta_j\})$ write $\tilde{q} = q - s$. Since $\tilde{p}$, $\tilde{p}_i$ with $i = 0, 1$ are harmonically extendable to Δ_j, we have $R(\tilde{p},\tilde{p}_i) = 0$ and hence obtain equations of the form $R(p,p_i) = R(\tilde{p},s) + R(s,\tilde{p}_i)$ for $i = 0,1$. Using $h+k = 1$ we infer that

$$(2)\quad D(p-p_{hk}) - \int_\beta p*dp$$

$$= (h-k)R(\tilde{p},s) - h^2R(\tilde{p}_0,s) + k^2R(\tilde{p}_1,s).$$

Let

$$s = \text{Re}\sum_{\nu=1}^{\infty} b_\nu^{(j)} z^{-\nu} + c^{(j)}\log|z|$$

and

$$p = \text{Re}\sum_{\nu=0}^{\infty} a_\nu^{(j)} z^{\nu} + s$$

in Δ_j where z is a local parameter which makes ζ_j correspond to 0. Let q^* denote a harmonic conjugate of q. By virtue of $\int_{\partial\Delta_j} s\, d\tilde{p}^* = -\int_{\partial\Delta_j}\tilde{p}^* ds$ we have

$$R(\tilde{p},s) = \sum_j \text{Im}\int_{\partial\Delta_j} (\tilde{p}+i\tilde{p}^*)(ds + i*ds) = C(p)$$

where

$$C(p) = \operatorname{Im} 2\pi i \sum_{j=1}^{n} (c^{(j)}a_0^{(j)} - \sum_{\nu=1}^{\infty} \nu a_\nu^{(j)} b_\nu^{(j)}).$$

Thus

$$(3) \quad D(p-p_{hk}) - \int_\beta p * dp = k^2 C(p_1) - h^2 C(p_0) + (h-k) C(p).$$

1C. Main Extremal Theorem. The above results may be extended to an arbitrary open Riemann surface R. Let p be a function on R, harmonic except for the singularity s, whose restriction to each regular subregion Ω of an exhaustion $\{\Omega\}$ of R is admissible. This means that p has vanishing flux across each P-dividing cycle of a regular partition P in case $k \neq 0$. The reasoning in 1B may be applied to each sufficiently large subregion Ω, and in the resulting equation (3) the directed limit $\Omega \to R$ may be taken. The principal functions on Ω tend uniformly on compacta to the corresponding principal functions on R by Theorem I.2.3B. Consequently the Dirichlet integral and the power series coefficients also converge. Thus (3) is true in this more general case.

Finally, we note that the results apply for any real h and k provided the admissible functions have the singularity $(h+k)s$. If $h+k \neq 0$ this assertion is obvious by the homogeneity of (3) when the functions p and p_{hk} are divided by $h+k$.

The case $h+k = 0$ requires slightly more proof. Then an admissible function q is harmonic on all of R. We write $D(q-p_{hk}) = D((q+p_1) - p_{h,k+1})$ and apply (2). This yields

$$D(q-p_{hk}) - \int_\beta q * dq = R(q,p_1) + (h-k-1) R(q+p_1,s)$$

$$-h^2 R(p_0,s) + (k+1)^2 R(p_1,s).$$

This is precisely (2) with p replaced by q since $R(q,p_1-s) = 0$ and the coefficient of $R(p_1,s)$ becomes k^2. Therefore the following theorem (Sario [9], [13], [14]) has been established.

Main Extremal Theorem. *Let R be an arbitrary Riemann surface; $\zeta_1, \cdots, \zeta_n$ points of R; s a harmonic function defined on a* *punctured neighborhood of each ζ_j and zero outside a compact subset of R; P a regular partition of the ideal boundary; p a harmonic function with the singularity $(h+k)s$ and, if $k\neq 0$, with vanishing flux across each P-dividing cycle. Let p_0 and p_1 be the associated principal functions. Let*

$$s = \operatorname{Re} \sum_{\nu=1}^{\infty} b_\nu^{(j)} z^{-\nu} + c^{(j)} \log |z|$$

and

$$p = \operatorname{Re} \sum_{\nu=0}^{\infty} a_\nu^{(j)} z^{\nu} + (h+k)s$$

near ζ_j, and set

$$C(p) = \operatorname{Im} 2\pi i \sum_j \left(c^{(j)} a_0^{(j)} - \sum_{\nu=1}^{\infty} \nu a_\nu^{(j)} b_\nu^{(j)}\right).$$

Then the functional $(h-k)C(p) + \int_\beta p * dp$ is minimized by p_{hk} and the value of the minimum is $h^2 C(p_0) - k^2 C(p_1)$:

$$(3') \qquad (h-k)C(p) + \int_\beta p * dp = h^2 C(p_0) - k^2 C(p_1) + D(p - p_{hk}).$$

2. Special cases

2A. The functions $p_0 + p_1$. We continue to use the notation of the Main Extremal Theorem. An important special case of that theorem is $h = k = \frac{1}{2}$. Since the Dirichlet integral is nonnegative we immediately obtain the

Corollary. *For any singularity s the function $\frac{1}{2}(p_0 + p_1)$ minimizes the functional $p \rightarrow \int_\beta p * dp$ among all admissible p. The value of the minimum is $\frac{1}{4}(C(p_0) - C(p_1))$.*

The last assertion is a consequence of the Main Extremal Theorem for the cases $(h,k) = (1,0)$ and $(0,1)$.

2B. Meromorphic and logarithmic poles. In case s has a single pole of the form $\operatorname{Re} e^{i\theta}(z-\zeta)^{-1}$ then $C(p) = -2\pi \operatorname{Re} a_1 e^{i\theta}$.

If s has only two logarithmic poles, $\log|z-\zeta_1|$ and $-\log|z-\zeta_2|$, then $C(p) = 2\pi \operatorname{Re}[a_0^{(1)} - a_0^{(2)}]$. Thus we have the

Corollary. *Among all admissible functions p with singularity $s = \operatorname{Re} e^{i\theta}(z-\zeta)^{-1}$, p_0 minimizes $\int_\beta p * dp - 2\pi \operatorname{Re} a_1 e^{i\theta}$ and p_1 minimizes $\int_\beta p * dp + 2\pi \operatorname{Re} a_1 e^{i\theta}$. For a singularity s with two logarithmic poles, $\log|z-\zeta_1|$ and $-\log|z-\zeta_2|$, p_0 minimizes $\int_\beta p * dp + 2\pi \operatorname{Re}[a_0^{(1)} - a_0^{(2)}]$ and p_1 minimizes $\int_\beta p * dp - 2\pi \times \operatorname{Re}[a_0^{(1)} - a_0^{(2)}]$.*

In certain contexts it is natural to consider only those admissible functions p for which $\int_\beta p * dp \leq 0$. This condition is of course satisfied for p_0, p_1. On a planar Riemann surface the real parts of univalent functions also have this property (see 2.1C). We obtain:

For admissible functions p with $\int_\beta p * dp \leq 0$, the inequalities

$$C(p_0) \leq C(p) \leq C(p_1)$$

hold.

In particular, $\operatorname{Re} a_1 e^{i\theta}$ is maximized by p_0 and minimized by p_1, in case $s = \operatorname{Re} e^{i\theta}(z-\zeta)^{-1}$. For a singularity $\log|z-\zeta_1|$ and $-\log|z-\zeta_2|$ the maximum and minimum of $\operatorname{Re}(a_0^{(2)} - a_0^{(1)})$ are attained for p_0 and p_1.

2C. Regular principal functions. We now consider the case $h+k = 0$. Any harmonic function will be admissible if it has vanishing flux across all P-dividing cycles.

Applying the Main Extremal Theorem to a pair of logarithmic poles at ζ_1, ζ_2 leads to a solution of the problem of minimizing the "span" $u(\zeta_1) - u(\zeta_2)$ among functions u.

Corollary. *Given $\zeta_1, \zeta_2 \in R$ the expression*

$$D(u) + 4\pi u(\zeta_1)$$

is minimized, among all admissible functions which vanish at ζ_2, by the difference $p_0 - p_1$ of the principal functions corresponding to the singularities $\log|z-\zeta_1|$ and $-\log|z-\zeta_2|$. For the singularity $\operatorname{Re}(z-\zeta)^{-1}$ the function $p_0 - p_1$ minimizes

$$D(u) - 4\pi \left(\frac{\partial u}{\partial x}\right)_{z=\zeta}.$$

For both types of singularities the minimum is $-D(p_0-p_1)$ *and the deviation from it is* $D(u-(p_0-p_1))$.

Here the value of the minimum is obtained by choosing $p \equiv 0$ in (3′).

§2. CONFORMAL MAPPING

In this section we restrict our attention to planar Riemann surfaces R. Then principal functions can be used to form meromorphic functions on R. The classical canonical mappings may be constructed in this way, and the extremal properties just established allow simple proofs of their geometric mapping properties.

1. Parallel slit mappings

1A. Principal meromorphic functions. Throughout this section R will be a planar Riemann surface and the L_1 normal operator will be defined with respect to the canonical partition Q.

Let $\zeta \in R$ and let z be a local parameter near ζ which maps ζ to the origin. Denote by s^θ the singularity function on $R-\{\zeta\}$ equal to $\operatorname{Re} e^{i\theta}/z$ near ζ and identically 0 near the ideal boundary of R. We let $p_0{}^\theta$, $p_1{}^\theta$ be the corresponding principal functions. They satisfy

$$\int_\gamma *dp_0{}^\theta = \int_\gamma *dp_1{}^\theta = 0 \tag{4}$$

for every dividing cycle γ on $R-\{\zeta\}$. Every cycle on $R-\{\zeta\}$ is dividing since R is planar. Thus (4) implies that there exist harmonic functions $q_j{}^\theta$ $(j=0,1)$ such that $P_j{}^\theta = p_j{}^\theta + iq_j{}^\theta$ is meromorphic on R. These functions are determined only to within an additive constant. To obtain uniqueness we normalize them so that $P_j{}^\theta(z) - e^{i\theta}z^{-1}$ tends to 0 as $z \to \zeta$.

Let us show that $P_0{}^\theta = -iP_1{}^{\theta+\pi/2}$. If R is a compact bordered surface then $q_0{}^\theta + p_1{}^{\theta+\pi/2}$ is harmonic on R. On each boundary contour it is constant and has vanishing flux. It is therefore constant on all of R, as may be seen by evaluating its Dirichlet integral. By Theorem I.2.3B these functions on an open surface R are limits of corresponding functions on compact bordered subregions. Thus $q_0^\theta + p_1^{\theta+\pi/2}$ is constant on an arbitrary R. This implies that $P_0^\theta + iP_1^{\theta+\pi/2}$ is a constant which according to our normalization must be 0. Therefore we may restrict our attention to P_0^θ in studying the mapping properties of these principal meromorphic functions.

Write $P_0 = P_0^0$, $P_1 = P_1^0$. In the same way one can establish the formula

$$P_0^\theta = P_0 \cos \theta + iP_1 \sin \theta \tag{5}$$

which will be of use later.

1B. Horizontal slit mapping. Assume R is the interior of a compact bordered surface. Then it is easily seen that P_0^θ maps R univalently onto a horizontal slit region. That is, the image is the complement in the extended plane of a finite number of horizontal slits. Indeed, P_0^θ extends holomorphically onto each border contour. There its imaginary part remains constant and its real part varies between a finite minimum and maximum. Thus the image of each contour is a horizontal slit.

Suppose w_0 in the image plane does not lie on any of these slits. Then the winding number of the slits about w_0 is 0. By the argument princple, the number of points in R which are mapped to w_0 is the same as the number of poles of P_0^θ, namely one. Thus P_0^θ is a univalent map of R onto a horizontal slit region.

1C. For an arbitrary open surface R we may conclude immediately that P_0^θ, as a nondegenerate limit of univalent functions, remains univalent. That its image is a slit region will follow from extremal properties of principal functions.

Lemma. *Among all univalent meromorphic functions on R with an expansion of the form*

$$\frac{e^{i\theta}}{z} + az + \cdots \tag{6}$$

near ζ, the function P_0^θ maximizes Re $ae^{i\theta}$.

This lemma follows directly from 1.2B, once it is observed that $\int_\beta p * dp \leq 0$ whenever p is the real part of a univalent meromorphic function P on R with a simple pole. For if β_Ω is the oriented boundary of a subregion $\Omega \subset R$ which contains the pole of P then $P(\beta_\Omega)$ is the negatively oriented boundary of A, the complement of $P(\Omega)$. If P maps into the (u,v)-plane we have

$$(7) \qquad \int_{\beta_\Omega} p * dp = \int_{P(\beta_\Omega)} udv = -\iint_A dudv \leq 0.$$

1D. Let E be any component of the complement of $P_0^\theta(R)$. If E is neither a point nor a slit, we shall obtain a contradiction to the above lemma by considering $\varphi \circ P_0^\theta$ where φ is a conformal mapping of the complement of E onto a slit region.

First observe that if f is a biholomorphic mapping of a Riemann surface S onto another surface S' and if p is a principal function on S' then $p \circ f$ is a principal function on S. Now let S be the complement of E. By the Riemann mapping theorem, if E is not a point S is conformally equivalent to the interior of a compact bordered surface. In 1B we saw that a principal function with a simple pole in the interior of such a surface maps it onto a horizontal slit region. Thus there is an L_0-principal function φ on S, with an expansion $\varphi(w) = w + bw^{-1} + \cdots$ near ∞, which maps S univalently onto the complement of a horizontal slit. From Lemma 1C it follows that Re $b \geq 0$. Furthermore, equality holds only if $\varphi(w) = w$, since the identity function w is admissible.

Let

$$(8) \qquad P_0^\theta(z) = \frac{e^{i\theta}}{z} + a_\theta z + \cdots$$

near ζ. Then $\varphi \circ P_0^\theta(z) = e^{i\theta} z^{-1} + (a_\theta + be^{-i\theta})z + \cdots$. By Lemma 1C Re $a_\theta e^{i\theta} \geq$ Re$(a_\theta e^{i\theta} + b)$, whence Re $b = 0$. Therefore $\varphi(w) = w$, and consequently E is a horizontal slit. Thus we have proved most of the

Theorem. *The principal function P_0^θ maps R biholomorphically into the extended plane. The complement of $P_0^\theta(R)$ has area zero and its components consist of points or horizontal slits.*

Making use of (5) it follows that P_1 maps R biholomorphically onto a region whose complementary components are vertical slits or points.

1E. Extremal property. Let P be a univalent meromorphic function on R with the development (6) near ζ. Let the expansion of P_0^θ be given by (8). If $p = \operatorname{Re} P$, the Main Extremal Theorem shows that

$$(9) \qquad 2\pi \operatorname{Re}(ae^{i\theta}) - \int_\beta p * dp = 2\pi \operatorname{Re}(a_\theta e^{i\theta}) - D(p - p_0^\theta).$$

Let A_P denote the area of the complement of $P(R)$. The next result (Sario [16]) completes the proof of Theorem 1D and characterizes P_0^θ by an extremal property. Its proof follows immediately from (7) and (9).

Theorem. *Among all univalent P with the development* (6), *the functional*

$$P \to 2\pi \operatorname{Re}(ae^{i\theta}) + A_P$$

is maximized by P_0^θ. The value of the maximum is $2\pi \operatorname{Re}(a_\theta e^{i\theta})$ and the deviation from it is $D(p - p_0^\theta)$.

2. Mapping by $P_0 + P_1$

2A. Compact bordered surfaces. In 1 we found that the functions P_0, P_1 were univalent and mapped R onto a horizontal and vertical slit region, respectively. The function $P_0 + P_1$ is also univalent. A discussion of the geometric properties due to Oikawa-Suita [1] of this mapping for surfaces of arbitrary connectivity must wait until Chapter III when we have the method of extremal length at our disposal. We shall then make use of the following result (Grunsky [1], Schiffer [1], Sario [16]) which can be established by the present methods.

Theorem. *Let $\bar{R}$ be a compact bordered Riemann surface with interior R. The function $P_0 + P_1$ gives a univalent mapping of R onto a region bounded by convex analytic curves.*

The proof requires a closer study of the behavior of P_0^θ on the boundary β of R. The convexity property can then be derived and the univalency deduced from it.

2B. Boundary behavior of P_0^θ. Let β_k be any boundary contour of $\bar{R}$ and let σ_k be the horizontal slit $P_0^\theta(\beta_k)$. For any point

w in the plane consider $\int_{\beta_k} f^{-1}df$ where $f = P_0^\theta - w$. By the Extended Residue Theorem (e.g. Ahlfors [7], pp. 247 ff.) the principal value of this integral is $2\pi i(n(w) - n(\infty))$ where $n(w)$ is the number of times P_0^θ takes the value w in $\bar{R}$. Multiplicities are counted as usual for w-points in R and with half their ordinary values if they lie on the boundary. We have seen that $\int_{\beta_k} f^{-1}df = 0$ if $w \notin \sigma_k$. If $w \in \sigma_k$ the integral reduces to $\int_{\beta_k} d \log P_0^\theta$. Thus it is zero in all cases and $n(w) = n(\infty) = 1$. It follows that f has either a simple zero in R, two simple zeros on some β_k, or a double zero on some β_k. Consequently $p_0^\theta|\beta_k$ takes every value between its maximum and minimum at two distinct points.

2C. Convexity of the boundary. On β_k we have $p_0^\theta = \operatorname{Re} P_0^\theta = \operatorname{Re}[(P_0+P_1)e^{i\theta}] +$ const. This shows that the line $\operatorname{Re}(we^{i\theta}) =$ const. has at most two points of intersection with the curve $w = (P_0+P_1)|\beta_k$. Thus the image γ_k of β_k under P_0+P_1 is convex.

We shall now show that each γ_k is an analytic curve, i.e. $(P_0+P_1)' \neq 0$ on β_k. We have seen that each of P_0' and P_1' has two simple zeros on β_k. These zeros must occur at distinct points. For suppose that $P_0' = P_1' = 0$ at a point b of β_k. By (5) this would imply that the first two derivatives of P_0^θ vanish at b for an appropriate choice of θ. Indeed, P_0'' is real and P_1'' is imaginary so there is a θ such that at b, $P_0''/P_1'' = -i \tan \theta$. However, P_0^θ takes every value with multiplicity ≤ 2. Thus P_0', P_1' cannot vanish simultaneously. It follows that $(P_0+P_1)'$ cannot vanish on β_k, for P_0' is real and P_1' is imaginary along β_k.

We now compute the winding number of γ_k about a point inside it by investigating the way the tangent turns as a point traces β_k. The tangent to γ_k has slope $\operatorname{Im} F$ where $F = P_1'/P_0'$.

Note that $F \neq 0, \infty$ in R and that $F(\zeta) = 1$ where ζ is the pole of P_0 and P_1. On each β_k, F has two simple poles but is otherwise imaginary. At the two points where P_0^θ has vanishing derivative we see from (5) that $F = i \cot \theta$. Thus F takes every imaginary value at least twice. For a fixed θ apply the Extended Residue Theorem to $F - i \cot \theta$. There are exactly two poles and at least two zeros on each β_k. In R there are no poles. Hence there are exactly two zeros on each β_k and none in R. We conclude that $\operatorname{Re} F > 0$ in R and consequently $\operatorname{Im} F$ decreases as β_k is traced in the positive direction. Therefore each γ_k is an analytic convex curve traced exactly once in the direction of decreasing slope.

To prove that P_0+P_1 is univalent in R, choose a point w inside of m curves γ_k. The winding number of $(P_0+P_1)(\partial R)$ about w is $-m$. This integer represents $n(w)-n(\infty)$, and therefore $m = -1$. The proof of Theorem 2A is complete.

2D. Extremal property of P_0+P_1. Corollary 1.2A gives an extremal property of P_0+P_1. The functional which is minimized has a geometric interpretation according to (7). As in 1E, let A_P denote the area of the complement of $P(R)$, when P is a univalent function on R. Then $\frac{1}{2}(P_0+P_1)$ maximizes A_P among all univalent meromorphic functions on R with an expansion $(z-\zeta)^{-1}+a(z-\zeta)+\cdots$ near ζ. The value of the maximum is, according to the Main Extremal Theorem,

$$A_{(P_0+P_1)/2} = \tfrac{1}{4}[C(p_1)-C(p_0)] = \tfrac{1}{2}\pi \operatorname{Re}[a(P_0)-a(P_1)].$$

We shall see below that $a(P_0)-a(P_1)$ is real. This will prove the following extremal property (Grunsky [1], Schiffer [1], Lokki [1], Lehto [1], Sario [13], [16]):

Theorem. *The function $\frac{1}{2}(P_0+P_1)$ maximizes A_P among all univalent mappings with a development $(z-\zeta)^{-1}+a(z-\zeta)+\cdots$ near ζ. The value of the maximum is*

$$A_{(P_0+P_1)/2} = \tfrac{1}{2}\pi\,[a(P_0)-a(P_1)].$$

2E. Extremal property of P_0-P_1. To prove that $a(P_0)-a(P_1)$ is ≥ 0 consider $Q = P_0-P_1$. Corollary 1.2C shows that Q minimizes $D(U)-4\pi \operatorname{Re} a(U)$ among all analytic functions U on R with a development $a_0+a(U)(z-\zeta)+\cdots$ near ζ. Now compare Q with $Qe^{i\theta}$. The minimum property becomes $\operatorname{Re} a(Q) \geq \operatorname{Re}[a(Q)e^{i\theta}]$. This holds for all θ and consequently $a(Q)\geq 0$. The proof of Theorem 2D is herewith complete.

The extremal property of Q is given more precisely by

$$D(U)-4\pi a(U) = -D(Q)+D(U-Q).$$

Thus the value of the minimum can be written as $-2\pi a(Q) = -D(Q)$. From the above theorem we have $2\pi a(Q) = 4A_{(P_0+P_1)/2} = A_{P_0+P_1}$. This establishes the following result (Sario [13]):

Theorem. *The analytic function* $Q = P_0 - P_1$ *minimizes*

$$D(U) - 4\pi \operatorname{Re} a(U)$$

among all analytic functions U *on* R *with* $U = a_0 + a(z-\zeta) + \cdots$ *near* ζ. *The value of the minimum is*

$$-2\pi a(Q) = -A_{P_0+P_1}.$$

As a consequence, $D(U)$ is minimized by $Q/a(Q)$ among analytic functions U with $U'(\zeta) = 1$.

3. Circular and radial slit planes

3A. Principal meromorphic functions F_0, F_1. Fix two points ζ_1, ζ_2 on the planar surface R. On $R - \{\zeta_1, \zeta_2\}$ choose a singularity function s of the form $\log|z-\zeta_1|$, $-\log|z-\zeta_2|$ near ζ_1, ζ_2 respectively and $\equiv 0$ near the ideal boundary of R. As in I.2B, principal functions p_0, p_1 may be formed with this singularity. In the case of p_1, we choose the L_1-operator which corresponds to the canonical partition of the ideal boundary of R. The admissible functions are now those which have simple logarithmic poles at ζ_1, ζ_2 with conjugate periods of 2π, -2π there. The flux is zero across every cycle which separates $\{\zeta_1, \zeta_2\}$ from the ideal boundary of R. Since R is planar this means that the flux across any cycle is an integral multiple of 2π.

If p is admissible and $\xi \in R - \{\zeta_2\}$ then

$$F(\xi) = \exp \int_{\zeta_1}^{\xi} dp + i * dp$$

is a well-defined holomorphic function on $R - \{\zeta_2\}$, independent of the path from ζ_1 to ξ. It extends meromorphically to all of R with a simple pole at ζ_2. Let $c = c(F)$ be the residue of F at ζ_2. Let F_0, F_1 correspond to the principal functions p_0, p_1.

3B. Mapping and extremal properties. If R is the interior of a compact bordered surface $\bar{R}$ then F_0, F_1 extend holomorphically to $\partial\bar{R}$. On any contour $|F_1|$ and $\arg F_0$ are constant. By the argument principle F_1 maps R univalently onto the extended plane less a finite number of concentric circular slits. Similarly F_0 is a univalent map onto a region bounded by a finite number of radial slits.

To determine the mapping properties for an arbitrary R we first note the extremal properties of F_0, F_1. In the corollary of 1.2B, the quantities $a_0^{(1)}$, $a_0^{(2)}$, and $\int_\beta p*dp$ can be expressed in terms of F. The normalization means $a_0^{(1)} = 0$, $a_0^{(2)} = \log|c(F)|$. If F is univalent the integral $\int_\beta p*dp = \int_\beta \log|F| d \arg F$ can be interpreted as the logarithmic area of the complement of $F(R)$, denoted $A_{\log}(F)$. The extremal property can be restated as follows (Sario [13]):

Theorem. *Among all normalized univalent F the expression*

$$2\pi \log|c(F)| + A_{\log}(F)$$

is maximized by the radial slit mapping F_0, and the expression

$$2\pi \log|c(F)| - A_{\log}(F)$$

is minimized by the circular slit mapping F_1.

To see that the complements of $F_0(R)$ and $F_1(R)$ consist of radial and circular slits (or points) respectively, one may proceed as in 1D, making use of the above extremal property. The details will be omitted.

§3. REPRODUCING DIFFERENTIALS

In this section we apply principal functions to the study of harmonic differentials. First the Hilbert space structure is introduced for the space of such differentials, and some important subspaces are defined.

Any continuous linear functional on a Hilbert space can be represented as an inner product with a fixed element of the space. We call the fixed element the reproducing differential for that functional. This notion also applies to any closed subspace and we are led to consider reproducing differentials for subspaces of harmonic differentials. In several cases we construct these reproducing differentials using principal functions. These results will be used in §§4, 5.

Reproducing differentials for analytic functions were first treated by Bergman [1] in the plane, and then generalized to Riemann surfaces and to various function spaces by Ahlfors [6], Goldstein [1], Rodin [1], and Weill [1]. The method of construction in 2F goes back to Sario [3], where the first proof was given of the existence of nonzero harmonic differentials of finite norm on arbitrary nonplanar Riemann surfaces.

1. The Hilbert space Γ_h

1A. Harmonic differentials. Recall (0.2.3D) that a first-order differential ω is said to be harmonic if $d\omega = d*\omega = 0$. Equivalently, ω is harmonic on the Riemann surface R if, in each parametric disk contained in R, ω is the differential of a complex-valued function whose real and imaginary parts are harmonic. Such functions will be called complex harmonic.

The norm $||\omega||$ of a harmonic differential ω is defined by

$$||\omega||^2 = \iint_R \omega \wedge *\bar{\omega}$$

where the integral is the limit as $\Omega \to R$ of similar integrals taken over exhausting subregions Ω. Since the integrand is nonnegative this integral is finite or $+\infty$. If ω is exact, $\omega = df$ say, then $||df||^2 = D(f)$. The set of harmonic differentials of finite norm on R forms a complex vector space which we denote by $\Gamma_h(R)$, or, if reference to the surface R seems unnecessary, by Γ_h. The norm on Γ_h comes from the inner product

$$(\omega_1, \omega_2) = \iint_R \omega_1 \wedge *\bar{\omega}_2$$

by setting $||\omega||^2 = (\omega, \omega)$. With these definitions Γ_h becomes a complex inner product vector space.

An inner product vector space which is, by means of its norm, a complete metric space is called a Hilbert space.

Theorem. *The vector space Γ_h is a Hilbert space.*

1B. For the proof we must show that every Cauchy sequence from Γ_h converges. Let $\{\omega_n\}$ be such a sequence. Consider an open cover of the Riemann surface R by parametric neighborhoods V_α. Then each ω_n is exact in V_α, say $\omega_n = du_{n\alpha}$ in V_α. We may normalize $u_{n\alpha}$ by picking a point $\zeta_\alpha \in V_\alpha$ and requiring that $u_{n\alpha}(\zeta_\alpha) = 0$. Now

$$||\omega_n - \omega_m|| \geq ||\omega_n - \omega_m||_{V_\alpha} = D_{V_\alpha}(u_{n\alpha} - u_{m\alpha})$$

where the subscript V_α indicates that the corresponding integral is taken over the subsurface V_α. Since $||\omega_n - \omega_m|| \to 0$, the above inequality implies the existence of a complex harmonic function u_α in V_α such that $u_{n\alpha} \to u_\alpha$ uniformly on compact subsets of V_α (see 0.2.3G). Since $u_{n\alpha} - u_{n\beta}$ is constant in $V_\alpha \cap V_\beta$ the same is true for the limit as $n \to \infty$. That is, setting $\omega = du_\alpha$ in V_α defines a harmonic differential on R. The proof will be complete if we show that $||\omega - \omega_n|| \to 0$.

The sequence $\{||\omega_n||\}$ is necessarily bounded. The convergence, $||du_\alpha - du_{n\alpha}||_{V_\alpha} \to 0$ as $n \to \infty$, established above implies that $||\omega - \omega_n||_\Omega \to 0$ as $n \to \infty$ for any relatively compact $\Omega \subset R$. Thus from $||\omega||_\Omega \leq ||\omega - \omega_n||_\Omega + ||\omega_n||$ we see that $||\omega|| < \infty$. Using the triangle inequality again and the fact that $||\cdot||_\Omega$ is an increasing function of Ω it is easy to verify that

$$(10) \quad ||\omega - \omega_n|| \leq ||\omega - \omega_n||_\Omega + ||\omega_N - \omega_n|| + ||\omega_N||_{R-\Omega} + ||\omega||_{R-\Omega}.$$

Choose N so that the second term on the right is arbitrarily small for all $n \geq N$. Now choose Ω so that the third and fourth terms are also small. Letting $n \to \infty$ we conclude that $\lim_n ||\omega - \omega_n|| = 0$.

1C. Subspaces of Γ_h. A differential is said to be analytic if it is locally of the form du where u is a complex analytic function. Denote by Γ_a the subset of Γ_h consisting of analytic differentials. Let $\bar{\Gamma}_a = \{\bar{\omega} | \omega \in \Gamma_a\}$. Then Γ_a and $\bar{\Gamma}_a$ are orthogonal closed subspaces of Γ_h. The identity $\omega = \frac{1}{2}(\omega + i*\omega) + \frac{1}{2}\overline{(\bar{\omega} + i*\bar{\omega})}$ shows that Γ_a and $\bar{\Gamma}_a$ span Γ_h. Thus they give an orthogonal direct sum decomposition of Γ_h which we symbolize by

$$\Gamma_h = \Gamma_a \oplus \bar{\Gamma}_a.$$

Another closed subspace of Γ_h is Γ_{he}, the space of exact differentials in Γ_h. Let Γ_{he}^* consist of all $*\omega$ such that $\omega \in \Gamma_{he}$. We denote the orthogonal complement of Γ_{he}^{*} by Γ_{ho}. Thus by definition

$$\Gamma_h = \Gamma_{he}^{*} \oplus \Gamma_{ho}.$$

Let P be a regular partition of the ideal boundary of R. Denote by $(P)\,\Gamma_{hse}$ the closed subspace of Γ_h consisting of those ω which satisfy

$$\int_{\gamma} \omega = 0 \tag{11}$$

for all P-dividing cycles γ. Set $(P)\,\Gamma_{ase} = (P)\,\Gamma_{hse} \cap \Gamma_a$. If P is the identity partition I, then (11) is automatic by Stokes' formula. For the special case $P = Q$, the canonical partition, the notation is simplified to Γ_{hse} and Γ_{ase}, and the elements are referred to as *semiexact differentials*. The orthogonal complement of $(P)\,\Gamma_{hse}^{*}$ is denoted by $(P)\,\Gamma_{hm}$:

$$\Gamma_h = (P)\,\Gamma_{hm}^{*} \oplus (P)\,\Gamma_{hse}. \tag{12}$$

$(Q)\,\Gamma_{hm}$ is called the space of *harmonic measures*, often denoted simply by Γ_{hm}. Observe that $(I)\,\Gamma_{hm} = \{0\}$.

For an account of other subspaces we refer to Ahlfors-Sario [1, pp. 265–331].

2. Reproducing differentials

2A. Basic properties. Let Γ_x be a closed subspace of Γ_h. Suppose c is a 1-chain on R and ψ_c a differential in Γ_x with the property

$$(\omega, \psi_c) = \int_c \omega$$

for all $\omega \in \Gamma_x$. Under these conditions ψ_c is called the *Γ_x-reproducing differential for c.* If such a differential exists it is clearly unique.

Let ζ be a point on the Riemann surface R and $z = x + iy$ a local coordinate system centered at ζ. An element $\psi_\zeta \in \Gamma_x$ with the property

$$(\omega, \psi_\zeta) = \frac{\partial^n u}{\partial x^n}(0)$$

for all $\omega \in \Gamma_x$, where $\omega = du$ in terms of the parameter z, is called a Γ_x-reproducing differential for nth derivatives at ζ. It is uniquely determined by the uniformizer z and the positive integer n.

2B. Construction of ψ_ζ. The point ζ corresponds to the origin of a local coordinate system z which maps a parametric neighborhood V of ζ into $\{|z|<1\}$. For $R-\{\zeta\}$ consider the singularity functions s and t defined by

$$s = \operatorname{Re}\frac{1}{z^n}, \qquad t = \operatorname{Im}\frac{1}{z^n}$$

near ζ and by the constant zero in a boundary neighborhood of R. Let L be the normal operator consisting of $(P)L_1$ in the boundary neighborhood of R and the Dirichlet operator in V. The Main Existence Theorem yields principal harmonic functions, to be denoted by p_{Ps} and p_{Pt}. They have $(P)L_1$ behavior near the ideal boundary of R and have the singularities $\operatorname{Re} z^{-n}$ and $\operatorname{Im} z^{-n}$ at ζ. If the operator L_0 is used in place of $(P)L_1$, the corresponding principal functions are denoted by p_{0s} and p_{0t}.

2C. Consider the differential $dp_{Is} + *dp_{It}$ where I is the identity partition. The functions p_{Is} and p_{It} have finite Dirichlet integrals over the complement of any neighborhood of ζ while dp_{Is} and $-*dp_{It}$ have the same singular parts at ζ. Consequently $dp_{Is} + *dp_{It}$ extends harmonically to ζ and is a differential in Γ_h. We shall prove (Rodin [4]):

Theorem. *The differential $dp_{Is} + *dp_{It}$ reproduces for the space Γ_h. Explicitly, we have*

$$(\omega, dp_{Is} + *dp_{It}) = -\frac{2\pi}{(n-1)!}\frac{\partial^n}{\partial x^n}u(0)$$

where $\omega = du(z)$ near ζ.

2D. The proof of this theorem makes use of the following

Lemma. *Let p be the $(P)L_1$ principal function for a singularity which is identically zero near the ideal boundary. Then $\int_\beta p\omega = 0$ for all $\omega \in (P)\Gamma_{hse}$.*

The symbolic integral over β stands for $\lim_{\Omega \to R} \int_{\partial\Omega} p\omega$, where $\{\Omega\}$ is an exhaustion of R by regular subregions. The proof is identical to that of (c) of Theorem I.3.2B if in that theorem we replace $*df$ by ω and the partition by P. To avoid repetition we omit the details.

2E. Proof of the theorem. In proving the above theorem we may assume that ω is real. Recall that V is a parametric disk about ζ with local parameter z. Let $\partial V = \alpha$. In order to simplify the notation we shall temporarily designate the functions p_{I_s} and p_{I_t} by p and q respectively. Furthermore, if f is a harmonic function let f^* be its conjugate. Note that $*df = df^*$. In Cl V let $\omega = du(z)$ and $p = \operatorname{Re} z^{-n} + v(z)$ where u and v are harmonic in $\{|z| \leq 1\}$. We then compute

$$(\omega, dp + *dq) = \iint_R \omega \wedge (dp^* - dq)$$

$$= \iint_{R-V} \omega \wedge dp^* - \omega \wedge dq + \iint_V \omega \wedge (dp^* - dq).$$

The integral over $R-V$ is the limit as $\Omega \to R$ of integrals over $\Omega - V$. We apply Stokes' formula to these integrals and, after simplification, find

$$(\omega, dp + *dq) = \int_\beta (p * \omega + q\omega) - \int_\alpha (p * \omega + p^* \omega).$$

The integral along β vanishes by the lemma. Hence

$$(\omega, dp + *dq) = -\int_{|z|=1} (\operatorname{Re} z^{-n} + v)\, du^* + (\operatorname{Im} z^{-n} + v^*)\, du$$

$$= \operatorname{Re} \int (u^* - iu)\, d(z^{-n}) - \int v du^* - u dv^*$$

$$= \operatorname{Re} ni \int \frac{u + iu^*}{z^{n+1}}\, dz$$

$$= \frac{-2\pi}{(n-1)!} \frac{\partial^n}{\partial x^n} u(0).$$

2F. We now seek an expression for the Γ_h-reproducing differential for a chain in terms of principal functions. Let V be a parametric disk on R and c a 1-simplex contained in V. In terms of the parameter z which maps V onto the unit disk we define a singularity function $\sigma = \log|(z-\zeta_2)/(z-\zeta_1)|$ where $\partial c = \zeta_2 - \zeta_1$. We set $\sigma \equiv 0$ outside of a compact set containing V. The corresponding $(P)L_1$-principal function is $p_{P\sigma}$. Let τ be the singularity function $\tau = \arg(z-\zeta_2)/(z-\zeta_1)$ in $V-c$ and $\tau \equiv 0$ near the ideal boundary of R. On the surface $R-c$ we choose the normal operator which is composed of $(P)L_1$ for a boundary neighborhood of R and of the Dirichlet operator for $V-c$. This Dirichlet operator maps a continuous function on ∂V into the restriction to $V-c$ of the harmonic function in V with these boundary values. The direct sum of these operators yields a function $p_{P\tau}$ harmonic on $R-c$. Since $d\tau$ extends harmonically to $c-\{\zeta_1,\zeta_2\}$, the differential $dp_{P\tau}$ can be extended harmonically to all of $R-\{\zeta_1,\zeta_2\}$. We shall continue to denote the extension by $dp_{P\tau}$, even though it is not exact. If c is an arbitrary 1-chain it is homologous to a finite sum $\sum n_i c_i$ where each c_i is a 1-simplex contained in a parametric disk and each n_i is an integer. We extend the definitions of $dp_{P\sigma}$ and $dp_{P\tau}$ to arbitrary c by letting $dp_{P\sigma} = \sum n_i dp_{P\sigma_i}$ and similarly for $dp_{P\tau}$. That these differentials are well defined will follow from the next theorem (Rodin [4]):

Theorem. *The differential $dp_{I\sigma} + {*dp_{I\tau}}$ reproduces for Γ_h. Specifically, if σ and τ correspond to a 1-chain c, then*

$$(\omega, dp_{I\sigma} + {*dp_{I\tau}}) = 2\pi \int_c \omega$$

for all $\omega \in \Gamma_h$.

2G. Because of the linearity it suffices to prove the theorem for the case that c is a 1-simplex contained in a parametric disk $V: \{|z|<1\}$ and $\partial c = \zeta_2 - \zeta_1$. We shall shorten the notation and write p and q for $p_{I\sigma}$ and $p_{I\tau}$ respectively. As in 2E it can be shown that

$$(\omega, dp + {*dq}) = -\int_\alpha (p * \omega + p^* \omega)$$

where $\partial V = \alpha$. In V let $\omega = du$ and $p = \log|(z-\zeta_2)/(z-\zeta_1)| + v$, where u and v are harmonic. Let α_1 and α_2 be disjoint circles in V with centers ζ_1 and ζ_2 respectively. After applying Stokes' formula one obtains

$$(\omega,dp+*dq) = \int_{\alpha_2} u*d\log|z-\zeta_2| - \int_{\alpha_1} u*d\log|z-\zeta_1|.$$

By the mean value formula the last two integrals reduce to $2\pi(u(\zeta_2)-u(\zeta_1))$. We have shown that $(\omega,dp+*dq) = 2\pi\int_{\zeta_1}^{\zeta_2}du = 2\pi\int_c\omega$. This completes the proof of the theorem.

3. Orthogonal projections

3A. The space Γ_a. The preceding two theorems give the existence of reproducing differentials for Γ_h and express them explicitly in terms of principal functions. By orthogonal projections the same can be done for certain closed subspaces of Γ_h. That is, if ψ is a Γ_h-reproducing differential, $\Gamma_h = \Gamma_x \oplus \Gamma_y$ an orthogonal decomposition of Γ_h, $\psi = \psi_x+\psi_y$ a resolution of ψ with $\psi_x \in \Gamma_x$, $\psi_y \in \Gamma_y$, then one easily sees that ψ_x is a Γ_x-reproducing differential.

From the decomposition $\Gamma_h = \Gamma_a \oplus \bar{\Gamma}_a$ and the corresponding resolution $\psi = \frac{1}{2}(\psi+i*\psi)+\frac{1}{2}(\overline{\bar{\psi}+i*\bar{\psi}})$ we obtain the

Corollary. *The differentials*

$$\frac{-(n-1)!}{4\pi}(dp_{I_s}+*dp_{I_t}+i(*dp_{I_s}-dp_{I_t}))$$

and

$$\frac{1}{4\pi}(dp_{I_\sigma}+*dp_{I_\tau}+i(*dp_{I_\sigma}-dp_{I_\tau}))$$

are the reproducing kernels for Γ_a.

3B. The spaces Γ_{hm} and $\Gamma^*_{hse}\cap\Gamma_{ho}$. Consider the decompositions $\Gamma_h = \Gamma^*_{he}\oplus\Gamma_{ho}$ and $\Gamma_h = (P)\Gamma^*_{hse}\oplus(P)\Gamma_{hm}$.

Since $\Gamma_{he}\subset(P)\Gamma_{hse}$ it follows that $(P)\Gamma_{hm}\subset\Gamma_{ho}$. Consequently there is a three-way orthogonal decomposition of Γ_h

$$\Gamma_h = \Gamma^*_{he}\oplus(P)\Gamma_{hm}\oplus(P)\Gamma^*_{hse}\cap\Gamma_{ho}. \tag{13}$$

We shall show that the identity

$$(14)\qquad -(dp_{Is}+*dp_{It}) = *(dp_{0t}-dp_{It})+(dp_{Ps}-dp_{Is})-(*dp_{0t}+dp_{Ps})$$

represents a resolution into components corresponding to the decomposition (13).

The first term on the right side of (14) is evidently in Γ^*_{he}. In order to show that the second term $(dp_{Ps}-dp_{Is})$ is in $(P)\Gamma_{hm}$ it will be sufficient to prove that it is orthogonal to $(P)\Gamma^*_{hse}$. Let $*\omega \in (P)\Gamma^*_{hse}$. Then

$$(dp_{Ps}-dp_{Is}, *\omega) = -\iint_R (dp_{Ps}-dp_{Is}) \wedge \bar{\omega} = \int_\beta p_{Is}\bar{\omega} - \int_\beta p_{Ps}\bar{\omega}.$$

Each of the integrals over β vanishes by Lemma 2D.

Finally, we wish to prove that $*dp_{0t}+dp_{Ps}$ is in $(P)\Gamma^*_{hse} \cap \Gamma_{ho}$. It is obviously an element of $(P)\Gamma^*_{hse}$. It remains to show that if $*\omega \in \Gamma^*_{he}$ then $(*dp_{0t}+dp_{Ps}, *\omega) = 0$. Setting $\omega = du$ we have

$$\begin{aligned}(*dp_{0t}+dp_{Ps}, *\omega) &= \iint_R d\bar{u} \wedge (*dp_{0t}+dp_{Ps}) \\ &= \int_\beta \bar{u} * dp_{0t} + \int_\beta \bar{u}\, dp_{Ps} \\ &= \lim_{\Omega\to R} \int_{\partial\Omega} \bar{u}(*dp_{0t} - *dp_{0t\Omega}) - \int_\beta p_{Ps} d\bar{u}\end{aligned}$$

where as usual $p_{0t\Omega}$ refers to the principal function on the bordered subregion Ω; it has vanishing normal derivative along $\partial\Omega$. By Lemma 2D the last integral over β vanishes. The Schwarz inequality provides us with an estimate

$$(15)\qquad \int_{\partial\Omega} \bar{u}(*dp_{0t} - *dp_{0t\Omega}) \leq \|d\bar{u}\| \cdot \|dp_{0t}-dp_{0t\Omega}\|_\Omega.$$

In I.3.2B we saw that $\|p_{0\Omega}-p_0\|_\Omega \to 0$ as $\Omega \to R$. Thus the resolution in (14) corresponds to the decomposition in (13). The same reasoning applies to the singularities σ and τ. We obtain the following consequences.

Corollary. *The differentials*

$$\frac{(n-1)!}{2\pi}(dp_{Ps}-dp_{Is})$$

and

$$\frac{1}{2\pi}(dp_{I\sigma}-dp_{P\sigma})$$

are the reproducing differentials for the space $(P)\,\Gamma_{hm}$.

Corollary. *The differentials*

$$\frac{-(n-1)!}{2\pi}(dp_{Ps}+*dp_{0t})$$

and

$$\frac{1}{2\pi}(dp_{P\sigma}+*dp_{0\tau})$$

are the reproducing differentials for $(P)\,\Gamma^*_{hse}\cap\Gamma_{ho}$.

3C. The space Γ_{ho}. Since $(I)\,\Gamma^*_{hse}\cap\Gamma_{ho}=\Gamma_{ho}$ the last result yields:

Corollary. *The differentials*

$$\frac{-(n-1)!}{2\pi}(dp_{Is}+*dp_{0t})$$

and

$$\frac{1}{2\pi}(dp_{I\sigma}+*dp_{0\tau})$$

are reproducing differentials for Γ_{ho}.

3D. The spaces Γ_{he} and $\Gamma^*_{hse} \cap \Gamma_{he}$. The space Γ_h has the orthogonal decomposition

$$\Gamma_h = (P)\,\Gamma_{hm} \oplus \Gamma^*_{ho} \oplus (P)\,\Gamma^*_{hse} \cap \Gamma_{he},$$

as may be seen from (12), the definition of Γ_{ho}, and the fact that $\Gamma_{ho} \subset (P)\,\Gamma_{hse}$. Since we are in possession of reproducing differentials for $(P)\,\Gamma_{hm}$ and Γ^*_{ho} (see 3B) we obtain the next result immediately.

Corollary. *The differentials*

$$\frac{(n-1)!}{2\pi}\,(dp_{0s} - dp_{Ps})$$

and

$$\frac{1}{2\pi}\,(dp_{P\sigma} - dp_{0\sigma})$$

are the reproducing kernels for $(P)\,\Gamma^*_{hse} \cap \Gamma_{he}$.

Taking P to be the identity partition in the above corollary gives the reproducers for Γ_{he}.

Corollary. *The differentials*

$$\frac{(n-1)!}{2\pi}\,(dp_{0s} - dp_{Is})$$

and

$$\frac{1}{2\pi}\,(dp_{I\sigma} - dp_{0\sigma})$$

are the reproducing kernels for Γ_{he}.

3E. The space Γ_{hse}. The kernel for $(P)\,\Gamma_{hse}$ can be found from the identity

$$-(dp_{Is} + *dp_{It}) = (*dp_{Pt} - *dp_{It}) - (dp_{Is} + *dp_{Pt})$$

and the orthogonal decomposition (12).

Corollary. *The differentials*

$$\frac{-(n-1)!}{2\pi}(dp_{Is}+*dp_{Pt})$$

and

$$\frac{1}{2\pi}(dp_{I\sigma}+*dp_{P\tau})$$

are the reproducing kernels for $(P)\,\Gamma_{hse}$.

3F. The space Γ_{ase}. Since $(P)\,\Gamma_{hm}\subset(P)\,\Gamma_{hse}$, (12) implies $(P)\,\Gamma_{hse} = (P)\,\Gamma_{hm}\oplus(P)\,\Gamma_{hse}\cap(P)\,\Gamma_{hse}^*$. This leads to an orthogonal decomposition

$$\Gamma_h = (P)\,\Gamma_{hm}\oplus(P)\,\Gamma_{hm}^*\oplus(P)\,\Gamma_{ase}\oplus(P)\,\bar{\Gamma}_{ase}$$

which allows us to project the Γ_h-reproducers into $(P)\,\Gamma_{ase}$.

Corollary. *The differentials*

$$\frac{-(n-1)!}{4\pi}(dp_{Ps}+*dp_{Pt}+i(*dp_{Ps}-dp_{Pt}))$$

and

$$\frac{1}{4\pi}(dp_{P\sigma}+*dp_{P\tau}+i(*dp_{P\sigma}-dp_{P\tau}))$$

are the reproducing differentials for $(P)\,\Gamma_{ase}$.

§4. INTERPOLATION PROBLEMS

Solvability of interpolation problems on Riemann surfaces can be characterized in terms of linear independence of certain sets of differentials of principal functions. It will be seen that this independence is assured if the ideal boundary is sufficiently "strong."

Interpolation problems on Riemann surfaces were first studied in Sario [14]. See also Ahlfors [2].

1. Generalities

1A. Let R be a Riemann surface and Γ_x a closed subspace of Γ_h. Let $\{\zeta_i\}_{i\epsilon I}$ be a discrete closed point set on R, z_i a local coordinate system centered at ζ_i, and let $\{c_j\}_{j\epsilon J}$ be a collection of 1-chains on R. We shall consider the following interpolation problem: Find a differential ω in Γ_x of minimum norm whose integrals $\int_{c_j}\omega$ with $j \in J$ take prescribed values; moreover if we let $\omega = du_i$ at ζ_i then some of the derivatives $\partial^n u_i/\partial x_i^n$ at ζ_i are to take preassigned values.

This interpolation problem may be reformulated in terms of reproducing differentials. A prescribed value of $\int_c \omega$ means a prescribed value of (ω,ψ_c) where ψ_c is the Γ_x-reproducing differential for c. Similarly, prescribing a value of $\partial^n u_i/\partial x_i^n$ at ζ_i is equivalent to requiring a fixed value for an inner product with a certain Γ_x-reproducing differential. *Thus an interpolation problem amounts to seeking a differential of minimum norm with prescribed inner products with a given collection of reproducing differentials.*

1B. There exists as yet no satisfactory theory concerning infinite interpolation problems, that is, problems in which an infinite number of periods or derivatives are prescribed.

Consider the finite problem of finding a differential ω of minimum norm with any prescribed set of values $c_1,\cdots,c_n$ for its inner products $(\omega,\psi_i)_{i=1}^n$ with a given finite set of reproducing differentials $\psi_1,\cdots,\psi_n$. Let us first show that this problem has a solution for every set of prescribed values $c_1,\cdots,c_n$ if and only if the differentials $\psi_1,\cdots,\psi_n$ are linearly independent.

Let Ψ be the linear subspace of Γ_h spanned by $\psi_1,\cdots,\psi_n$. By orthogonal projection into Ψ we see that if there is a solution for one set of prescribed values $c_1,\cdots,c_n$, then there is a unique solution for these same values which lies in Ψ and has minimum norm.

Set $(\psi_i,\psi_j) = a_{ij}$ and let T be the matrix (a_{ij}). Then the vector $(x_1,\cdots,x_n)T$ has its ith component $(\sum x_j\psi_j,\psi_i)$. Thus we seek a vector $(x_1,\cdots,x_n)$ with $(x_1,\cdots,x_n)T = (c_1,\cdots,c_n)$. Consequently, the interpolation problem is solvable for any set $c_1,\cdots,c_n$ of prescribed inner products if and only if T is nonsingular. And T is nonsingular if and only if $\psi_1,\cdots,\psi_n$ are linearly independent.

2. Bordered surfaces

2A. Reflection of differentials. Let R be the interior of a bordered Riemann surface $\bar{R}$. Denote by $\hat{R}$ the double of $\bar{R}$ and let $\zeta \rightarrow \zeta^*$ be the natural antiholomorphic involutory self-mapping of $\hat{R}$ which leaves the border β of $\bar{R}$ pointwise fixed. We do not require $\bar{R}$ to be compact.

Let p be a principal function on R corresponding to the L_0 or $(P)L_1$ normal operator. It is clear that dp can be extended to all of $\hat{R}$ as a harmonic differential with singularities. We wish to obtain explicit expressions for the singularities of the extended differential.

The following facts are fundamental. Suppose u and v are harmonic in the half disk $\{z \,|\, |z| \leq 1 \text{ and } y \geq 0\}$ and u is constant on the real diameter $\{-1 \leq x \leq 1\}$ while v has vanishing normal derivative there. Define

$$\tilde{u}(z) = \begin{cases} u(z) & \text{if } \operatorname{Im} z > 0, \\ -u(z) & \text{if } \operatorname{Im} z < 0, \end{cases}$$

$$\tilde{v}(z) = \begin{cases} v(z) & \text{if } \operatorname{Im} z > 0, \\ v(\bar{z}) & \text{if } \operatorname{Im} z < 0. \end{cases}$$

Then $d\tilde{u}$ and $d\tilde{v}$ can be extended to all of $\{|z| < 1\}$ as harmonic differentials and agree with du and dv in the upper half disk.

Now suppose p is a $(P)L_1$-principal function on R with singularity s at a point $\zeta_0 \in R$. Extend dp to $\hat{R} - \bar{R}$ by defining $\tilde{p}(\zeta^*) = -p(\zeta)$ for $\zeta \in R - \{\zeta_0\}$ and setting $dp = d\tilde{p}$ in $\hat{R} - \bar{R}$. Then dp is harmonic on $\hat{R} - \beta - \{\zeta_0, \zeta_0^*\}$. Since p is constant on each component of β the above considerations show that dp can be extended to become a harmonic differential on $\hat{R} - \{\zeta_0, \zeta_0^*\}$. In a neighborhood V^* of ζ_0^*, the differential dp is of the form $d(-s(\zeta^*) + u(\zeta^*))$ for $\zeta \in V$ where u is a harmonic function in V^*. For brevity we may describe this by stating that the reflected differential has singularity $-ds$ at ζ_0^*.

In a similar manner it can be shown that the differential of an L_0-principal function p with singularity s at ζ_0 can be reflected to a harmonic differential on $R - \{\zeta_0, \zeta_0^*\}$ with singularity ds at ζ_0^*.

2B. Interpolation on bordered surfaces. We now consider interpolation problems for several subspaces Γ_x of Γ_h. It will be convenient to classify such problems according as periods are prescribed or not along a set of cycles.

Problem A. On a Riemann surface R there is given a set of 1-chains $\{c_1,\cdots,c_m\}$, a sequence of points $\{\zeta_1,\cdots,\zeta_n\}$, a function $f:\{1,\cdots,n\}\rightarrow\{1,2,3,\cdots\}$, and a local parameter $z_j = x_j+iy_j$ centered at ζ_j $(j = 1,\cdots,n)$. To each set of complex numbers $\{a_1,\cdots,a_m,b_1,\cdots,b_n\}$ find a differential ω in Γ_x of minimum norm such that

$$\int_{c_k} \omega = a_k \qquad (k = 1,\cdots,m),$$

$$\frac{\partial^{f(j)}}{\partial x^{f(j)}} u_j(0) = b_j,$$

for $\omega = du_j$ near ζ_j $(j = 1,\cdots,n)$. The conditions are assumed to be consistent, that is, the chains $\{c_1,\cdots,c_m\}$ are homologously independent modulo dividing cycles and the integers $f(j)$ and $f(i)$ are distinct if $\zeta_j = \zeta_i$, $j\neq i$.

A restricted form of Problem A is obtained by requiring that the chains $c_1,\cdots,c_m$ be linearly independent modulo cycles, that is, that no nontrivial linear combination of $c_1,\cdots,c_m$ can be a cycle. We shall refer to this modified problem as *Problem B*. The reason for this distinction is that a priori Problem A cannot be solved for any subspace of Γ_{he}. However it is possible that Problem B might have a solution for such a subspace.

Theorem. *Let R be the interior of a bordered Riemann surface $\bar{R}$, compact or not, with a nonempty border. Problem A is solvable for the subspaces Γ_h, Γ_a, $(P)\,\Gamma_{hse}$, and $(P)\,\Gamma_{ase}$. Problem B can be solved for $(P)\,\Gamma^*_{hse}\cap\Gamma_{he}$ and Γ_{he}.*

2C. To prove the theorem we first reformulate it in terms of prescribing the inner product with a set of appropriate reproducing differentials. Denote those differentials which correspond to the chains by $\psi_{c_1},\cdots,\psi_{c_m}$, and those which correspond to derivatives by $\varphi_1,\cdots,\varphi_n$. According to 1B it must be shown that the set

$\{\psi_{c_1},\cdots,\psi_{c_m},\varphi_1,\cdots,\varphi_n\}$ is linearly independent. Assume a linear relation

$$\sum \alpha_i\psi_{c_i} + \sum \beta_j\varphi_j = 0. \tag{16}$$

The expressions for these reproducing kernels in terms of principal functions show that they extend harmonically, except for possible singularities, to the double $\hat{R}$, as described in 2A. In fact, a typical reproducing kernel for the spaces mentioned in the theorem is of the form $dp + *dq$ where p and q have singularities s and s^* respectively, and the same L_0- or L_1-behavior on the border of $\bar{R}$. Consequently, $dp + *dq$ has singularity $\pm 2ds$ on $\hat{R}$. In the case of the φ_j the singularities are $s = \operatorname{Re}(z-\zeta_j)^n$ and they are distinct. By the uniqueness of harmonic continuation, (16) can be interpreted as an equation on $\hat{R}$, and we see that no cancellation is possible among the φ_j. Thus each β_j must be zero.

The relation (16) reduces to $\sum \alpha_i\psi_{c_i} = 0$, the ψ_{c_i} being regarded as differentials on $\hat{R}$. Let $\partial c_i = b_i - a_i$. Then the differential $\sum \alpha_i\psi_{c_i}$ or $-\sum \alpha_i\psi_{c_i}$ has logarithmic poles of order α_i at b_i^* and $-\alpha_i$ at a_i^*. Thus $\sum \alpha_i(b_i^* - a_i^*)$ is the zero element of the 0-dimensional homology group over the field $\mathbf{C}$ of complex numbers, and the same must be true of $\sum \alpha_i(b_i - a_i)$.

In the case of Problem B the assumption that no nontrivial integral combination of $c_1,\cdots,c_m$ can be a cycle means that the elements $(b_1-a_1),\cdots,(b_m-a_m)$ are independent over the set $\mathbf{Z}$ of integers. This implies that they are independent over $\mathbf{C}$. In fact they are obviously independent over the rationals. Hence

$$b_i - a_i = \sum_{k=1}^{N} q_{ik}e_k \qquad (i = 1,\cdots,m)$$

where $e_1,\cdots,e_N$ are distinct points of R and the rational matrix (q_{ik}) has rank m. The rank of this matrix does not change if we consider its elements as belonging to the larger field $\mathbf{C}$. Thus if $\sum \alpha_i(b_i - a_i) = 0$ for $\alpha_i \in \mathbf{C}$ then each $\alpha_i = 0$.

Now consider Problem A. We have just seen that a nontrivial relation among $(b_1-a_1),\cdots,(b_m-a_m)$ over $\mathbf{C}$ implies that a nontrivial combination of $c_1,\cdots,c_m$ over $\mathbf{Z}$ is a cycle. By hypothesis this cycle is nondividing. To complete the proof we may assume that each $\alpha_i \in \mathbf{Z}$, that $\gamma = \sum \alpha_i c_i$ is a nondividing cycle, and derive that each $\alpha_i = 0$.

Since γ is nondividing there is a cycle c' such that the intersection number $c' \times \gamma \neq 0$ where $c' \times \gamma$ can be interpreted as the number of times γ crosses c' from left to right (0.1.3C). Now $\sum \alpha_i \psi_{c_i} = \psi_\gamma$ by the reproducing property. We shall show that $(2\pi)^{-1} \int_\delta * \psi_\gamma = \delta \times \gamma$ for any cycle δ. It will then follow that each α_i must vanish.

2D. For the subspaces under consideration at this point ψ_γ is of the form $\sum_\nu * dp_{\tau_\nu}$ where p_{τ_ν} is the principal function defined in 3.2F for an arc, say γ_ν, contained in a parametric disk. The terms dp_{σ_ν} cancel out since we are assuming γ is a cycle. By virtue of linearity we need only prove that

$$\int_\delta dp_{\tau_\nu} = -2\pi(\delta \times \gamma_\nu). \tag{17}$$

Since dp_{τ_ν} is exact outside of the parametric disk we may assume that δ is contained in this disk. In terms of the local parameter in this disk dp_{τ_ν} is of the form

$$d\left(\arg \frac{z-\zeta_2}{z-\zeta_1} + u(z)\right)$$

where u is harmonic and $\partial \gamma_\nu = \zeta_2 - \zeta_1$. Thus formula (17) holds provided

$$\int_\delta d \arg \frac{z-\zeta_2}{z-\zeta_1} = -2\pi(\delta \times \gamma_\nu).$$

This last equation is geometrically evident from the interpretation of $\delta \times \gamma_\nu$ as the number of times γ_ν crosses δ from left to right. A proof could be based on simplicial approximation.

2E. If R has finite genus we may apply Theorem 2B for the space Γ_{ase} and require all periods to be zero. We obtain:

Corollary. *Let R be the interior of a bordered Riemann surface $\bar{R}$ of finite genus, whose border is nonempty. At a finite number of* *points there are prescribed a finite number of Taylor series coefficients with respect to fixed local coordinates. There is a unique analytic function which has these coefficients and whose Dirichlet integral is minimum and finite.*

3. Open regions

3A. Corollary 2E shows that interpolation problems are solvable on surfaces whose ideal boundary is fairly strong. In this same vein we have the following result (Sario [14]):

Theorem. *Let R be a plane region containing ∞ such that complement of R has positive area. Given any finite set of Taylor coefficients at a finite number of points there is a unique analytic function which has these coefficients and whose Dirichlet integral is minimum and finite. Furthermore, this function is bounded.*

This theorem can be generalized in part to an open Riemann surface R of finite genus which carries nonconstant Dirichlet-finite analytic functions (Nakai [5]).

3B. The theorem answers a question of the same form as *Problem B of* 2B for the space $\Gamma_{ae}(R)$. We must prove that any finite consistent set of reproducing differentials in Γ_{ae} is linearly independent. Let the set be $\psi_1,\cdots,\psi_N$. Since R is planar $\Gamma_{ase} = \Gamma_{ae}$. Corollary 3F then shows that each ψ_j is of the form $\omega_j + i * \omega_j$ where $\omega_j = dp_{Qs_j} + * dp_{Qt_j}$ and s_j, t_j are singularities of the type $\operatorname{Re}(z-\zeta)^{-n}$ and $\operatorname{Im}(z-\zeta)^{-n}$ $(n \geq 1)$ respectively, or else of the type $\log|(z-\zeta_2)/(z-\zeta_1)|$ and $\arg[(z-\zeta_2)/(z-\zeta_1)]$.

We first observe that $*dp_{Qt_j} = -dp_{0s_j}$. This is true on any planar surface because $*dp_{Qt_j}$ is exact on $W-\{\zeta\}$ or $W-\{\zeta_1,\zeta_2\}$, depending on the type of the singularity t_j. By Theorem I.2.3B, $p_{Qt_j} = \lim_{\Omega\to R} p_{Qt_j\Omega}$ where $p_{Qt_j\Omega}$ is constant on each component of $\partial\Omega$. Consequently the normal derivative $\partial p^*_{Qt_j\Omega}/\partial n = \partial p_{Qt_j\Omega}/\partial s$ is identically zero on $\partial\Omega$. Now p_{0s_j} is also a limit of functions $p_{0s_j\Omega}$ whose normal derivatives vanish along $\partial\Omega$ according to the same theorem. Thus p_{0s_j} and $-p^*_{Qt_j}$ are L_0-principal functions with the same singularity. They must therefore differ by a constant. Similarly $*dp_{Qs_j} = dp_{0t_j}$.

3C. We have shown that each ψ_j is of the form $dq_{s_j}+i*dq_{s_j}$ where $q_{s_j} = p_{Qs_j}-p_{0s_j}$. Suppose a linear relation $\sum w_j\psi_j = 0$ is given, where $w_j = u_j+iv_j$ is a complex number. This is equivalent to a real linear relation $\sum u_jdq_{s_j}-v_j*dq_{s_j} = 0$, or taking account of the identities above,

$$\text{(18)} \qquad \sum (u_jdp_{Qs_j}+v_jdp_{Qt_j})-(u_jdp_{0s_j}+v_jdp_{0t_j}) = 0.$$

For singularities of the type $s_j+it_j = (z-\zeta)^{-n}$ it is seen immediately that $u_jdp_{Qs_j}+v_jdp_{Qt_j} = dp_{Qr_j}$ where $r_j = \text{Re}\ \bar{w}_j(z-\zeta)^{-n}$.

For singularities of the other type we let c_j be a cycle with $\partial c_j = \zeta_2-\zeta_1$. Then on $R-c_j$ we may form the principal function p_{Qr_j} with singularity $r_j = \text{Re}\ \bar{w}_j \log[(z-\zeta_2)/(z-\zeta_1)]$ exactly as in 3.2F. The differential dp_{Qr_j} is harmonic on $R-\{\zeta_1,\zeta_2\}$. Equation (18) becomes

$$\text{(19)} \qquad \sum dp_{Qr_j}-dp_{0r_j} = 0.$$

Let r be the singularity function which is equal to each r_j in the domain of r_j. Form dp_{Qr} and dp_{0r}.

3D. We have $dp_{Qr} = \sum dp_{Qr_j}$ and $dp_{0r} = \sum dp_{0r_j}$. Both dp_{Qr} and dp_{0r} are exact on $R-\cup c_j$. Let p be any harmonic function on $R-\cup c_j$ such that $p_{Qr}-p$ extends to a regular harmonic function on R with a finite Dirichlet integral. Assume also that $d(p_{Qr}-p) \in \Gamma^*_{hse}(R)$. Let $\{\Delta_j\}$ be disjoint parametric disks with $c_j \subset \Delta_j$.

In the following computations we make use of relations of the form $\int_\beta p*dp_0 = 0$. They follow from Theorem I.3.2B. Note also that $\int_\beta *dp = 0$.

The validity of equation (1) of 1.1B is not affected if p, p_0, p_1 are not defined on $\cup c_j$. We therefore have

$$\text{(20)} \qquad D(p-\tfrac{1}{2}(p_{0r}+p_{Qr}))-\int_\beta p*dp$$

$$= -\frac{1}{2}\int_\beta p_{0r}*dp+p*dp_{Qr}+\frac{1}{4}\int_\beta p_{0r}*dp_{Qr}.$$

The first integral on the right is independent of p. In fact

$$\int_\beta p_{0r} * dp = \int_\beta (p_{0r} - p_{Qr}) * d(p - p_{0r}),$$

$$\int_\beta p * dp_{Qr} = \int_\beta (p - p_{Qr}) * d(p_{Qr} - p_{0r})$$

$$= \int_\beta (p_{Qr} - p_{0r}) * d(p - p_{Qr}),$$

so (20) yields

$$D(p - \tfrac{1}{2}(p_{0r} + p_{Qr})) = \int_\beta p * dp - \frac{1}{2} \int_\beta (p_{0r} - p_{Qr}) * d(p_{Qr} - p_{0r})$$

$$+ \frac{1}{4} \int_\beta (p_{0r} - p_{Qr}) * d(p_{Qr} - p_{0r}).$$

Since $D(p_{0r} - p_{Qr}) = \int_\beta (p_{0r} - p_{Qr}) * d(p_{0r} - p_{Qr}) = 0$ by (19), we finally arrive at the conclusion that

$$\int_\beta p * dp \geq 0 \tag{21}$$

for all admissible functions p.

3E. Recall that the singularity function r was locally of the form Re $\bar{w}_j(z - \zeta)^{-n}$ or Re $\bar{w}_j \log[(z - \zeta_2)/(z - \zeta_1)]$. In particular, if r is regular then each $w_j = 0$, the relation $\sum w_j \psi_j = 0$ is trivial, and the independence of the reproducing differentials is established. If r is not regular a contradiction to (21) can be derived. Since R is a region on the Riemann sphere $\mathbf{P}$ containing ∞ we can easily find a harmonic function p with possible poles, single-valued on $\mathbf{P} - \cup c_j$, and such that $dp - dp_{Qr} \in \Gamma^*_{hse}$. If r is not regular p is not constant and the zeros of dp are isolated. Therefore

$$\iint_{\mathbf{P}-R} dp \wedge * dp = D_{\mathbf{P}-R}(p) > 0$$

since $\mathbf{P} - R$ has positive area. The contradiction follows from $D_{\mathbf{P}-R}(p) = \int_{-\beta} p * dp$.

The proof has also shown that the minimizing function is a linear combination of principal functions. Hence it is necessarily bounded and the proof of Theorem 3A is complete.

§5. THE THEOREMS OF RIEMANN-ROCH AND ABEL

The zeros and poles of meromorphic functions on closed surfaces cannot be arbitrarily prescribed. On noncompact surfaces, however, any set of isolated singularities is the singular part of some meromorphic function (Behnke-Stein [1]). A unity can be found in these situations if regularity conditions at the ideal boundary are imposed on all functions. The regularity of principal functions will be required, and for such functions some classical theorems can be extended to an arbitrary Riemann surface, compact or not.

References related to this section are made here to Kusunoki [3], Ahlfors [5], Rodin [1], Royden [7], Sainouchi [1], Kobori-Sainouchi [1], and Watanabe [1], among others.

1. Riemann-Roch type theorems

1A. A *divisor* on a Riemann surface R is a 0-chain, that is, a finite formal sum of points with integral coefficients. It is said to be *integral* if all the coefficients are nonnegative.

Suppose f is a meromorphic function on R, ζ a point of R, and $f = \sum_{-\infty}^{\infty} c_n(z-\zeta)^n$ in a neighborhood of ζ. Let $v_\zeta(f)$ be the smallest index n such that $c_n \neq 0$. The *divisor of* f is defined to be $\sum_{\zeta \in R} v_\zeta(f)\zeta$. The function f is said to be a *multiple* of the divisor $\mathbf{D}$ if $\sum v_\zeta(f)\zeta - \mathbf{D}$ is integral. The divisor of a meromorphic differential α is defined in the same way by setting $v_\zeta(\alpha) = v_\zeta(f)$ if $\alpha = f(z)dz$ near ζ.

Let $\mathbf{D}$ be a divisor on a Riemann surface R and let $\mathbf{D} = \mathbf{B} - \mathbf{A}$ where $\mathbf{A}$ and $\mathbf{B}$ are disjoint integral divisors. Suppose that $\mathbf{A} = \sum m_j a_j$ and $\mathbf{B} = \sum n_k b_k$, with the a_j and b_k points of R and the m_j and n_k positive integers. For each j and k let Δ_j and Δ_k' be closed parametric disks with centers at a_j and b_k respectively. We assume that these disks are mutually disjoint. Let $\Delta = \bigcup \Delta_j$ and $\Delta' = \bigcup \Delta_k'$.

Let $S = s+it$ be a meromorphic function in Δ' and a multiple of the divisor $-\mathbf{B}$. Define $S \equiv 0$ in a boundary neighborhood of R and form the differential $dF_S = dp_{Ps}+i\,dp_{Pt}$. Let $\mathcal{B}$ denote the complex vector space consisting of all such dF_S. The dimension of $\mathcal{B}$ is equal to the degree of $\mathbf{B}$, i.e. $\dim \mathcal{B} = \deg \mathbf{B} = \sum n_k$.

1B. Suppose ω is a harmonic differential except for analytic singularities. This means that locally $\omega = f\,dz+du$ where f is meromorphic and u a complex-valued harmonic function. Then ω can be written in a unique way as the sum of a meromorphic differential and the complex conjugate of a meromorphic differential. In fact, set $\varphi = \frac{1}{2}(\omega+i*\omega)$, $\psi = \frac{1}{2}(\bar{\omega}+i*\bar{\omega})$. Then $\omega = \varphi+\bar{\psi}$, and local considerations show that ψ is actually holomorphic.

Apply this decomposition to $dF_S = \varphi_S+\bar{\psi}_S$. Then $\bar{\psi}_S = \frac{1}{2}(dF_S-i*dF_S)$ is in $(P)\,\bar{\Gamma}_{ase}$. According to Corollary 3.3F ψ_S has certain reproducing properties. The following lemma is a generalization of that fact. For a divisor $\mathbf{E}$ let $(P)\,\Gamma_{ase}[\mathbf{E}]$ denote the vector space of meromorphic differentials which are multiples of $\mathbf{E}$, are square integrable near the ideal boundary of R, and have vanishing periods along each P-dividing cycle outside of some compact set.

Lemma. *Let* $\alpha \in (P)\,\Gamma_{ase}[-\mathbf{A}]$. *Then*

$$(\alpha,\psi_S)_{R-\Delta} = i\int_{\partial\Delta'} S\alpha + i\int_{\partial\Delta} F_S\alpha. \tag{22}$$

The left-hand side is to be understood as a Cauchy limit. That is, consider the subdisk of radius r contained in Δ_k', and let Δ'' be the union of these for all k. Then this inner product is the limit of inner products taken over $R-\Delta-\Delta''$ as $r\to 0$. Note that

$$(\alpha,\psi_S)_{R-\Delta-\Delta''} = (dF_S,\bar{\alpha})_{R-\Delta-\Delta''} = -i \lim_{\Omega\to R} \int_{\partial(\Omega-\Delta-\Delta'')} F_S\alpha.$$

A modification of the proof of Lemma 3.2D shows that $\int_{\partial\Omega} F_S\alpha \to 0$ as $\Omega\to R$. In Δ' we have $F_S = S+u$ where S is meromorphic and u is harmonic. When the disks in Δ'' shrink to points we obtain (22).

We shall make use of the following algebraic facts. Let U and V be vector spaces over a field K. A bilinear mapping $T: U \times V \to K$ is called a pairing of U and V. The left kernel U_0 is the space $\{u \in U | T(u,V) = 0\}$, and the right kernel V_0 is the space $\{v \in V | T(U,v) = 0\}$. If one of the quotient spaces U/U_0 or V/V_0 is finite dimensional then there is an isomorphism $U/U_0 \simeq V/V_0$.

Consider the pairing $T: \mathfrak{B} \times (P)\Gamma_{ase}[-\mathbf{A}] \to \mathbf{C}$ defined by $T(dF_S, \alpha) = \int_{\partial\Delta'} S\alpha$.

Suppose dF_S is in the left kernel of this pairing. From (22) we have $(\alpha, \psi_S)_{R-\Delta} = i\int_{\partial\Delta} F_S\alpha$ for all $\alpha \in (P)\Gamma_{ase}[-\mathbf{A}]$. We may replace α by ψ_S, which is regular in Δ. Letting the disks of Δ shrink, we obtain $\|\psi_S\|_R = 0$. Hence $\psi_S = 0$ and dF_S is meromorphic. We also find that

$$\int_{\partial\Delta} F_S\alpha = 0. \tag{23}$$

We apply (23) to differentials $\alpha \in (P)\Gamma_{ase}[-\mathbf{A}]$ with singularities of the form

$$\frac{dz}{z-a_j} - \frac{dz}{z-a_{j'}}, \qquad \frac{dz}{(z-a_j)^{\mu}} \qquad (2 \leq \mu \leq m_j).$$

That such α exist is easily seen by using principal functions. For example, $\alpha = dp + i * dp$ where p is the $(P)L_1$-principal function for the singularity $\log(|z-a_j|/|z-a_{j'}|)$ or $\operatorname{Re}(z-a_j)^{1-\mu}$. For these choices of α (23) shows that F_S, suitably normalized by an additive constant, is a multiple of the divisor $-\mathbf{D}$. Conversely, if dF_S is meromorphic then $\psi_S = 0$ and $T(dF_S, \alpha) = \int_{\partial\Delta'} S\alpha = -\int_{\partial\Delta} F_S\alpha$. The differential α has a pole of order at most m_j at a_j, and $F_S^{(k)}(a_j) = 0$ $(k = 1, \cdots, m_j - 1)$ if F_S is a multiple of $\mathbf{A}$. Hence $\int_{\partial\Delta} F_S\alpha = 0$ by the Cauchy integral formula. Thus dF_S is in the left kernel $\mathfrak{B}_0$ if and only if the function F_S is meromorphic and can be normalized so as to be a multiple of $-\mathbf{D}$.

A differential $\alpha \in (P)\Gamma_{ase}[-\mathbf{A}]$ is in the right kernel if and only if $\int_{\partial\Delta'} S\alpha = 0$ for all S which are multiples of $-\mathbf{B}$. Appropriate choices for S show that α must be a multiple of $\mathbf{B}$, i.e. $\alpha \in (P)\Gamma_{ase}[\mathbf{D}]$.

Since $\mathfrak{B}$ is finite dimensional we have

$$\mathfrak{B}/\mathfrak{B}_0 \simeq (P)\,\Gamma_{ase}[-\mathbf{A}]/(P)\,\Gamma_{ase}[\mathbf{D}]$$

or

$$(24) \qquad \dim \mathfrak{B}_0 = \dim \mathfrak{B} - \dim (P)\,\Gamma_{ase}[-\mathbf{A}]/(P)\,\Gamma_{ase}[\mathbf{D}].$$

Let $(P)M$ be the complex vector space of meromorphic functions on R whose real and imaginary parts have $(P)L_1$ behavior near the ideal boundary. Denote by $(P)M[\mathbf{E}]$, where $\mathbf{E}$ is a divisor, the subspace of $(P)M$ consisting of functions which are multiples of $\mathbf{E}$. The homomorphism $d\colon (P)M[-\mathbf{D}] \to \mathfrak{B}_0 : F_S \to dF_S$ is surjective. The kernel is $\mathbf{C}$ if $\deg \mathbf{A} = 0$, and it is zero if $\deg \mathbf{A} \neq 0$. From (24) we obtain (Rodin [1])

$$(25) \quad \dim (P)M[-\mathbf{D}] = \deg \mathbf{B} + 1 - \min(1, \deg \mathbf{A})$$
$$- \dim (P)\,\Gamma_{ase}[-\mathbf{A}]/(P)\,\Gamma_{ase}[\mathbf{D}].$$

Theorem. *Let* $\mathbf{A}$ *and* $\mathbf{B}$ *be disjoint integral divisors on a Riemann surface* R *and let* $\mathbf{D} = \mathbf{B} - \mathbf{A}$. *Then* (25) *is valid for any regular partition* P *of the ideal boundary of* R.

1C. The classical case. Suppose now that R is a compact surface of genus g. Let $A_1, B_1, \cdots, A_g, B_g$ be a canonical homology basis for the 1-cycles on R (see 0.1.3D). Consider the vector space homomorphism $\Gamma_h \to \mathbf{C}^{2g}$ defined by associating to each harmonic differential ω its $2g$ periods $(\int_{A_1}\omega, \int_{B_1}\omega, \cdots, \int_{B_g}\omega)$. This is actually an isomorphism since a harmonic differential in the kernel would be exact and, by the maximum principle, any harmonic function defined on a compact surface is constant. The mapping is surjective since, for example, the Γ_h reproducing differential for A_i has periods 0 over all the basis cycles except B_i, where the period is nonzero (see 4.2C, 2D). Thus Γ_h has dimension $2g$.

The direct sum $\Gamma_h = \Gamma_a \oplus \bar{\Gamma}_a$ shows that Γ_a therefore has dimension g. The dimension of $\Gamma_a[-\mathbf{A}]$, $\mathbf{A} \neq 0$, can be computed as follows. Let $\{\alpha_1, \cdots, \alpha_g\}$ be a basis for Γ_a. Let $a_1, \cdots, a_J$ be the distinct points of $\mathbf{A}$ which have nonzero coefficients. Let c_j be a

1-chain with $\partial c_j = a_j - a_1$ ($j = 2,\cdots,J$). Let $S_{k_j} = (z-a_j)^{1-k_j}$ if $k_j \geq 2$. Then $\{dp_{\sigma(c_j)} + i * dp_{\sigma(c_j)}\}_{2\leq j\leq k}$ and $\{\varphi_{S k_j}\}_{2\leq k_j\leq m_j}$ together with $\{\alpha_1,\cdots,\alpha_g\}$ form a basis for $\Gamma_a[-\mathbf{A}]$ of $g+\deg \mathbf{A}-1$ elements.

Since the P-semiexactness conditions are vacuous (25) becomes

$$\text{(26)} \qquad \dim M[-\mathbf{D}] = \deg \mathbf{D} - g + 1 + \dim \Gamma_a[\mathbf{D}],$$

which is the classical Riemann-Roch theorem.

1D. An application to conformal mapping. We illustrate the Riemann-Roch theorem by solving a classical conformal mapping problem (Bieberbach [1]). Let R be a planar compact bordered Riemann surface. We wish to find an analytic function on R which represents it as an n-sheeted unit disk where n is the connectivity of R.

More precisely, let $\beta_1,\cdots,\beta_n$ be the oriented contours of R. We seek a function f, analytic on R, with the property that each $f(\beta_i)$ is the unit circle $\{|w| = 1\}$ traversed once in the positive direction.

To obtain a solution form the double $\hat{R}$ and pick a point $\zeta_i \in \beta_i$ for each $i = 1,\cdots,n$. Note that $\hat{R}$ has genus $n-1$. Thus from the Riemann-Roch theorem (26) for the closed surface $\hat{R}$ and the divisor $\mathbf{D} = \zeta_1+\zeta_2+\cdots+\zeta_n$ we conclude that $\dim M[-\mathbf{D}] \geq 2$. Thus there exists a nonconstant meromorphic function F on $\hat{R}$ whose only singularities are possible simple poles at the ζ_i's.

Since $F(z)$ can be replaced by $F(z)+\bar{F}(z^*)$, where $z \to z^*$ is the natural antiholomorphic involution, we see that F may be assumed to be real on each β_i. We shall show that F maps R onto an n-sheeted half-plane. Thus the desired function f may be obtained by composing F with a linear transformation. It suffices then to prove the following assertions:

(a) F has a simple pole at each ζ_i.
(b) Each $F(\beta_i)$ traverses the extended real axis exactly once.
(c) $\operatorname{Im} F$ is of constant sign on the interior of R.

To prove (a) we may assume F has simple poles at $\zeta_1,\cdots,\zeta_k$ ($1 \leq k \leq n$) and then show $k = n$. Note that F takes every real value on each $\beta_1,\cdots,\beta_k$. Note also that F takes any value exactly

k times on $\hat{R}$. The last statement obviously follows from the general fact that a meromorphic function h on a closed Riemann surface must have the same number of zeros as poles. And that fact, in turn, can be proved easily by applying Stokes' theorem (Residue Theorem) to the closed differential dh/h. Now our function F takes some real value on β_{k+1}, if $k<n$. We have seen that this value is also assumed by F on $\beta_1,\cdots,\beta_k$. Thus F assumes this value $k+1$ times, a contradiction unless $k = n$.

Assertion (b) is now immediate. Indeed, F takes every value n times, and $F(\beta_i)$ covers the extended real axis for $1 \leq i \leq n$.

Since F takes every real value n times on the border $\beta_1 \cup \cdots \cup \beta_n$ it cannot be real anywhere on the interior of R. Hence (c) is proved.

1E. We have now established most of the following result (Bieberbach [1]):

Theorem. *Any n-connected bordered plane region R can be mapped biholomorphically onto an n-sheeted covering, with branch points, of the unit disk in such a way that each contour of R covers the unit circle once. Moreover, the n pre-images of a given point on the unit circle may be prescribed at will. The total order of the branch points is $2n-2$.*

The last statement still requires proof. Denote the mapping function by f. Apply (26) to the divisor $\mathbf{D}$ of the extended meromorphic differential df on $\hat{R}$. The dimensions of $M[-\mathbf{D}]$, $\Gamma_a[\mathbf{D}]$ can be evaluated explicitly. Indeed, the map $h \rightarrow h\, df$ is an isomorphism of $M[-\mathbf{D}]$ onto Γ_a. Hence $\dim M[-\mathbf{D}] = g$. Similarly the isomorphism $\alpha \rightarrow \alpha/df$ of $\Gamma_a[\mathbf{D}]$ into the space of meromorphic functions has only regular functions in its image. These are the constants, and hence $\dim \Gamma_a[\mathbf{D}] = 1$. Thus (26) reduces to

$$\deg \mathbf{D} = 2g-2.$$

We remark for future reference that the reasoning leading to this equation was valid for the divisor of any meromorphic differential on any closed surface of genus g.

Since $g = n-1$ on $\hat{R}$, and since the only singularities of df on

$\hat{R}$ are n double poles, it follows that df must have zeros of total order $2(n-1)-2+2n = 4n-4$. The total order of the branch points is the total order of the zeros of df on R. In the existence proof (1D) we saw that f is one-to-one on each β_i, and has symmetry on $\hat{R}$. Hence df has $2n-2$ zeros on R, and none on the boundary.

For other proofs of this theorem and its generalizations we refer to the bibliography in Mizumoto [2]. For the relation of such mappings to extremal problems see Ahlfors [1, 2] and Royden [8].

2. Abel's theorem

2A. Suppose ω is a harmonic differential on R and is exact outside of a compact subset. Then it makes sense to say that the multiple-valued function $\int^z \omega$ has $(P)L_1$-behavior near the ideal boundary of R. If f is a meromorphic function on R, $d\log f$ is exact outside of a compact subset, and $\int^z d\log f$ has $(P)L_1$-behavior near the ideal boundary of R, then we say f is *quasi-rational* and denote this by $f \in \exp(P)L_1$.

In 4.2F we introduced differentials $dp_{P\sigma}$ and $dp_{P\tau}$ associated with a 1-chain c. It will be convenient to use the notation $dF_{Pc} = dp_{P\sigma} + i\, dp_{P\tau}$. We observe the useful formula

$$\int_\delta dF_{Pc} = -2\pi i(\delta \times c) \tag{27}$$

which holds for any cycle δ. The proof follows from (17). We shall establish the following extension of Abel's theorem (Ahlfors [5], Rodin [1]):

Theorem. *A necessary and sufficient condition for a divisor to be the divisor of a function in* $\exp(P)L_1$ *is that it be the boundary of a 1-chain c with the property*

$$\int_c \omega = 0 \tag{28}$$

for all $\omega \in (P)\Gamma_{ase}$.

2B. In the decomposition of dF_{Pc} into the sum of its meromorphic and anti-analytic parts, $dF_{Pc} = \varphi + \bar{\psi}$, Corollary 3.3F identifies the differential ψ as the reproducing differential for $(P)\,\Gamma_{ase}$. Therefore (28) is valid if and only if dF_{Pc} is meromorphic. Equation (27) shows that

$$f(z) = \exp \int_{z_0}^{z} dF_{Pc}$$

is single-valued. Evidently it is in $\exp(P)L_1$ provided dF_{Pc} is meromorphic. Since f has divisor ∂c the sufficiency of (28) is proved.

2C. For the converse suppose $f \in \exp(P)L_1$ and ∂c is the divisor of f. Then $d \log f - dF_{Pc}$ is regular and exact on the complement of a compact subset of R. The nonzero periods of $d \log f$ and dF_{Pc} are integral multiples of $2\pi i$ and correspond to a finite set of nondividing cycles. By (27) it is possible to find a cycle, call it c', such that $dF_{Pc'}$ has the same periods as $d \log f - dF_{Pc}$. Let $c'' = c + c'$. Then $dF_{Pc''} = dF_{Pc} + dF_{Pc'}$ has the same singularities as f since c' is a cycle. Hence $d \log f - dF_{Pc''}$ is regular and exact on all of R. We may therefore write it as dQ where Q is a harmonic function whose real and imaginary parts have $(P)L_1$-behavior. Such a function must be a constant. Hence $dF_{Pc''} = d \log f$ is meromorphic. This implies that $\psi = \frac{1}{2} d\bar{F}_{Pc''} + i * d\bar{F}_{Pc''})$ is zero.

Since ψ is the reproducing differential for c'' for $(P)\,\Gamma_{ase}$ except for a multiplicative constant it follows that (28) is valid for all $\omega \in (P)\,\Gamma_{ase}$.

§6. EXTREMAL LENGTH

A family of curves on a Riemann surface forms a geometric configuration to which a numerical conformal invariant, called extremal length, can be associated. Extremal length can often be determined or estimated from metric properties of the configuration, and may be considered to be a geometric quantity.

It is of interest to relate analytic invariants to this geometric invariant, an aspect which will be treated in this section and in Chapter IV. However, extremal length also provides a powerful tool for investigating purely analytical problems as will be seen in Chapter III. In many cases the metric determining extremal length is derived from principal functions.

The material in this section contains work of Accola [2], Ahlfors-Beurling [1], Blatter [1], Fuglede [1], Marden [2], Marden-Rodin [1]-[3], Ohtsuka [1], and Rodin [2], among others.

1. Fundamentals

1A. Linear densities. Basically, we would like to consider all possible Riemannian metrics ds on a Riemann surface R which are compatible with the conformal structure of R. This means that at each point of R there is a local parameter, say z, such that in terms of it $ds = \rho(z)|dz|$ with ρ a positive continuous function. However, it is necessary to allow more general metrics by requiring only that ρ be a nonnegative real-valued Borel measurable function. In that case ds is called a *linear density*. Thus a linear density associates with each local parameter a nonnegative Borel measurable function such that if ρ_1, ρ_2 are associated with local parameters z, w then $\rho_1(z) = \rho_2(w(z))|w'(z)|$ wherever $w = w(z)$. The last condition is expressed by saying that $\rho(z)|dz|$ is invariant.

If c is a finite or countable union of arcs on R, with or without end points, then c will be called a *curve*. If ds is a linear density and c is a curve then ds is measurable along c and $\int_c ds$ may be defined if we allow ∞ as a possible value. Let F be a family of curves on R and define the ds-length of F to be

$$L(F,ds) = \inf\left\{\int_c ds \,\middle|\, c \in F\right\}.$$

Given a linear density ds we also consider the area element $d\sigma$ associated with it. If $ds = \rho(z)|dz|$ in terms of z then $d\sigma = \rho^2(z)dxdy$ in this same coordinate system. The area of R in this metric is

$$A(R,ds) = \iint_R d\sigma.$$

Other notations for these quantities are $L(F,ds) = L(F,\rho)$ and $A(R,ds) = A(\rho) = A(ds)$.

1B. Extremal length. Suppose F is a family of curves on R. We obtain a conformal invariant by considering

$$\lambda(F) = \sup \frac{L^2(F,ds)}{A(R,ds)}$$

where the supremum is taken with respect to all linear densities ds on R for which $L^2(F,ds)/A(R,ds)$ is not an indeterminate expression. This invariant is called the extremal length of F. It may be denoted by $\lambda_R(F)$ as well.

We shall use the fact that $\lambda_R(F) = \lambda_S(F)$ if R is contained in a Riemann surface S. To prove this first observe that any density on R can be extended to S by defining it to be $\equiv 0$ on $S-R$. This shows that $\lambda_R(F) \leq \lambda_S(F)$. On the other hand, an admissible density on S can be restricted to R and this will not decrease the value L^2/A. Thus $\lambda_R(F) \geq \lambda_S(F)$ and our assertion is valid.

1C. Extremal and conjugate extremal distance. Let R be the interior of a compact bordered Riemann surface $\bar{R}$, and let the contours of $\bar{R}$ be partitioned into sets α^0, α^1, γ^0, γ^1 with α^0, α^1 nonempty.

In addition to $\bar{R}$, we shall use another compactification R_* of R. The topological space R_* is defined as the quotient space of $\bar{R}$ obtained by identifying all points of $\partial\bar{R}$ which belong to the same contour. Thus R_* may be realized by shrinking each contour of $\bar{R}$ to a point. The sets of contours α^0, α^1, γ^0, γ^1 of $\bar{R}$ correspond to finite point sets $\alpha_*{}^0$, $\alpha_*{}^1$, $\gamma_*{}^0$, $\gamma_*{}^1$ of R_*.

Let F be the class of arcs in $R_* - \gamma_*{}^0$ which go from α^0 to α^1. Let F^* consist of all c such that c is a sum of closed curves in $R_* - \gamma_*{}^1$ and c separates α^0 from α^1.

A suggestive example of these definitions is obtained by taking $\bar{R}$ to be an annulus with circular slits γ^1, radial slits γ^0, inner circumference α^0, and outer circumference α^1 as in Figure 1. The curve drawn with short dashes is in F, the one drawn with longer dashes is in F^*.

Although F, F^* were defined as classes of curves on R_*, they naturally determine classes of curves on R by restriction. In this sense we may consider $\lambda(F)$, $\lambda(F^*)$ and ask how they are related to each other and to analytic invariants of the configuration.

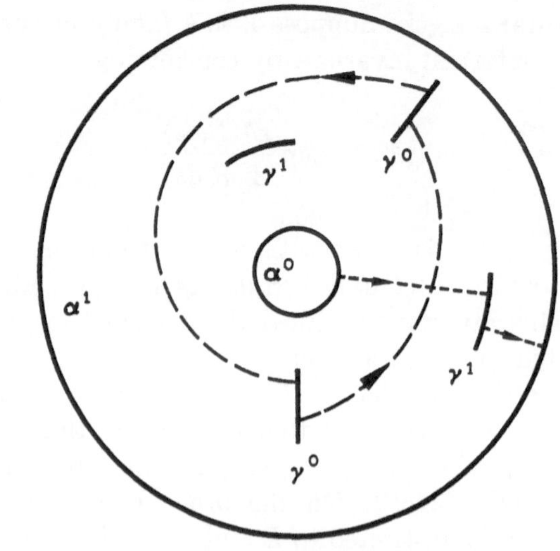

Figure 1

In the special case $\gamma^1 = \varnothing$, $\lambda(F)$ is called the *extremal distance* between α^0 and α^1, and $\lambda(F^*)$ is the *conjugate extremal distance.*

1D. Making use of the Main Existence Theorem I.1.2B we construct the harmonic function u which is 0 on α^0, 1 on α^1, and satisfies $u = L_0 u$ near each γ^0 and $u = L_1 u$ near each γ^1. Thus $\partial u/\partial n = 0$ along γ^0 and u is constant and has flux zero across each contour of γ^1. Let $||du||^2$ denote the Dirichlet integral of u.

Theorem. *On a compact bordered Riemann surface*

$$\lambda(F^*) = \frac{1}{\lambda(F)} = ||du||^2.$$

1E. To begin the proof we estimate the extremal lengths using $ds = |du + i * du|$ as a linear density. The ds-area of R is $A(ds) = \iint_R (u_x^2 + u_y^2)\, dxdy = ||du||^2$. To estimate the ds-length of F we use $ds \geq |du|$ and note that the variation of u along $c \in F$ is at least $u(\alpha^1) - u(\alpha^0) = 1$. Thus $L(F, ds) \geq 1$. For F^* use $ds \geq |*du|$ and note that if $c \in F^*$ then the ds-length of c is not less than the flux of u across c, namely $||du||^2$. Hence $L(F^*, ds) \geq ||du||^2$. This leads to the inequalities

$$\lambda(F) \geq ||du||^{-2} \geq \frac{1}{\lambda(F^*)}. \tag{29}$$

For estimates in the opposite direction we consider the level curves of u and of a harmonic conjugate function u^*.

1F. Level curves. We shall find a parametrization l_s of the level curves of u^* such that $l_s \in F$ for almost all s. Orient the contours of α^0, α^1, γ^0, γ^1 so that R lies to the left. Fix a point ζ on α^0 as the origin and determine a route which traverses α^0 exactly once in the positive direction, beginning and ending at ζ. Let u^* denote a local harmonic conjugate of u such that $u^*(\zeta) = 0$. By analytic continuation define $u^*(z)$ for each z on this route. Then since u^* is strictly decreasing along this route, it takes on each value s with $0 \geq s \geq -||du||^2$ exactly once.

For each such s a single level curve $\bar{l}_s$ of u^* leaves α^0. Orient each $\bar{l}_s$ so that u increases along the positive direction. We now define two complementary subsets S and C of the interval $-||du||^2 < s \leq 0$.

If the connected level curve $\bar{l}_s$ passes through a critical point of u in $\bar{R}$ put s in C.

If s is not already in C then $\bar{l}_s$ ends at either α^1 or γ^1. If it ends at α^1 put s in S and write l_s instead of $\bar{l}_s$.

Suppose $\bar{l}_s$ ends on some contour γ_i^1 of γ^1. Let ζ_s denote the end point of $\bar{l}_s$ and follow γ_i^1 from ζ_s in the direction of increasing $*du$. When γ_i^1 is traced in this manner the arc $\bar{l}_s$ and hence R lie to the left. Let ζ_s' be the first point for which $\int *du = 0$, the integral being taken over the arc on γ_i^1 from ζ_s to ζ_s'.

If ζ_s' is a critical point of u, put s in C. Otherwise the level curve $\bar{l}_s'$ of u^* passing through ζ_s', when oriented in the direction of increasing u, begins at ζ_s' and travels into R. We refer to $\bar{l}_s'$ as the continuation of $\bar{l}_s$.

If $\bar{l}_s'$ passes through a critical point of u, put s in C. Otherwise repeat the above procedure. Note that $\bar{l}_s'$ cannot return to γ_i^1 without passing through a critical point.

Thus each s with $-||du||^2 < s \leq 0$ is either in C or there is an l_s which is a finite union of connected level curves of u^* obtained by successive continuations beginning with $\bar{l}_s$, and which runs from α^0 to α^1. Each such l_s belongs to F. The reader may also check that $l_s \cap l_t = \emptyset$ if s, t are distinct elements of S. Furthermore, C is finite.

1G. Except at the finite number of critical points $u+iv$ can be used as a local parameter where v is any local harmonic conjugate of u. If $\rho|dz|$ is any linear density and $s \in S$ we use the Schwarz inequality to obtain

$$\int_{l_s} \rho|dz| = \int_{l_s} \rho\, du \leq \left(\int_{l_s} \rho^2\, du \int_{l_s} du\right)^{1/2} = \left(\int_{l_s} \rho^2\, du\right)^{1/2}.$$

Integration over $s \in S$ gives

$$\left(\inf_S \int_{l_s} \rho|dz|\right)^2 \int_S ds \leq \int_S \int_{l_s} \rho^2\, du ds = A(\rho|dz|),$$

and hence $\lambda(F) \leq ||du||^{-2}$. From this and (29) we obtain one equality in Theorem 1D.

Now consider the level curve, not necessarily connected, of $u = c$ for $0<c<1$. Each such curve l_c belongs to F^*. Orient l_c in the direction of increasing $*du$. Repeating the above argument with the Schwarz inequality gives

$$\int_{l_c} \rho|dz| = \int_{l_c} \rho\, dv \leq \left(\int_{l_c} \rho^2\, dv \int_{l_c} dv\right)^{1/2} = ||du|| \left(\int_{l_c} \rho^2\, dv\right)^{1/2},$$

$$\left(\inf_S \int_{l_c} \rho|dz|\right)^2 \int_0^1 dc \leq ||du||^2 A(\rho|dz|),$$

and hence $\lambda(F^*) \leq ||du||^2$. From (29) we deduce $\lambda(F^*) = ||du||^2$, which completes the proof.

1H. An application. As an immediate corollary of Theorem 1D we obtain:

Theorem. *On an annulus $r \leq |z| \leq R$, the extremal length of all curves which separate the inner and outer contours is $2\pi/\log(R/r)$. The extremal length of the family of open arcs which tend to the inner contour in one direction and the outer contour in the other direction is $(2\pi)^{-1}\log(R/r)$.*

1I. Let R be a rectangle $0<x<a$, $0<y<b$, and let F be the family of open arcs in R which tend to a side of length a in one direction and to the other side of length a in the opposite direction. Estimating $\lambda(F)$ with the Euclidean metric $|dz|$ yields $\lambda(F)\geq b/a$.

To obtain the opposite quantity consider any linear density ρ in R. For $0<t<a$ let l_t be the line segment $\{\operatorname{Re} z = t, 0<\operatorname{Im} z<b\}$. Then

$$\int_{l_t}\rho|dz| = \int_{l_t}\rho\,dy \leq \left(\int_{l_t}\rho^2\,dy\int_{l_t}dy\right)^{1/2} = \left(b\int_{l_t}\rho^2\,dy\right)^{1/2}.$$

Hence

$$aL^2(F,\rho)\leq bA(\rho),$$

and we obtain $\lambda(F)\leq b/a$. Therefore we have proved:

Theorem. *In a rectangle with sides of length a and b, let F be the family of open arcs which tend to a side of length a in one direction and tend to the opposite side in the opposite direction. Then $\lambda(F) = b/a$.*

1J. A geometric inequality. As a consequence of the above theorem we see that if F^* denotes the family of open arcs which tend to a side of length b in one direction and to the opposite side in the opposite direction, then $\lambda(F) = 1/\lambda(F^*)$. Since extremal length is unchanged by conformal mapping we can restate this fact for arbitrary quadrilaterals.

Let R now denote the inside of a closed Jordan curve β in the plane. Let there be given four distinct points on β. Then R may be called a quadrilateral and the families F, F^* may be defined in the obvious way. Now let a be the infimum of the Euclidean lengths of all curves joining one pair of opposite sides of R, and let b be the corresponding quantity for the other pair of opposite sides. Let us prove the following purely geometric result of Loewner (Besicovitch [1]):

Theorem. *The Euclidean area A of the quadrilateral R is no smaller than ab.*

We have $a^2/A\leq\lambda(F)$, $b^2/A\leq\lambda(F^*)$. Since $\lambda(F) = 1/\lambda(F^*)$ we obtain $a^2/A\leq A/b^2$, and the conclusion follows at once.

2. Infinite extremal length

2A. Basic properties. A family of curves F with $\lambda(F) = \infty$ is in many ways analogous to a set of measure zero. We shall derive some useful properties of such families and consider some examples.

Lemma. *Let $\{F_n\}$ be a finite or countable collection of families of curves, and let G be a family of curves with $G \subset \cup F_n$. Then*

$$\frac{1}{\lambda(G)} \leq \sum \frac{1}{\lambda(F_n)}.$$

For each n we may choose a density $\rho_n|dz|$ such that $L(\rho_n, F_n) = 1$ and $1/A(\rho_n)$ is arbitrarily close to $\lambda(F_n)$. Let $\rho|dz|$ be defined by $\rho = \sup \rho_n$. Then $L(\rho, G) \geq 1$, hence

$$\frac{1}{\lambda(G)} \leq \frac{A(\rho)}{L^2(\rho, G)} \leq A(\rho) = \iint \rho^2 dxdy \leq \iint \sum \rho_n^2 dxdy = \sum A(\rho_n).$$

Since $\sum A(\rho_n)$ can be made arbitrarily close to $\sum 1/\lambda(F_n)$ the proof is complete.

As an application we obtain the

Corollary. *Let $F = G \cup H$ where F, G, H are families of curves and $\lambda(G) = \infty$. Then $\lambda(F) = \lambda(H)$.*

2B. We say a statement is true for *almost every* curve in a family F if the subfamily of curves for which the statement is false has extremal length infinity. The following result is very useful (Fuglede [1]):

Lemma. *Let F be a family of curves and $\{ds_n\}$ a sequence of linear densities. If* $\lim A(ds_n) = 0$ *then there is a subsequence* $\{ds_{n_i}\}$ *such that*

$$\lim \int_c ds_{n_i} = 0$$

for almost every $c \in F$.

For the proof we may assume, by passing to a subsequence, that $A(ds_n) < 2^{-3n}$. Set

$$B_n = \left\{c \in F \,\middle|\, \int_c ds_n > 2^{-n}\right\},$$

$$C_n = \bigcup_{i=n}^{\infty} B_i,$$

$$D = \bigcap_{i=1}^{\infty} C_i.$$

For any n we have $\lambda(D)^{-1} \leq \lambda(C_n)^{-1}$ since $D \subset C_n$ and $\lambda(C_n)^{-1} \leq \sum_{i=n}^{\infty} \lambda(B_i)^{-1}$ by Lemma 2A. Using ds_n as a competing density for estimating $\lambda(B_n)$ we obtain $\sum_{i=n}^{\infty} \lambda(B_i)^{-1} \leq \sum_{i=n}^{\infty} 2^{-i} = 2^{1-n} \to 0$. Thus $\lambda(D) = \infty$. If $\limsup \int_c ds_n > 0$ then c belongs to infinitely many B_n, hence to every C_n, and a fortiori $c \in D$. Thus $\int_c ds_n \to 0$ for every $c \in F - D$.

2C. Examples. By considering some examples we can see that if F is a family of curves on a Riemann surface which tend to "part" of the ideal boundary, then the condition $\lambda(F) = \infty$ means that that part is very "small."

Let ζ be a point on a surface R and let F be the family of all curves which run from a fixed compact subset $A \subset R - \{\zeta\}$ to ζ. It is easy to see that $\lambda(F) = \infty$. In fact, let Δ be a disk with $\zeta \in \Delta$, $\Delta \subset R - A$. Let G be the family of curves in Δ running from the circumference to ζ. Clearly $L(G, ds) \leq L(F, ds)$ for any ds, and therefore $\lambda(G) \leq \lambda(F)$. That $\lambda(G) = \infty$ can be seen directly from Theorem 1H, but we shall consider the following more general example with ζ replaced by a set of points.

Let E be a closed point set in a disk Δ and let F consist of all curves running from $\partial\Delta$ to E. Then $\lambda(F) = \infty$ if and only if E has capacity zero. The definition of capacity zero may be stated in the following form. Let $\{\Omega_n\}$ be an exhaustion of $\Delta - E$ with $\partial\Delta \subset \partial\Omega_n$. Let u_n be the harmonic function in Ω_n with values 0 on $\partial\Delta$ and 1 on the remaining contours β_n of $\partial\Omega_n$. Then $\{u_n\}$ forms a decreasing sequence of harmonic functions. Therefore it converges uniformly on compacta to a harmonic limit function u. If $u \equiv 0$ then E is said to be of capacity zero.

Let F_n be the family of curves in Ω_n from $\partial\Delta$ to β_n. By Theorem 1D we have $\lambda(F_n) = ||du_n||^{-2}$. Clearly $\lambda(F) \geq \lambda(F_n)$ so $\lambda(F) = \infty$ if E has capacity zero. The converse is also true. We shall not give a proof here since a more general result will be proved in III. 2.4B.

A related result is the following: Let R be an annulus and let E be a closed set on one contour. Let F be the family of curves which tend to E in one direction and to the other contour in the opposite direction. Then $\lambda(F) = \infty$ if and only if E is of capacity zero.

2D. Level curves. Let R be a Riemann surface on which there exists a nonconstant harmonic function u with a finite Dirichlet integral. Let c_t be a level curve $u = t$. Suppose T is a measurable set of real numbers such that for each $t \in T$ the curve c_t is nonempty and does not pass through a critical point of u.

Theorem. *If the family* $\{c_t | t \in T\}$ *has infinite extremal length then* T *has measure zero.*

For the proof suppose $ds = \rho|dz|$ is any linear density. Except at the critical points of u we may use $dz = du + i * du$ as a local parameter. Suppose the curves c_t are oriented so that $*du > 0$ along them. Then

$$\int_{c_t} \rho|dz| = \int_{c_t} \rho * du \leq \left(\int_{c_t} \rho^2 * du \int_{c_t} * du\right)^{1/2},$$

and hence $L(\{c_t\}, ds) \leq (M_t \int_{c_t} \rho^2 * du)^{1/2}$ where $\int_T M_t \, dt \leq ||du||^2 < \infty$. On integrating over T we obtain

$$\mu(T) L(\{c_t\}, ds) \leq ||du|| A^{1/2}(ds)$$

where $\mu(T)$ is the measure of T. Therefore, $\mu(T) \neq 0$ would imply $\lambda(\{c_t\}) \leq ||du||^2/\mu^2(T) < \infty$.

2E. Relative homology. We have already had occasion to consider curves which tend to the ideal boundary in a certain manner. For a unified treatment we now introduce the notion of relative chains and weak homology.

Let c_i^n denote a singular n-simplex on a surface R ($n = 0,1,2$). A formal sum $c^n = \sum_{i=1}^{\infty} x_i c_i^n$, where x_i is an integer, will be called a *relative n-chain* provided that for each compact subset $K \subset R$ the set of indices i for which $x_i \neq 0$ and $c_i^n \cap K \neq \emptyset$ is finite. The boundary homomorphism ∂ applied to a relative n-chain yields a relative $(n-1)$-chain. A relative n-chain c^n is a *relative cycle* if $\partial c^n = 0$, and it is a *relative boundary* if $c^n = \partial c^{n+1}$ for some relative $(n+1)$-chain. Note that an open arc on R which tends to the ideal boundary in both directions can be considered as a relative 1-cycle. Two relative chains are *weakly homologous* if their difference is a relative boundary.

2F. Integration. Let $\omega = a\,dx + b\,dy$ be a first order differential on R, and $c = \sum x_i c_i$ a relative 1-chain. We define $\int_c \omega$ as $\sum x_i \int_{c_i} \omega$ provided the series is absolutely convergent. If ds is a linear density then $\int_c ds$ is to be $\sum |x_i| \int_{c_i} ds$. It is always nonnegative and may be infinite. In particular, $\langle \omega \rangle = (|a|^2 + |b|^2)^{1/2} |dz|$ is a linear density if the coefficients of ω are Borel measurable. If $\int_c \langle \omega \rangle < \infty$ then $\int_c \omega$ exists and

$$\left| \int_c \omega \right| \leq \int_c \langle \omega \rangle.$$

2G. Principal functions. Let p_0, p_1 be the principal functions on R corresponding to a singularity function s and to the identity partition I. Let $E = \{\zeta_1, \cdots, \zeta_m\}$ be the singular points of s. If R is the interior of a compact bordered surface $\bar{R}$ then $\int dp_1 = 0$ for any arc on $\bar{R} - E$ with end points on $\partial \bar{R}$. Similarly, if c is an arc on $\bar{R} - E$ such that c together with part of $\partial \bar{R}$ bound a region which does not meet E, then $\int_c * dp_0 = 0$. These properties serve to characterize the boundary behavior of p_0 and p_1. The generalization of these properties to an arbitrary R is given in the following

Theorem. *Let p_0, p_1 and E be defined as above. Then $\int_c dp_1 = 0$ for almost every relative cycle c on R which does not meet E, and $\int_d * dp_0 = 0$ for almost every relative boundary d of the form ∂d^2 with $d^2 \cap E = \emptyset$.*

2H. In Marden-Rodin [1] an example is given to show that the above theorem would be false if "almost every" were replaced by "every rectifiable." To prove the theorem consider an exhaustion $\{\Omega_n\}$ of R with $E \subset \Omega_1$. Let p_{0n}, p_{1n} denote the principal functions for Ω_n with the same singularities as p_0, p_1. We know that $\lim_{n\to\infty} \|dp_i - dp_{in}\|_{\Omega_n} = 0$ for $i = 0, 1$ (see I.3.2B). Consider the linear densities $ds_{in} = \langle dp_i - dp_{in}\rangle = |\operatorname{grad}(p_i - p_{in})||dz|$ where we define $dp_{in} \equiv 0$ on $R - \Omega_n$. The above remark shows that $\lim_{n\to\infty} A(ds_{in}) = 0$. By Lemma 2B we may assume that a subsequence of $\{ds_{in}\}$ has been selected so that, without changing notation for the subsequence, we have

$$\lim_{n\to\infty} \int_c ds_{1n} = 0,$$

$$\lim_{n\to\infty} \int_d ds_{0n} = 0,$$

for almost every relative cycle c and relative boundary d of $R - E$.

Since for almost every such c, d we have

$$0 \leq \left| \int_c dp_1 - dp_{1n} \right| \leq \int_c ds_{1n} \to 0$$

and

$$0 \leq \left| \int_d *dp_0 - *dp_{0n} \right| \leq \int_d ds_{0n} \to 0,$$

the proof will be complete if we show that $\int_c dp_{1n} = 0$, $\int_d *dp_{0n} = 0$.

2I. It is geometrically evident that $\int_d *dp_{0n} = 0$ if d is a relative boundary. For a careful proof let $d = \partial d^2$ where $d^2 = \sum x_i d_i^2$ is a relative 2-chain, $d^2 \cap E = \emptyset$. There are only finitely many 2-simplices d_j^2 whose boundary intersects Ω_n; assume they are $d_1^2, \cdots, d_N^2$. Let $d' = \sum_1^N x_j \partial d_j^2$. Then $d - d' = \sum_{N+1}^\infty x_j \partial d_j^2$ lies in $R - \Omega_n$, and clearly $\int_{d-d'} *dp_{0n} = 0$. Now d' is an ordinary chain which is homologous to zero on $R - E$, and therefore $\int_{d'} *dp_{0n} = 0$. Thus $\int_d *dp_{0n} = 0$ as claimed. In the same way one may show that $\int_c dp_{1n} = 0$ if c is a relative cycle. This completes the proof of Theorem 2G.

3. Extremal length of homology classes

3A. Generalized homology. Let Γ_x be a closed subspace of Γ_h and let c_1, c_2 be ordinary (finite) cycles on R. We say c_1 and c_2 are Γ_x-homologous, denoted by $c_1 = c_2 \pmod{\Gamma_x}$, if $\int_{c_1-c_2}\omega = 0$ for all $\omega \in \Gamma_x$. The equivalence class of a cycle c will be denoted by c^x.

Lemma. *Let c be a cycle and let $\lambda(c^x)$ be the extremal length of all cycles Γ_x-homologous to c. Then $\lambda(c^x) \geq ||\psi||^2$ where ψ is the Γ_x-reproducing differential.*

3B. For the proof consider the linear density $ds = |\psi + i*\psi|$. Then $A(ds) = ||\psi||^2$. If $\delta \in c^x$ then

$$\int_\delta ds \geq \int_\delta |\psi| \geq \left|\int_\delta \psi\right| = \left|\int_c \psi\right| = (\psi,\psi).$$

Thus $L(c^x, ds) \geq ||\psi||^2$ and we have

$$\lambda(c^x) \geq \frac{L^2(c^x,ds)}{A(ds)} \geq ||\psi||^2.$$

3C. Compact surfaces. The following result contains a partial converse to Lemma 3A. Let P be a regular partition of the boundary of R. Denote by $\psi_P(c)$ the $(P)\,\Gamma_{hse}$-reproducing differential for c, and by c^P the $(P)\,\Gamma_{hse}$-homology class of c.

Lemma. *Let R be a closed Riemann surface or the interior of a compact bordered surface $\bar{R}$. Then $\lambda(c^P) = ||\psi_P(c)||^2$.*

The proof will make use of the representation $2\pi\,\psi_P(c) = *dp_{P\tau}$ established in 3.3E. Note that $dp_{P\sigma} = 0$ since c is a cycle. Also of importance will be the fact that $*\psi_P(c)$ has integral periods $\int_\delta *\psi_P(c) = \delta \times c$ for any cycle δ (cf. (17) of 4.2D).

3D. For the proof of Lemma 3C we denote the sets of contours of $\partial\bar{R}$ determined by P by $\beta_1, \cdots, \beta_n$. Let V be the set of points in $\bar{R}$ which can be joined to a given point on β_1 by an arc δ for which $\int_\delta *\psi_P(c)$ is an integer. If R is closed let the given point be anywhere on R. Then V is a closed set which is locally the level curve of a harmonic function. Let $R - V = R_1 \cup \cdots \cup R_m$ be a decomposition into components. Each R_ν is the interior of a compact bordered surface $\bar{\bar{R}}_\nu$ whose border can be realized by the curves V and $\partial \bar{R}$. In terms of this realization we shall refer to boundary points of $\bar{\bar{R}}_\nu$ as points in V or $\partial \bar{R}$.

Observe that $*\psi_P(c)$ is exact on each R_ν. Indeed, if $*\psi_P(c)$ had a nonzero period along a cycle in R_ν this period would necessarily be an integer. But then we could construct an arc δ from β_1 to an interior point of R_ν such that $\int_\delta *\psi_P(c)$ is an integer.

Thus $*\psi_P(c)$ is the differential of a harmonic function p_ν in R_ν. Now $\partial\bar{\bar{R}}_\nu$ consists of points of V where p_ν is constant by definition of V, or of points of $\partial\bar{R}$ where p_ν is constant by the L_1-behavior of $p_{P\tau}$. We may adjust p_ν by an additive constant so that the smallest such constant on a part of V is zero. Let α_ν be the collection of those boundary components of $\bar{\bar{R}}_\nu$ on which $p_\nu = 0$, let σ_ν consist of those on which $p_\nu = 1$, and let γ_ν contain the remaining ones. Consider α_ν, σ_ν, and γ_ν as curves on $\bar{R}$ and orient them so that $\partial\bar{\bar{R}}_\nu = \alpha_\nu + \sigma_\nu + \gamma_\nu$. The points of α_ν and σ_ν belong to V, those of γ_ν belong to $\partial\bar{R} - V$.

Let us show that if γ_ν contains a point t of some β_k, then it must contain all of β_k. For $t_1 \in \beta_1$, $\int_{t_1}^{t} *\psi_P(c)$ is not an integer and since $*\psi = -(2\pi)^{-1}\, dp_{P\tau}$ and $p_{P\tau}$ is constant along β_k it follows that β_k has a connected neighborhood disjoint from V. This neighborhood must be in R_ν, hence $\beta_k \subset \gamma_\nu$.

Let $\omega \in (P)\,\Gamma_{hse}$ and assume that ω extends harmonically to $\bar{R}$. Then

$$\int_c \omega = (\omega, \psi_P(c)) = \sum_\nu (\omega, *dp_\nu).$$

By Stokes' theorem

$$(\omega, *dp_\nu) = \iint_{R_\nu} dp_\nu \wedge \omega = \int_{\sigma_\nu} \omega + \int_{\gamma_\nu} p_\nu \omega.$$

We have seen that γ_ν is a union of P-parts $\beta_{\nu 1}, \cdots, \beta_{\nu k}$ on each of which p_ν is a constant. Since ω has zero period over any P-dividing cycle we obtain

$$(30) \qquad \int_c \omega = \int_{\Sigma\sigma_\nu} \omega, \quad \text{all} \quad \omega \in (P)\,\Gamma_{hse}.$$

The function p_ν has boundary values 0 on α_ν, 1 on σ_ν, and constants $k_{\nu\mu}$ on $\beta_{\nu\mu}$, those P-parts of $\partial\bar{R}$ which make up γ_ν. These constants must satisfy $0<k_{\nu\mu}<1$ in order for the flux condition $\int_{\beta_{\nu\mu}} *dp_\nu = 0$ to hold. Consequently for $t\in(0,1)$ the level curves $\sigma_{\nu t} = p_\nu^{-1}(t)$ are compact and $(P)\,\Gamma_{hse}$-homologous to σ_ν, except for the finite number of values $t = k_{\nu u}$. Let $\sigma(t) = \sum_\nu \sigma_{\nu t}$.

Consider a linear density ds on R. With the exception of a finite number of critical points, $\psi_P(c) + i*\psi_P(c) = dz$ can be used for a local coordinate system. Writing $ds = \rho|dz| = \rho|\psi + i\psi^*|$ we have for almost all $t\in(0,1)$

$$L^2(c^P,ds) \leq \left(\int_{\sigma(t)} \rho\psi_P(c)\right)^2 \leq \int_{\sigma(t)} \rho^2\psi_P(c) \int_{\sigma(t)} \psi_P(c)$$

$$= \|\psi_P(c)\|^2 \int_{\sigma(t)} \rho^2\psi_P(c).$$

On integrating over $t\in(0,1)$ we obtain

$$L^2(c^P,ds) \leq \|\psi_P(c)\|^2 A(ds). \tag{31}$$

The inequality opposite to (31) is given by Lemma 3A. The proof of Lemma 3C is herewith complete.

3E. Open surfaces. The most important special cases of Lemma 3C are obtained by taking P to be the canonical partition Q or the identity partition I. Then $(Q)\,\Gamma_{hse} = \Gamma_{hse}$ and $(I)\,\Gamma_{hse} = \Gamma_h$.

A cycle γ is Γ_{hse}-homologous to zero only if it is a dividing cycle. In fact, if γ is not a dividing cycle choose a cycle δ with $\delta\times\gamma\neq 0$. Then $\int_\gamma *\psi_Q(\delta)\neq 0$ and $*\psi_Q(\delta) = -(2\pi)^{-1}\,dp_{Qt}$ is in Γ_{hse}.

If γ is Γ_h-homologous to zero on a compact bordered surface $\bar{R}$ then γ is homologous to zero. The reasoning above shows that γ must be a dividing cycle. By considering du_i where u_i is the harmonic measure of a contour β_i of $\partial\bar{R}$, one can easily construct a differential in Γ_h with nonzero period over any dividing cycle which is not homologous to zero.

Two (finite) cycles are weakly homologous (6.2E) if their difference is a dividing cycle. Thus we have shown that c^Q is the weak homology class of c and, for compact bordered surfaces, c^I is the usual homology class of c. We shall prove (Accola [2], Blatter [1], Rodin [2]):

Theorem. *Let R be an arbitrary Riemann surface and c a cycle on R. The extremal length of the homology class of c is $\|\psi_I(c)\|^2$, the squared norm of the Γ_h-reproducer for c. The extremal length of the weak homology class of c is $\|\psi_Q(c)\|^2$ where $\psi_Q(c)$ is the Γ_{hse}-reproducer for c.*

3F. For the proof we let $\{\Omega\}$ be an exhaustion of R and $\psi_{P\Omega}(c)$ the $(P)\,\Gamma_{hse}(\Omega)$-reproducer for c. By Theorem I.2.3B and Corollary 3.3E we know that $\psi_P(c) = \lim_{\Omega\to R}\psi_{P\Omega}(c)$.

Let F_I (respectively F_Q) be the homology (respectively weak homology) class of c. Note that $\lambda(F)$ increases as F decreases. Hence for $P = I$ or Q

$$\lambda(F_P) \geq \lambda(c^P).$$

By Lemma 3A

$$\lambda(c^P) \geq \|\psi_P(c)\|^2 = \lim_{\Omega\to R} \|\psi_{P\Omega}(c)\|^2.$$

We also have

$$\|\psi_{P\Omega}(c)\|^2 = \lambda_\Omega(F_{P\Omega})$$

where $F_{P\Omega} = \{\gamma \in F_P | \gamma \subset \Omega\}$. By 1B, $\lambda_\Omega(F_{P\Omega}) = \lambda(F_{P\Omega})$. Thus

$$\lambda(F_P) \geq \|\psi_p(c)\|^2 \geq \lim_{\Omega\to R} \lambda(F_{P\Omega}) \geq \lambda(F_P).$$

Thus $\lambda(F_P) = \|\psi_p(c)\|^2$ and the theorem is proved.

3G. Γ_{ho}-reproducers. We shall prove a theorem analogous to 3E for the space Γ_{ho}. The Γ_{ho}-reproducer for a cycle c is of the form $\psi = (2\pi)^{-1} * dp_{0r}$ as shown by Corollary 3.3C. As before $*\psi$ has integral periods.

On an arbitrary Riemann surface R choose a point and let V be the equivalence class of that point as in 3D. Then $R - V$ is a union of components $\cup_{\nu=1}^{\infty} R_\nu$. It is possible to show that this union is finite although we shall not need that result here.

The form $*\psi$ is exact in each R_ν, say $*\psi = dp_\nu$. The relative boundary of R_ν consists of relative chains σ_ν on which we may take $p_\nu = 1$, and remaining relative chains α_ν on which $p_\nu = 0$. Orient them so that R_ν has oriented relative boundary $\alpha_\nu - \sigma_\nu$.

We claim that $\sum \sigma_\nu$, a relative cycle, is weakly homologous to c, i.e. $c - \sum \sigma_\nu$ is a relative boundary. To prove this it suffices to show that $\delta \times c = \delta \times \sum \sigma_\nu$ for every finite cycle δ. Let $\{\Omega\}$ be an exhaustion of R such that each Ω contains c and a given δ, and let $\psi_\Omega(\delta)$ be the $\Gamma_h(\Omega)$-reproducing differential for δ. Then $*\psi_\Omega(\delta) = (2\pi)^{-1}dp_{I_T\Omega}$ is zero along $\partial\Omega$. Hence $\delta \times c = \int_\delta *\psi = (*\psi, \psi_\Omega(\delta))_\Omega = -\sum \int_{\sigma_\nu \cap \Omega} *\psi_\Omega(\delta) = \delta \times \sum \sigma_\nu \cap \Omega = \delta \times \sum \sigma_\nu$.

For $t \in (0,1)$ let $\sigma_{\nu t}$ be the level line $p_\nu = t$. Then $\sigma_{\nu t}$ is weakly homologous to σ_ν since together they bound the open set where $t < p_\nu < 1$. Consequently each $c_t = \sum \sigma_{\nu t}$ is a relative cycle weakly homologous to c.

Let $ds = \rho|dz|$ be any linear density and consider $dz = \psi + i*\psi$ as a local variable on R less the countable number of zeros of ψ. By Theorems 2D and 2G we have $\int_{c_t - c}\psi = 0$ for almost all $t \in (0,1)$ and therefore $\int_{c_t}\psi = ||\psi||^2$ for almost all such t. We obtain

$$\int_{c_t} \rho|dz| \le \left|\int_{c_t} \rho\psi\right| \le \left(\int_{c_t} \rho^2\psi \int_{c_t} \psi\right)^{1/2}.$$

An integration over a.e. $t \in (0,1)$ leads to

$$L(\dot{c},ds) \le ||\psi|| \left(\int_0^1 \int_{c_t} \rho^2\psi\, dt\right)^{1/2} \le ||\psi|| A^{1/2}(ds)$$

where $\dot{c}$ is the family of relative cycles weakly homologous to c. We have proved that $\lambda(\dot{c}) \le ||\psi||^2$.

To obtain the opposite inequality consider now the linear density $ds = |\psi + i*\psi|$. Then $A(ds) = ||\psi||^2$. By Theorem 2G $\int_\delta \psi = \int_c \psi = ||\psi||^2$ for every $\delta \in \dot{c}$ except for a subfamily of infinite extremal length. By Corollary 2A we may ignore these exceptions in computing $\lambda(\dot{c})$. Hence

$$L(\dot{c},ds) = \inf_{\delta \in \dot{c}} \int_\delta |\psi + i*\psi| \ge \left|\int_\delta \psi\right| = ||\psi||^2,$$

and consequently we have $\lambda(\dot{c}) \ge L^2(\dot{c},ds)/A(ds) \ge ||\psi||^2$. Thus we have proved (Marden [2]) the

Theorem. *Let R be an arbitrary Riemann surface, c a finite cycle on R, $\dot{c}$ the family of relative cycles weakly homologous to c, and ψ the Γ_{h0}-reproducing differential for c. Then* $\lambda(\dot{c}) = ||\psi||^2$.

In Marden [2] this result is generalized to apply also to relative cycles c.

CHAPTER III
CAPACITY, STABILITY, AND EXTREMAL LENGTH

We begin by considering a harmonic function whose boundary behavior is a combination of L_0- and L_1-type behavior. This leads to a generalization of harmonic measure and capacity. In §2 we relate this function to some extremal length problems. The remaining sections treat applications to conformal mapping and stability problems.

In their original form, the capacity functions and stability problems were introduced in Sario [12], [15], and the extremal length in Ahlfors-Beurling [1] and Strebel [2]. The extensions and generalizations presented in this chapter were obtained in Marden-Rodin [2].

§1. GENERALIZED CAPACITY FUNCTIONS

We wish to construct a harmonic function u on an arbitrary Riemann surface which generalizes the function constructed in II.6.1D; i.e. u should be 0 on one part α^0 of the ideal boundary, 1 on another part α^1, have L_0-behavior on a third part γ^0, and L_1-behavior on a fourth part γ^1. It turns out that there are two equally natural ways of doing this so that the function u will not depend on the choice of an exhaustion of the surface. We also construct a harmonic function p with the same boundary behavior as that of u, but with a single logarithmic pole at an interior point of the surface.

1. Construction of u

1A. Induced partitions. Let R be an arbitrary Riemann surface and let the ideal boundary of R be partitioned into four disjoint sets α^0, α^1, γ^0, γ^1. Take a canonical exhaustion $\{\Omega\}$ of R.

Denote by $\bar{R}$ the Kerékjártó-Stoïlow compactification of R (see 0.1.2C).

For each Ω we wish to define the partition α_Ω^0, α_Ω^1, γ_Ω^0 γ_Ω^1 of $\partial\Omega$ "induced" by the partition α^0, α^1, γ^0, γ^1.

Consider the components of $\bar{R}-\Omega$ which contain a point of α^i ($i = 0,1$) and denote their union by A^i. We shall require that $A^0 \cap A^1 = \emptyset$. This can always be attained for sufficiently large Ω if we assume that α^0, α^1 are closed subsets of $\bar{R}$. We can then define α_Ω^i to be the union of those contours of $\partial\Omega$ which are part of ∂A^i. For the applications we have in mind it is natural to assume that α^0, α^1 are closed and also that they are nonempty. However, the definitions of γ_Ω^0, γ_Ω^1 are not so naturally determined.

A contour c of $\partial\Omega$ bounds a component A^c of $\bar{R}-\Omega$. Suppose $c \notin \alpha_\Omega^0 \cup \alpha_\Omega^1$. If all ideal boundary points in A^c belong to γ^0 we put c in γ_Ω^0. Similarly put c in γ_Ω^1 if all ideal points of A^c are in γ^1. If A^c contains points of both γ^0 and γ^1, we shall arbitrarily declare that c belongs to γ_Ω^1. It will be seen that the effect of this decision is to treat points of γ^0 which are limit points of γ^1 as if they were actually points of γ^1. It will therefore be convenient to assume at the outset the following properties of the partition α^0, α^1, γ^0, γ^1:

(A1) α^0, α^1 are nonempty closed subsets of $\bar{R}$.

(A2) $\alpha^0 \cup \alpha^1 \cup \gamma^1$ is a closed subset of $\bar{R}$.

In a later section (3.3E) we shall consider the effect of replacing A2 by

(A2′) $\alpha^0 \cup \alpha^1 \cup \gamma^0$ is a closed subset of $\bar{R}$.

1B. Consider now a countable canonical exhaustion $\{\Omega_n\}_{n=1}^{\infty}$ of R (see 0.2.1F). Although the existence of such an exhaustion is not essential for the following convergence proofs, we shall make use of one because it simplifies the notation and will, in fact, be needed for later work with extremal length.

Given Ω_n let $\tilde{\Omega}_n$ denote the possibly noncompact region obtained by adjoining to Ω_n the components of $R-\Omega_n$ whose borders are in $\gamma_n^0 \equiv \gamma_{\Omega_n}^0$. Orient the contours of $\tilde{\Omega}_n$ so that $\partial\tilde{\Omega}_n = \alpha_n^1 + \gamma_n^1 - \alpha_n^0$.

Let $\{\Psi_i\}_{i=1}^{\infty}$ be a canonical exhaustion of R and set $\Omega_{ni} =$

$\tilde{\Omega}_n \cap \Psi_i$ for sufficiently large i with $\partial\tilde{\Omega}_n \subset \Psi_i$. Choose the notation and orientation of the contours of Ω_{ni} so that $\partial\Omega_{ni} = \alpha_n^1 + \gamma_n^1 + \gamma_{ni}^0 - \alpha_n^0$. For sufficiently large i, γ_{ni}^0 is homologous to γ_n^0.

Define a harmonic function u_{ni} in Ω_{ni} by the following boundary conditions:

(B1) $u_{ni} = 0$ along α_n^0, $u_{ni} = 1$ along α_n^1.

(B2) $*du_{ni} = 0$ along γ_{ni}^0.

(B3) $du_{ni} = 0$ along each component c of γ_n^1 and $\int_c *du_{ni} = 0$.

1C. Convergence of u_{ni}. We shall show that $\lim_{i\to\infty} u_{ni} = u_n$ exists in the sense of the Dirichlet norm, and is independent of the exhaustion $\{\Omega_{ni}\}_i$ of $\tilde{\Omega}_n$.

For $j>i$ the equation

$$(du_{ni}, du_{nj})_{\Omega_{ni}} = \int_{\partial\Omega_{ni}} u_{nj} *du_{ni} = \int_{\alpha_{n1}} *du_{ni} = \|du_{ni}\|^2_{\Omega_{ni}}$$

implies that

$$\|du_{ni} - du_{nj}\|^2_{\Omega_{ni}} < \|du_{nj}\|^2_{\Omega_{nj}} - \|du_{ni}\|^2_{\Omega_{ni}}. \tag{1}$$

In particular, $\|du_{ni}\|^2_{\Omega_{ni}}$ is increasing with i.

Let h be the harmonic function in Ω_n defined by the boundary conditions $h = 0$ on α_n^0, $h = 1$ on $\partial\Omega_n - \alpha_n^0$. Then since

$$(du_{ni}, dh)_{\Omega_n} = \int_{\partial\Omega_n} h * du_{ni} = \|du_{ni}\|^2_{\Omega_{ni}}$$

we find that

$$\|du_{ni} - dh\|^2_{\Omega_n} \le \|dh\|^2_{\Omega_n} - \|du_{ni}\|^2_{\Omega_{ni}},$$

and therefore $\|du_{ni}\|_{\Omega_{ni}}$ is bounded in i. It follows from equation (1) that $\lim_{i\to\infty} u_{ni} = u_n$ exists. Furthermore

$$\lim_{i\to\infty} \|du_{ni}\|_{\Omega_{ni}} = \|du_n\|_{\tilde{\Omega}_n}.$$

To prove this last assertion set $A = \lim_{i\to\infty}||du_{ni}||^2_{\Omega_{ni}}$. Then (using a simplified notation) let $i\to\infty$, $i>j$, in the inequality

$$||du_{ni}-du_{nj}||_j^2 \leq ||du_{ni}||_i^2 - ||du_{nj}||_j^2$$

to obtain

$$||du_n - du_{nj}||_j^2 \leq A - ||du_{nj}||_j^2$$

and a fortiori $\lim_{j\to\infty}||du_n - du_{nj}||_j = 0$. Our assertion now follows upon letting $j\to\infty$ in the inequality

$$|\,||du_n - du_{nj}||_j - ||du_n||_j| \leq ||du_{nj}||_j \leq ||du_n - du_{nj}||_j + ||du_n||_j,$$

since $||du_n||_j \to ||du_n||_{\tilde{\Omega}_n}$.

Clearly u_n is independent of the particular exhaustion $\{\Omega_{ni}\}$ of $\tilde{\Omega}_n$ and u_n is 0 on α_n^0, 1 on α_n^1, and constant on each component c of γ_n^1 with $\int_c *du_n = 0$.

Next we show that $\lim du_n = \omega$ exists as an exact harmonic differential independently of the exhaustion $\{\Omega_n\}$. If $n>m$ then $\tilde{\Omega}_m \subset \tilde{\Omega}_n$,

$$\partial(\Omega_{ni}\cap\tilde{\Omega}_m) = \alpha_m^1 + \gamma_m^1 - \alpha_m^0 + \gamma_{ni}^0 \cap \tilde{\Omega}_m,$$

and γ_m^1, α_m^0, and α_m^1 are homologous respectively to components in $\gamma_n^1, \gamma_{ni}^0$; $\alpha_n^0, \gamma_n^1, \gamma_{ni}^0$; $\alpha_n^1, \gamma_n^1, \gamma_{ni}^0$. Hence, since

$$\lim_{i\to\infty} ||du_n - du_{ni}||_{\Omega_{ni}} = 0,$$

$$(du_m, du_n)_{\tilde{\Omega}_m} = \lim_{i\to\infty} (du_m, du_{ni})_{\Omega_{ni}\cap\tilde{\Omega}_m} = \lim_{i\to\infty} \int_{\alpha_m^1} *du_{ni} = ||du_n||^2_{\tilde{\Omega}_n}.$$

Therefore

$$(2) \qquad ||du_m - du_n||^2_{\tilde{\Omega}_m} \leq ||du_m||^2_{\tilde{\Omega}_m} - ||du_n||^2_{\tilde{\Omega}_n}$$

and we see that $||du_m||^2_{\tilde{\Omega}_m}$ decreases as m increases and $\lim_{m\to\infty} du_m$ exists. The harmonic limit differential is exact and we denote it by du or $du(\alpha^0, \alpha^1, \gamma^0, \gamma^1)$, u being unique only up to an additive constant. We also see that $\lim ||du_m||_{\tilde{\Omega}_m} = ||du||$. The question arises: Does $\lim u_n = u$ exist for suitable u? This is false if $||du|| = 0$, but turns out to be correct if $||du|| > 0$. The proof follows easily from later results (Theorems 1F and 2.2A). For the present we do not need this fact.

If $\{\Omega_n'\}$ is another exhaustion of R then given k there exist m and n such that $\Omega_k' \subset \Omega_m \subset \Omega_n'$. It follows that $\tilde{\Omega}_k' \subset \tilde{\Omega}_m \subset \tilde{\Omega}_n'$ and we see that indeed du is independent of our choice of the exhaustion $\{\Omega_n\}$.

We shall have occasion to use the following simple observation: For each n we can find an integer i_n such that

$$||du_n||_{\tilde{\Omega}_n} - ||du_{ni_n}||_{\Omega_{ni_n}} < \frac{1}{n}.$$

Thus on setting $h_n = u_{ni_n}$ and replacing Ω_{ni_n} by Ω_n we see that $\lim ||dh_n||_{\Omega_n} = ||du||$. By passage to a subsequence we may assume the Ω_n are nested and thus form an exhaustion.

1D. Dependence of $u(\alpha^0, \alpha^1, \gamma^0, \gamma^1)$ on α^0. In many applications we begin with a Riemann surface R whose ideal boundary is partitioned into α^1, γ^0, γ^1. Let A be a regular subregion of R with $\partial A = \tilde{\alpha}^0$. Denote the part of the ideal boundary of $R-A$ which corresponds to $\tilde{\alpha}^0$ by α^0. In this way we obtain a partition α^0, α^1, γ^0, γ^1 of the ideal boundary of $R-A$ and can construct

$$du(\alpha^0, \alpha^1, \gamma^0, \gamma^1)$$

on $R-A$. By *abus de langage* we shall not distinguish between $\tilde{\alpha}^0$ and α^0, but simply describe the above process as follows: Given some analytic curves α^0 in R such that α^0 does not separate R into two noncompact regions, form $du(\alpha^0, \alpha^1, \gamma^0, \gamma^1)$.

In the situation described above the exhaustion $\{\Omega_n\}$ used to construct du on $R-A$ may be chosen so that α^0 is a contour of each Ω_n. Now suppose there is another such collection of curves δ^0 disjoint from α^0. Choose a corresponding exhaustion $\{\Omega_n'\}$ with $\partial\Omega_n' = \alpha_n^1 + \gamma_n^1 + \gamma_n^0 - \delta^0$ and construct

$$du' = du(\delta^0, \alpha^1, \gamma^0, \gamma^1).$$

As a consequence of the observation made at the end of 1C we may assume that $\{\Omega_n\}$ and $\{\Omega_n'\}$ are chosen so that $\lim\|dh_n\| = \|du\|$, $\lim\|dh_n'\| = \|du'\|$ (see the notation used there) and $\alpha^0\subset\Omega_n'$, $\delta^0\subset\Omega_n$ for all n.

Setting $\Psi_n = \Omega_n\cap\Omega_n'$ we see that $\partial\Psi_n = \alpha_n^1+\gamma_n^1+\gamma_n^0-\delta^0-\alpha^0$ and

$$(dh_n,dh_n')_{\Psi_n} = \int_{\partial\Psi_n} h_n * dh_n' = \int_{\alpha_n^1} * dh_n' - \int_{\delta^0} h_n * dh_n'$$

$$= \|dh_n'\|^2_{\Omega_n'} - \int_{\delta^0} h_n * dh_n'.$$

Hence

$$\|dh_n - dh_n'\|^2_{\Psi_n} \le \|dh_n\|^2_{\Omega_n} - \|dh_n'\|^2_{\Omega_n'} + 2\int_{\delta^0} h_n * dh_n'.$$

We may conclude that if $\lim\|dh_n\|_{\Omega_n} = \|du\| = 0$ then $\lim h_n = 0$ and $\lim\int_{\delta^0} h_n * dh_n' = 0$, so that

$$\lim_{n\to\infty} \|dh_n - dh_n'\|^2_{\Psi_n} = -\lim_{n\to\infty} \|dh_n'\|^2_{\Omega_n'} = -\|du'\|^2.$$

Therefore $\|du'\| = 0$ and we have proved most of the following result (Marden-Rodin [2]):

Theorem. *If α^0 and δ^0 are as above then $du(\alpha^0, \alpha^1, \gamma^0, \gamma^1) = 0$ if and only if $du(\delta^0, \alpha^1, \gamma^0, \gamma^1) = 0$.*

We have shown that if α^0, δ^0 are disjoint, then the conclusion holds. If they are not disjoint choose σ^0 disjoint from each and compare du, du' with $du'' = du(\sigma^0, \alpha^1, \gamma^0, \gamma^1)$.

1E. The function u_{ni} on Ω_{ni} can be constructed by a normal operator (see (B1) through (B3) of 1B). It therefore satisfies the following maximum principle which will be useful later:

Lemma. *Let c be a union of analytic Jordan curves which divides Ω_{ni} into two regions A, B in such a way that ∂B consists only of c and contours in γ_n^0 or γ_n^1. Then for all $z\in B$*

$$0 < \min u_{ni}|c \le u_{ni}(z) \le \max u_{ni}|c < 1.$$

1F. Boundary behavior of *du*. Recall that $\bar{R}$ is the Kerékjártó-Stoïlow compactification of R and $(\alpha^0, \alpha^1, \gamma^0, \gamma^1)$ denotes a partition of the ideal boundary of R which satisfies (A1), (A2) of 1A. Using the standard notion of an arc in a topological space we may consider the family $\mathfrak{F}$ of all arcs in $\bar{R}-\gamma^0$ which have one end point on α^0 and the other on α^1. If we delete the points in $\beta = \bar{R}-R$ from an arc in $\mathfrak{F}$ we obtain a relative chain on R (see II.6.2E, 2F). If $c \in \mathfrak{F}$ and ω is a first order differential on R then $\int_c \omega$ shall mean $\int_{c \cap R} \omega$. In this way we may extend the notion of extremal length to families of arcs on $\bar{R}$, and thereby consider $\lambda(\mathfrak{F})$.

The next result (Marden-Rodin [2]) shows that if $\lambda(\mathfrak{F}) < \infty$ then the functions $u_n = u_n(\alpha^0, \alpha^1, \gamma^0, \gamma^1)$ of 1C converge to a limit function u. This function has values 0 on α^0, 1 on α^1, and constants C on each component of γ^1 in the following sense (cf. the definition of almost all in II.6.2B):

Theorem. *Assume* $\lambda(\mathfrak{F}) < \infty$. *Then:*

(a) *There exists a function* u, *independent of the exhaustion, such that* $\lim u_n = u$ *and*

$$\int_c du = 1$$

for almost all $c \in \mathfrak{F}$.

(b) *To each component* $\sigma \in \gamma^1$ *there corresponds a constant* C *with* $0 < C < 1$, *such that for almost all arcs* c *in* $\bar{R}-\gamma^0$ *with initial point* $c(0)$ *in* R *and end point* $c(1)$ *on* σ

$$\int_c du = C - u(c(0)).$$

1G. For the proof suppose $\lambda(\mathfrak{F}) < \infty$. By (A1) of 1A there is an arc δ in R separating α^0 from α^1. Therefore the family G of arcs in $\bar{R}-\gamma^0$ which travel from δ to α^0 must have finite extremal length. Consider the linear densities $ds_n = |\operatorname{grad}(u-u_n)||dz|$ where u_n is defined to be $\equiv 0$ in $R-\tilde{\Omega}_n$. Then $A(ds_n) \to 0$ so that by Lemma II.6.2B and II.6.2F we have

$$\lim_{n_i \to \infty} \left| \int_\tau du - du_{n_i} \right| \le \lim_{n_i \to \infty} \int_\tau |\operatorname{grad}(u-u_{n_i})||dz| = 0$$

for some subsequence $\{u_{n_i}\}$ and for almost every $\tau \in G$. Since

$\int_\tau du_{n_i} = -u_{n_i}(\tau(0))$ and since we could have begun with any subsequence of $\{u_n\}$ it follows that $\{u_n(\tau(0))\}$ converges for τ in a nonempty subset of G. Now $\int |du_n - du| \to 0$ on any arc in R. Therefore $\{u_n\}$ converges pointwise to $u = u(\alpha^0, \alpha^1, \gamma^0, \gamma^1)$, say.

Now consider a c as in part (b) of the theorem. As above we see that $\lim \int_c du - du_{n_i} = 0$ for almost all such c. Let σ_n be the component of γ_n^1 determined by $\sigma \in \gamma^1$. Then u_n takes a constant value on γ_n^1, say C_n. Since $\int_c du_{n_i} = C_{n_i} - u_{n_i}(c(0))$ for almost all c, and since $u_{n_i}(c(0)) \to u(c(0))$ we have

$$\int_c du = \lim_{n_i \to \infty} C_{n_i} - u(c(0)).$$

Thus $\{C_{n_i}\}$ converges and so does $\{C_n\}$. By the maximum principle and Lemma 1E we see that $0 < \lim C_n < 1$. This completes the proof of (b). The proof of the remaining statement in (a) is similar and shall be omitted.

2. Construction of p

2A. Construction of $p(\zeta, \alpha^1, \gamma^0, \gamma^1)$. Suppose now that we have a partition of the ideal boundary of R into α^1, γ^0, γ^1 with the following properties:

(A1) α^1 is nonempty and closed in $\bar{R}$.

(A2) $\alpha^1 \cup \gamma^1$ is closed in $\bar{R}$.

Let ζ be a point of R and z a local parameter in a neighborhood Δ of ζ which makes ζ correspond to $z = 0$. We shall construct a function $p = p(\zeta, \alpha^1, \gamma^0, \gamma^1)$ harmonic on $R - \{\zeta\}$ with singularity $\log |z|$ at ζ, and with the same boundary behavior near α^1, γ^0, γ^1 as $u(\alpha^0, \alpha^1, \gamma^0, \gamma^1)$.

As in 1A we start with an exhaustion $\{\Omega_n\}$ of R, but require $\zeta \in \Omega_n$ for each n. Partition the contours of Ω_n into α_n^1, γ_n^0, γ_n^1 according to the following rules. If a contour c bounds a region on $\bar{R} - \Omega_n$ which contains a point of α^1, put c into α_n^1. If c bounds a similar region which contains only points of γ^0 put c into γ_n^0. Let γ_n^1 denote the remaining contours.

Adjoin to Ω_n the components of $R - \Omega_n$ which are bounded by contours of γ_n^0 and thereby obtain a region $\tilde{\Omega}_n$ with $\partial \tilde{\Omega}_n = \alpha_n^1 +$

γ_n^1. Let $\{\Omega_{n_i}\}_{i=1}^{\infty}$ be an exhaustion of $\tilde{\Omega}_n$ with $\partial\Omega_{n_i} = \alpha_n^1+\gamma_n^1+\gamma_{ni}^0$ where γ_{ni}^0 is homologous to γ_n^0. By the Main Existence Theorem (I.1.2B) we construct the function p_{ni}, harmonic on $\Omega_{ni}-\{\zeta\}$, determined by the conditions

(a) $p_{ni}(z) - \log|z|$ is harmonic at $z = 0$,
(b) $p_{ni}|\alpha_n^1 = 1$,
(c) $*dp_{ni} = 0$ along γ_{ni}^0,
(d) p_{ni} has a constant value on any component σ of γ_n^1 and $\int_\sigma *dp_{ni} = 0$.

This construction by the Main Existence Theorem is similar to that of the auxiliary functions in I.3.1A. The constant 1 of (b) can be obtained by appropriately choosing the arbitrary additive constant.

First we show that $\lim_{i\to\infty}p_{ni} = p_n$ exists. Extend p_{n1} to $\tilde{\Omega}_n$ by defining it to be $\equiv 0$ in $\tilde{\Omega}_n-\Omega_{n1}$. Let us temporarily drop the subscript n. For $j>i$ we have

$$\begin{aligned}(d(p_j-p_1),d(p_i-p_1))_{\Omega_i} &= (d(p_j-p_1),d(p_i-p_1))_{\Omega_1}+(dp_j,dp_i)_{\Omega_i-\Omega_1}\\ &= \int_{\gamma_1^0}(p_j-p_1)*d(p_i-p_1)-\int_{\gamma_1^0}p_j*dp_i\\ &= -\int_{\gamma_1^0}p_1*dp_i = \|d(p_i-p_1)\|^2_{\Omega_i}.\end{aligned}$$

Hence

$$\begin{aligned}\|d(p_j-p_i)\|^2_{\Omega_i} &= \|d(p_j-p_1)-d(p_i-p_1)\|^2_{\Omega_i}\\ &\le \|d(p_j-p_1)\|^2_{\Omega_j}-\|d(p_i-p_1)\|^2_{\Omega_i}.\end{aligned}$$

Thus $\|d(p_i-p_1)\|_{\Omega_i}$ increases with i. It is also bounded. To see this let v be harmonic with the singularity $\log|z|$ at ζ, $\equiv 1$ on $\partial\Omega_1$, and $\equiv 0$ in $R-\Omega_1$. Then

$$\|d(p_i-v)\|^2_{\Omega_i} = \|d(v-p_1)\|^2_{\Omega_1}-\|d(p_i-p_1)\|^2_{\Omega_i}.$$

Hence $\|d(p_i-p_1)\|^2_{\Omega_i}$ is bounded. It follows that $\lim_{i\to\infty}p_{ni} = p_n$ exists in $\tilde{\Omega}_n$ and $\lim_{i\to\infty}\|d(p_n-p_{ni})\|^2_{\Omega_i} = 0$.

2B. We now consider the limit of p_n. We shall see that it is $-\infty$ or else a function p harmonic on $R-\{\zeta\}$.

Let $p_0 = 1+\log|z|$ in the neighborhood Δ of ζ and set $p_0\equiv 0$ on $R-\Delta$. For $n>m$ we have

$$(d(p_n-p_0), d(p_m-p_0))_{\tilde{\Omega}_m} = \lim_{i\to\infty} (d(p_{ni}-p_0), d(p_m-p_0))_{\Omega_{ni}\cap\tilde{\Omega}_m}$$

$$= \lim_{i\to\infty}\left[\int_{\alpha_m^1} *dp_{ni} - \int_{\partial\Delta} p_m *dp_{ni} + \int_{\partial\Delta} (p_m-1) *d(p_{ni}-p_0)\right]$$

$$= -\int_{\partial\Delta} (p_m-1) *dp_0 = ||d(p_m-p_0)||^2_{\tilde{\Omega}_m}.$$

Hence

$$||d(p_n-p_m)||^2_{\tilde{\Omega}_m} \leq ||d(p_n-p_0)||^2_{\tilde{\Omega}_n} - ||d(p_m-p_0)||^2_{\tilde{\Omega}_m}.$$

Since $||d(p_n-p_0)||^2_{\tilde{\Omega}n} = -\int_{\partial\Delta}(p_n-1) *dp_0$ we see that $\lim p_n = p$ where either $p\equiv -\infty$ or p is harmonic on $R-\{\zeta\}$ with singularity $\log|z|$ at ζ. As in 1C, p is independent of the exhaustion $\{\Omega_n\}$. We may write $p = p(\zeta, \alpha^1, \gamma^0, \gamma^1)$ to emphasize the quantities on which p depends.

2C. Capacity. Suppose p, p_n have expansions

$$p = \log|z|+k+o(1),$$

$$p_n = \log|z|+k_n+o(1),$$

at ζ where $o(1)\to 0$ as $|z|\to 0$. We shall show that $k_n>k_{n+1}$ for all n and $\lim k_n = k$.

Set $p_0 = 1+\log(|z|/r)$ in the disk $\Delta_r = \{|z|<r\}$, and $p_0\equiv 0$ on $R-\Delta_r$. We find that

$$||d(p_n-p_0)||^2_{\tilde{\Omega}n} = -\int_{\partial\Delta_r} (p_n-1) *dp_0$$

$$= -\int_0^{2\pi} (\log r+k_n-1+o(1))d\theta$$

$$= -2\pi\log r-2\pi k_n-2\pi+o(1).$$

Since $\|d(p_n - p_0)\|_{\bar{\Omega}_n}$ is increasing, k_n is decreasing. Moreover, from $\lim p_n = p$ we have $\lim k_n = k$. The quantity e^k will be called the *generalized capacity* of α^1. It depends on the local parameter z as well as the partition.

Choose j so near $-\infty$ that the level line α^0 on which $p = j$ bounds a relatively compact subregion A of R. Then

$$u(\alpha^0, \alpha^1, \gamma^0, \gamma^1) = (1-j)^{-1}[p(\zeta, \alpha^1, \gamma^0, \gamma^1) - j]$$

in $R - A$. We shall indicate the proof of this statement.

Let u_{ni} be the approximations to $u(\alpha^0,\alpha^1,\gamma^0,\gamma^1)$ with respect to the exhaustion $\Omega_n' = \Omega_n - A$ of $R - A$. Let p_{ni} be the approximations to $p(\zeta,\alpha^1,\gamma^0,\gamma^1)$ with respect to the exhaustion Ω_n as constructed in 2A. Then

$$\left\| d\left[u_{ni} - \frac{1}{1-j}(p_{ni} - j)\right]\right\|^2_{\Omega_{ni}'} = \frac{1}{1-j}\int_{\alpha^0} (p_{ni} - j) * d\left(u_{ni} - \frac{1}{1-j}p_{ni}\right).$$

Letting $i \to \infty$ and then $n \to \infty$ we obtain

$$\left\| d\left[u(\alpha^0,\alpha^1,\gamma^0,\gamma^1) - \frac{1}{1-j}(p(\zeta,\alpha^1,\gamma^0,\gamma^1) - j)\right]\right\|^2_{R-A} = 0$$

since convergence is uniform on α^0.

Using the results of 1D we arrive at the following statement (Marden-Rodin [2]):

Theorem. *The condition* $p(\zeta,\alpha^1,\gamma^0,\gamma^1) = -\infty$ *is independent of the point* $\zeta \in R$.

The boundary behavior of p may be determined exactly as in 1F (ff.).

3. A maximum problem

3A. Partition the ideal boundary β of R into α^0, δ^0, δ^1 where α^0 and δ^1 are closed and $\delta^0 \cup \delta^1$ is isolated from α^0. Let α range over the points of $\beta' = \delta^0 \cup \delta^1$. When α is chosen set $\gamma^0 = \delta^0 - \alpha$

and $\gamma^1 = \delta^1 - \alpha$. Indicate the dependence of $u(\alpha^0,\alpha,\gamma^0,\gamma^1)$ on α by the notation u_α. Thus $\alpha \to \|du_\alpha\|$ is a real-valued function on the compact subset β' of β. Then we have (Marden-Rodin [2]):

Theorem. *The function $\alpha \to \|du_\alpha\|$ is upper semicontinuous, and there is a component $\alpha \in \beta'$ which maximizes $\|du_\alpha\|$.*

For the proof let $a = \limsup_{\alpha \to \bar{\alpha}} \|du_\alpha\|$ where $\bar{\alpha} \in \beta'$. Take a sequence α_n such that $a = \lim_n \|du_{\alpha_n}\|$ and $\alpha_n \to \bar{\alpha}$.

Consider $u_{\bar{\alpha}}$ and let $\tilde{\Omega}_k$ be one of the regions used to define $u_{\bar{\alpha}}$. Let u_k be the function on $\tilde{\Omega}_k$ which approximates $u_{\bar{\alpha}}$. For large n the function u_k is also an approximation to u_{α_n}. From 1C we have, for sufficiently large n,

$$\|du_k\|_{\tilde{\Omega}_k} \geq \|du_{\alpha_n}\| \qquad \text{and} \qquad \|du_k\|_{\tilde{\Omega}_k} \geq a.$$

This implies that $\|du_{\bar{\alpha}}\| \geq a$. Hence $\|du_\alpha\|$ is an upper semicontinuous function of α. The remainder of the theorem follows from the general fact that an upper semicontinuous function on a compact set achieves a maximum.

3B. Boundary components of maximal capacity. There is also a corresponding theorem for $p(\zeta,\alpha,\gamma^0,\gamma^1)$. Let δ^0, δ^1 be a partition of β and suppose δ^1 is closed. Let α vary over the points of β, and set $\gamma^0 = \delta^0 - \alpha$, $\gamma^1 = \delta^1 - \alpha$. Write p_α for $p(\zeta,\alpha,\gamma^0,\gamma^1)$ where ζ is a fixed point of R. About ζ, p_α has the expansion

$$p_\alpha(z) = \log|z| + k(\alpha) + o(1).$$

Recall that $e^{k(\alpha)}$ is called the generalized capacity of α (with respect to the partition δ^0, δ^1). Then we have the following analogue (Marden-Rodin [2]) of Theorem 3A:

Theorem. *The function $\alpha \to k(\alpha)$ is upper semicontinuous, and there is an ideal boundary component of maximal capacity.*

Let $\bar{\alpha} \in \beta$ and set $a = \limsup_{\alpha \to \bar{\alpha}} k(\alpha)$. Take a sequence $\{\alpha_n\}$ from β with $\alpha_n \to \bar{\alpha}$, $\lim k(\alpha_n) = a$.

Consider an exhaustion $\{\tilde{\Omega}_j\}$ used to define $p_{\bar{\alpha}}$. Let p_j be the function in $\tilde{\Omega}_j$ which approximates $p_{\bar{\alpha}}$ there. For large n the function p_j is also an approximation to p_{α_n}. Hence $k(j) \geq k(\alpha_n)$ for sufficiently large n. Therefore $k(j) \geq a$ and so $k(\bar{\alpha}) \geq a$. This completes the proof.

Recall that $k(\alpha)$ depends on the local parameter z. If w is another parameter with $w = 0$ at ζ then the corresponding constant $k'(\alpha)$ satisfies

$$k'(\alpha) = k(\alpha) + \log \left| \frac{dz}{dw} \right|_{\zeta}.$$

Thus the component α which maximizes $k(\alpha)$ does not depend on z.

§2. EXTREMAL LENGTH

We begin with a rather delicate topological technique which allows us to generalize Theorem II.6.1D to arbitrary open Riemann surfaces. Next we establish a connection between extremal distance and the functions u and p, and exploit the connection to derive properties of the functions. In 4 we leave the generality of an arbitrary partition of β and consider the special cases when γ^0 or γ^1 is empty. This leads to two notions of capacity for an ideal boundary component.

The results in this section are mainly due to Marden-Rodin [2]. Capacities on Riemann surfaces were introduced in Sario [12], [15] and Strebel [2].

1. Continuity lemma

1A. Definition of $\mathfrak{F}$, $\mathfrak{F}^*$. Let $\bar{R}$ be the Kerékjártó-Stoïlow compactification of R. Consider a fixed partition $(\alpha^0,\alpha^1,\gamma^0,\gamma^1)$ of the ideal boundary β of R which satisfies (A1), (A2) of 1.1A. The family $\mathfrak{F} = \mathfrak{F}(\alpha^0,\alpha^1,\gamma^0,\gamma^1)$ was defined as consisting of all arcs on $\bar{R}-\gamma^0$ which have initial point on α^0 and terminal point on α^1.

We also consider the class $\mathfrak{F}^* = \mathfrak{F}^*(\alpha^0,\alpha^1,\gamma^0,\gamma^1)$ of curves c on $\bar{R}$ which (roughly speaking) separate α^0 and α^1. The closure on $\bar{R}$ of each such c should not meet $\alpha^0 \cup \alpha^1 \cup \gamma^1$. We would also like c to be oriented so that the regions of $\bar{R} - c$ which contains α^0 lie to the left of c. The property of such a c which we need in our proofs is the following: If Ω_n is any sufficiently large canonical subregion of R, then $c \cap \partial\Omega_n \subset \gamma_n^0$ and $(c \cap \Omega_n) - \alpha_n^0$ is homologous on Ω_n to a sum of full contours in γ_n^1 plus some arcs on contours in γ_n^0. To avoid a topological investigation of assumptions on c which will guarantee this property, we shall take it as our defining characteristic of $\mathfrak{F}^*$.

1B. Continuity lemma. Let $\{\Omega_n\}_{n=1}^{\infty}$ be a canonical exhaustion of R. Let $\bar{\Omega}_n$ denote the Kerékjártó-Stoïlow compactification of Ω_n. Recall that the partition of β induces a partition α_n^0, α_n^1, γ_n^0, γ_n^1 of $\partial\Omega_n$ (1.1A). Without changing notation, let α_n^0, α_n^1, γ_n^0, γ_n^1 also stand for the corresponding partition of the ideal boundary points of $\Omega_n \subset \bar{\Omega}_n$. It will always be clear from the context whether α_n^0, for example, refers to a contour or an ideal boundary point.

Let F_n be the class of curves in $\bar{\Omega}_n - \gamma_n^0$ which go from α_n^0 to α_n^1 possibly via some points in γ_n^1. (For the remainder of this section we use capital italic letters F, F^* when working on a compact bordered surface, and script letters $\mathfrak{F}$, $\mathfrak{F}^*$ for the noncompact case.) More precisely, $l \in F_n$ if and only if l is a continuous mapping into $\bar{\Omega}_n - \gamma_n^0$, the domain of l consists of a finite union of closed intervals $[a_0,a_1] \cup [a_2,a_3] \cup \cdots \cup [a_{j-1},a_j]$ with $a_0 < a_1 < \cdots < a_j$, $l(a_0) \in \alpha_n^0$, $l(a_j) \in \alpha_n^1$, and for odd $i<j$, $l(i)$ and $l(i+1)$ belong to the same component of γ_n^1. Our immediate aim is to prove the following continuity lemma:

Lemma. *For a canonical exhaustion of R we have*

$$\liminf_{n\to\infty} \lambda(F_n) \geq \lambda(\mathfrak{F}).$$

1C. Restatement. In 1E we shall define a family $\mathfrak{F}'$ of relative 1-chains on R such that

(a) $\text{rest}_R(\mathfrak{F}) \subset \mathfrak{F}'$,
(b) $\liminf \lambda(F_n) \geq \lambda(\mathfrak{F}')$,
(c) $\lambda(\mathfrak{F}') = \lambda(\mathfrak{F})$,

where $\text{rest}_R(\mathfrak{F}) = \{l \cap R \mid l \in \mathfrak{F}\}$.

Once this is done the proof of the lemma will be complete. Recall that $\tilde{\Omega}_n$ was obtained from Ω_n by attaching to it all components of $R - \Omega_n$ whose boundaries belong to γ_n^0. Let $\mathfrak{F}_n$ be the family of curves in $\tilde{\Omega}_n$ which go from α_n^0 to α_n^1 possibly via some γ_n^1's. More precisely, $l \in \mathfrak{F}_n$ if and only if l is a continuous mapping into the closure (with respect to R) of $\tilde{\Omega}_n$, the domain of l consists of a finite union of closed intervals $[a_0,a_1] \cup [a_2,a_3] \cup \cdots \cup [a_{j-1},a_j]$ with $a_0 < a_1 < \cdots < a_j$, $l(a_0) \in \alpha_n^0$, $l(a_j) \in \alpha_n^1$, and for all odd $i<j$,

$l(i)$ and $l(i+1)$ belong to the same part of γ_n^1. Then $F_n \subset \mathfrak{F}_n$ and $\lambda(F_n) \geq \lambda(\mathfrak{F}_n)$. Hence instead of (b) it suffices to prove

(b′) $\liminf \lambda(\mathfrak{F}_n) \geq \lambda(\mathfrak{F}')$.

We wish to replace (b′) by another condition. Take any $x < \lambda(\mathfrak{F}')$ and choose a linear density $\rho|dz|$ such that $L^2(\mathfrak{F}',\rho) > x$ and $A(\rho) = 1$. To prove (b′) it is sufficient to show that $\liminf_n L^2(\mathfrak{F}_n,\rho) \geq x$. If this inequality failed to hold there would exist a subsequence along which $L^2(\mathfrak{F}_n,\rho)$ had a limit $y < x$. Let $L(l,\rho) = \int_l \rho|dz|$. If we can prove that for each $\varepsilon > 0$ there is an $l(\varepsilon) \in \mathfrak{F}'$ satisfying

$$L^2(l(\varepsilon),\rho) \leq y + 7\varepsilon$$

then we would have the desired contradiction since

$$y < x < L^2(\mathfrak{F}',\rho) \leq L^2(l(\varepsilon),\rho) \leq y + 7\varepsilon.$$

Therefore we may replace (b') by:

(b″) Given a density $\rho|dz|$ on R with $L^2(\mathfrak{F}_n,\rho) \to y$ as $n \to \infty$. Then for each $\varepsilon > 0$ there is an $l(\varepsilon) \in \mathfrak{F}$ (to be defined according to (a), (c)) such that

$$L^2(l(\varepsilon),\rho) \leq y + 7\varepsilon. \tag{3}$$

1D. Notation and terminology. Given $l_N \in \mathfrak{F}_N$ and $n \leq N$, we wish to define a type of restriction of l_N to $\tilde{\Omega}_n$, to be denoted by $l_N || \tilde{\Omega}_n$, with the property that $l_N || \tilde{\Omega}_n \in \mathfrak{F}_n$.

There is a greatest t for which $l_N(t) \in \alpha_n^0$; call it t_1. Let t_2 be the smallest t for which $l_N(t) \in \partial\tilde{\Omega}_n$ and also $t > t_1$. Then $l_n(t_2)$ is on some contour of γ_n^1 (or possibly α_n^1); call that contour c_2. Set t_3 equal to the greatest t for which $l_N(t) \in c_2$. We continue in this manner and obtain an even number of *stopping times* $t_1 < \cdots < t_k$, a sequence of *stopping points* $l_N(t_1), \cdots, l_N(t_k)$, and a *contour sequence* $\alpha_n^0 = c_1, \cdots, c_{(k+2)/2} = \alpha_n^1$ of distinct contours on $\partial\tilde{\Omega}_n$ such that $l_N(t_j) \in c_{[(j+2)/2]}$ $(j = 1, \cdots, k)$. Define $l_N || \tilde{\Omega}_n$ to be the restriction of l_N to $[t_1,t_2] \cup [t_3\, t_4] \cup \cdots \cup [t_{k-1},t_k]$.

1E. Definition of $\mathfrak{F}'$. A 1-chain l' on R will belong to $\mathfrak{F}'$ if either $l' = l \cap R$ for some $l \in \mathfrak{F}$, or if l' is a continuous map of an open dense subset of (0,1) into R such that:

(E1) If t_0 is not in the domain dom l' of l' and $0<t_0<1$ then there exist sequences $\{r_n\}$, $\{s_n\}$ in dom l' such that $r_n \nearrow t_0$, $s_n \searrow t_0$, and a point $* \in \gamma^1$ with $l'(r_n) \to *$, $l'(s_n) \to *$. If $t_0 = 0$ (resp. 1) we require only a sequence $\{s_n\}$ (resp. $\{r_n\}$) from dom l' with $s_n \searrow 0$ (resp. $r_n \nearrow 1$) and $l'(s_n) \to \alpha^0$ (resp. $l'(r_n) \to \alpha^1$).

(E2) There is a canonical exhaustion $\{\Omega_n\}$ such that $l' || \tilde{\Omega}_N \in \mathfrak{F}_N$ for each $N \geq 1$.

(E3) If $t \in$ dom l' then there exists an N such that $t \in$ dom $l'|\tilde{\Omega}_n$ for all $n \geq N$.

1F. Proof of (a) and (b''). We shall show that $\mathfrak{F}'$ satisfies (a), (b'') of 1C. Condition (a) is part of the definition of $\mathfrak{F}'$. To prove (b'') suppose an $\varepsilon > 0$ is given. By passage to a subsequence of $\{l_n\}$ we may assume that

$$|L^2(\mathfrak{F}_n, \rho) - y| < \frac{\varepsilon}{2^n} \qquad (n \geq 1). \tag{4}$$

Whenever a subsequence of $\{l_n\}$ is extracted and the notation is unchanged we tacitly agree that $\{\Omega_n\}$ shall refer to the corresponding subsequence of $\{n\}$.

Choose $l_n \in \mathfrak{F}_n$ such that

$$|L^2(l_n, \rho) - y| < \frac{\varepsilon}{2^n} \qquad (n \geq 1). \tag{5}$$

A subsequence of $\{l_n\}$, after some modification, will be used to construct $l(\varepsilon)$.

The first step is to find a subsequence $\{l_{n_i}\}$ of $\{l_n\}$ such that all $l_{n_i} || \tilde{\Omega}_N$ $(n_i \geq N)$ have the same contour sequence on $\partial\tilde{\Omega}_N$. Since there are only a finite number of possible contour sequences on $\partial\tilde{\Omega}_1$ we may select a first subsequence of $\{l_n\}$, all elements of which have the same contour sequence on $\partial\tilde{\Omega}_1$. By induction we obtain for each N a subsequence of the preceding one, all elements of which follow a common contour sequence on $\partial\tilde{\Omega}_N$. The diagonal process yields a subsequence with the desired property. We shall not change notation, but continue to designate this subsequence by $\{l_n\}$. Observe that (5) continues to hold.

The next step will be to modify each l_n so that not only will all $l_n || \tilde{\Omega}_N$ $(n \geq N)$ follow the same contour sequence, but also $l_n || \tilde{\Omega}_{n-1}$ and $l_{n-1} || \tilde{\Omega}_{n-1}$ will have the same sequence of stopping points on $\partial \tilde{\Omega}_{n-1}$. To do this we use the diagonal process to find a preliminary subsequence, again denoted by $\{l_n\}$, with the following property. Suppose l_N has k stopping points on $\partial \tilde{\Omega}_N$. Then for each $i \leq k$ the ith stopping point ζ_n of $l_n || \tilde{\Omega}_N$ $(n \geq N)$ gives rise to a convergent sequence of points $\{\zeta_n\}$ on a contour of $\partial \tilde{\Omega}_N$. Around the limit point of this sequence we put a topological disk the circumference of which has very small ρ-length. The actual length will be determined below. Note, however, that it can be required to be arbitrarily small. Indeed, the extremal length of all Jordan arcs in a punctured disk which surround a fixed point is zero (see II.6.1H), and hence for any $\rho|dz|$ there is such an arc of arbitrarily small ρ-length.

For each N we have as many disks on $\partial \tilde{\Omega}_N$ as there are stopping points for $l_n || \tilde{\Omega}_N$ (any $n \geq N$). Choose the circumferences of these disks so small that their total ρ-length is $< 2^{-N}\varepsilon$. By the diagonal process we can achieve a situation in which each stopping point of $l_n || \tilde{\Omega}_N$ on $\partial \tilde{\Omega}_N$ is inside its appropriate disk for all n, N with $n \geq N$. For each disk pick a point in the intersection of its circumference and the corresponding contour; call such a point a distinguished stopping point. By a modification of l_n we mean the result of replacing part of its path inside a disk by a path on the circumference of the disk. Now modify l_1 so that all its stopping points are distinguished; in general, modify l_n so that the stopping points of $l_n || \tilde{\Omega}_{n-1}$ on $\partial \tilde{\Omega}_{n-1}$ and $l_n || \tilde{\Omega}_n$ on $\partial \tilde{\Omega}_n$ are distinguished. Denote the modified sequence again by $\{l_n\}$. As a result of modification the ρ-length of l_n has been increased by no more than $2^{-n+1}\varepsilon + 2^{-n}\varepsilon$. For the present sequence $\{l_n\}$ equation (5) must be changed to

$$(5') \qquad |L(l_n, \rho) - y| < \frac{4\varepsilon}{2^n}.$$

These modifications can be accomplished so that each new l_n remains in $\mathfrak{F}_n$.

By induction, for each n reparametrize l_n so that dom $l_n || \tilde{\Omega}_{n-1} =$ dom l_{n-1}. Then dom l_{n-1} consists of a finite number of closed intervals $[t_1,t_2] \cup [t_3,t_4] \cup \cdots \cup [t_{k-1},t_k]$ and

$$l_{n-1}(t_i) = l_n(t_i) \quad (1 \le i \le k).$$

1G. We are now ready to construct $l(\varepsilon)$. On dom l_1 set $l(\varepsilon) = l_1$. In general, if $l(\varepsilon)$ has been defined on dom l_{n-1} set $l(\varepsilon) = l_n$ on dom $l_n -$ dom l_{n-1}. Then $l(\varepsilon)$ is a continuous 1-chain on R. Its domain is an open subset of $(0,1)$ which, by reparametrization, may be assumed to be dense.

To estimate the ρ-length of $l(\varepsilon)$ note that the ρ-length of l_n restricted to dom $l_n -$ dom l_{n-1} is $< (6\varepsilon/2^n)^{1/2}$. Indeed, $L^2(l_n,\rho) < y + 4\varepsilon/2^n$ by (5′) and since $l_n|$ dom $l_{n-1} = l_n||\tilde{\Omega}_{n-1} \in \mathfrak{F}_{n-1}$,

$$L^2(l_n||\tilde{\Omega}_{n-1},\rho) \ge L^2(\mathfrak{F}_{n-1},\rho) > y - \frac{\varepsilon}{2^{n-1}}$$

by (4). Hence

$$L^2(l(\varepsilon),\rho) < L^2(l_1,\rho) + \sum \frac{6\varepsilon}{2^n} < y + 7\varepsilon.$$

It remains to show that $l(\varepsilon) \in \mathfrak{F}'$. We can satisfy (E1) as follows. Suppose $t_0 \notin$ dom $l(\varepsilon)$ and $t_0 \ne 0$, 1. Consider the stopping times $t_1,\cdots,t_k$ of $l(\varepsilon)$ on $\partial\tilde{\Omega}_n$. For n sufficiently large t_0 is between two stopping times which correspond to stopping points on a common contour c_n of $\partial\tilde{\Omega}_n$. We cannot assert that these contours tend to a single point of $\beta = \bar{R} - R$. However, there is a subsequence which does have a limit point, say $* \in \beta$. The corresponding stopping times yield $\{r_n\}$, $\{s_n\}$. The cases $t_0 = 0$, 1 can be handled similarly. The checking of (E2), (E3) will be omitted.

1H. Proof of (c). If a 1-chain $l' \in \mathfrak{F}'$ can be extended continuously to $[0,1]$ with values in $\bar{R}$ then the extension will automatically be an arc in $\mathfrak{F}$. For each Ω_n we consider annular regions A_{ni} around each contour of $\partial\tilde{\Omega}_n$. We shall show that if no such annulus is crossed infinitely often by l' then l' can be extended continuously to [0,1]. This will prove (c) because the extremal length of a family of 1-chains, each of which crosses some A_{n_i} infinitely often, is ∞.

Given $t_0 \in \operatorname{dom} l'$, $t_0 \neq 0, 1$. Let $\{r_n\}$, $\{s_n\}$, $*$ be as in (E1). It suffices to prove that

$$\lim_{\substack{t \to t_0 \\ t \in \operatorname{dom} l'}} l'(t) = *. \tag{6}$$

Let G be a neighborhood of $*$ on $\bar{R}$. We wish to find a neighborhood of t_0 whose l'-image is in G, and for this we may assume that G is a component of $\bar{R}-\tilde{\Omega}_N$ for some N. Let A be the annular region around ∂G chosen above; for definiteness assume A and G have intersection ∂G. Now $l'(r_n)$, $l'(s_n) \in G$ for sufficiently large n. If (6) failed we could find, for some G, a sequence $u_n \in \operatorname{dom} l'$ with $u_n \to t_0$ and $l'(u_n) \notin G \cup A$. A subsequence of $\{u_n\}$ is monotonic, hence suppose $u_n \nearrow t_0$. Choose r_{i_1} and $u_{i_1} > r_{i_1}$ and $\tilde{\Omega}_{m_1} \supset \Omega_N$ so that r_{i_1}, $u_{i_1} \in \operatorname{dom} l' || \tilde{\Omega}_{m_1}$. Since $l' || \tilde{\Omega}_{m_1} \in \mathfrak{F}_{m_1}$ there is a crossing of A within (r_{i_1}, u_{i_1}). Next choose r_{i_2}, u_{i_2}, $\tilde{\Omega}_{m_2}$ such that $r_{i_1} < u_{i_1} < r_{i_2} < u_{i_2}$, and r_{i_2}, $r_{i_1} \in \operatorname{dom} l' || \tilde{\Omega}_{m_2}$. There must be a crossing of A within (r_{i_2}, u_{i_2}). In this way we see that l' crosses A infinitely many times. The cases $t_0 = 0, 1$ can be treated similarly.

2. Extremal and conjugate extremal distance

2A. The definitions of $\mathfrak{F}$, $\mathfrak{F}^*$ were given in 1A and $du = du(\alpha^0, \alpha^1, \gamma^0, \gamma^1)$ was constructed in 1.1C. We can now complete the generalization (Marden-Rodin [2]; cf. Strebel [2], Reich [1]) of Theorem II.6.1D.

Theorem. *On an arbitrary Riemann surface*

$$\lambda(\mathfrak{F}) = \frac{1}{\lambda(\mathfrak{F}^*)} = ||du||^{-2}.$$

We shall refer to $\lambda(\mathfrak{F})$ and $\lambda(\mathfrak{F}^*)$ respectively as the extremal distance and conjugate extremal distance between α^0 and α^1 with respect to the partition α^0, α^1, γ^0, γ^1.

2B. We first prove that

$$\lambda(\mathfrak{F}) = ||du||^{-2}. \tag{7}$$

With the notation introduced earlier we have, by Theorem II.6.1D,

$$\lambda(F_i) = ||du_{ni}||_{\Omega_{ni}}^{-2}.$$

We estimate $\lambda(\mathfrak{F}_n)$ using $ds = |du_n + i*du_n| = |\operatorname{grad} u_n||dz|$ and find

$$||du_n||_{\tilde{\Omega}_n}^{-2} = \left(\int_0^1 du_n\right)^2 ||du_n||_{\tilde{\Omega}_n}^{-2} \leq L^2(\mathfrak{F}_n, ds) A^{-1}(\tilde{\Omega}_n, ds).$$

Hence for all i

$$||du_n||_{\tilde{\Omega}_n}^{-2} \leq \lambda(\mathfrak{F}_n) \leq \lambda(F_i) = ||du_{ni}||_{\Omega_{ni}}^{-2}.$$

Consequently $\lambda(\mathfrak{F}_n) = ||du_n||_{\tilde{\Omega}_n}^{-2}$.

Since every curve in $\mathfrak{F}$ contains a curve in $\mathfrak{F}_n$, $\lambda(\mathfrak{F}) \geq \lambda(\mathfrak{F}_n)$. On the other hand we infer from the continuity lemma 1B that $||du||^{-2} = \lim \lambda(\mathfrak{F}_n) \geq \lambda(\mathfrak{F})$. The proof of (7) is now complete.

2C. Proof of $\lambda(\mathfrak{F}^*) = ||du||^2$. Let $\mathfrak{F}_n^*$ denote the class $\mathfrak{F}^*$ for the surface $\tilde{\Omega}_n$, and F_i^* the class $\mathfrak{F}^*$ for the surface Ω_{ni}. Since every curve in $\mathfrak{F}_n^*$ contains one in F_i^* for all i, we find by Theorem II.6.1D that

$$||du_{ni}||_{\Omega_{ni}}^2 = \lambda(F_i^*) \leq \lambda(\mathfrak{F}_n^*)$$

and therefore $||du_n||_{\tilde{\Omega}_n}^2 \leq \lambda(\mathfrak{F}_n^*)$.

All but a finite number of level curves $u_n = k$ in $\tilde{\Omega}_n$ contain a curve in $\mathfrak{F}_n^*$. By the Schwarz inequality we have

$$L(\mathfrak{F}_n^*, \rho) \leq \int_{u_n=k} \rho|dz| \leq \left[\int_{u_n=k} \rho^2 * du_n\right]^{1/2} ||du_n||_{\tilde{\Omega}_n}$$

as in II.6.1G. Integration for $k \in (0,1)$ gives $\lambda(\mathfrak{F}_n^*) \leq ||du_n||_{\tilde{\Omega}_n}^2$, and we obtain $\lambda(\mathfrak{F}_n^*) = ||du_n||_{\tilde{\Omega}}^2$.

Since every curve in $\mathfrak{F}_n^*$ is a curve in $\mathfrak{F}^*$, $\lambda(\mathfrak{F}^*) \leq \lambda(\mathfrak{F}_n^*)$ for all n. Therefore $\lambda(\mathfrak{F}^*) \leq ||du||^2$. It remains to prove that $||du||^2 \leq \lambda(\mathfrak{F}^*)$.

Recall the remark at the end of 1.1C: There is an exhaustion Ω_n of R such that $||du - dh_n||_{\Omega_n} \to 0$ as $n \to \infty$. Here h_n is the harmonic function which is 0 on α_n^0, 1 on α_n^1, has L_0-behavior on γ_n^0, and L_1-behavior on each part of γ_n^1. Note that if $c \in \mathfrak{F}^*$ then

$$\int_c *dh_n \geq \int_{\alpha_n^0} *dh_n = ||dh_n||^2 \tag{8}$$

because of the defining property of $\mathfrak{F}^*$.

Set $h_n \equiv 0$ in $R - \Omega_n$ and consider the densities $ds_n = |\operatorname{grad}(u - h_n)||dz|$. Then $A(ds_n) \to 0$ and it follows that there is a subsequence such that

$$\int_c |*du - *dh_{n_i}| \leq \int_c ds_{n_i} \to 0$$

for almost all $c \in \mathfrak{F}^*$. Therefore we have

$$\int_c *du = \lim ||dh_{n_i}||^2 = ||du||^2 \tag{9}$$

for almost all c.

Now consider the linear density $ds = |du + i*du|$. By (9) $\int_c ds \geq \int_c *du = ||du||^2$, and thus $||du||^2 = A(ds)$. Therefore $L^2(\mathfrak{F}^*, ds) \geq ||du||^2 A(ds)$ and it follows that $\lambda(\mathfrak{F}^*) \geq ||du||^2$. This completes the proof of Theorem 2A.

3. Properties of u and p

3A. Uniqueness of du. By means of Theorem 2A, extremal length considerations yield information about u and p. We begin with a uniqueness theorem (Marden-Rodin [2]) for $u = u(\alpha^0, \alpha^1, \gamma^0, \gamma^1)$.

Theorem. *If h is harmonic on R and $\int_c dh \geq 1$ for almost all $c \in \mathfrak{F}$ then $||dh|| \geq ||du||$. In the case $\lambda(\mathfrak{F}) < \infty$, $||dh|| = ||du||$ only if $dh = du$.*

If h is harmonic on R and $\int_c *dh \geq ||dh||^2$ for almost all $c \in \mathfrak{F}^*$ then $||dh|| \leq ||du||$. Equality of these norms implies $dh = du$.

Consider the first statement of the theorem. Let $ds_1 = |dh + i*dh|$. By hypothesis there is a subclass $\mathfrak{F}_1 \subset \mathfrak{F}$ such that

$L(\mathfrak{F}_1, ds_1) \geq 1$ and $\lambda(\mathfrak{F} - \mathfrak{F}_1) = \infty$. If $||dh|| < ||du||$ then we would have

$$\lambda(\mathfrak{F}) = \lambda(\mathfrak{F}_1) \geq \frac{L^2(\mathfrak{F}_1, ds_1)}{A(ds_1)} > ||du||^{-2} = \lambda(\mathfrak{F}),$$

a contradiction. Therefore $||dh|| \geq ||du||$. If equality holds, ds_1 is an *extremal metric* for $\mathfrak{F}_1$. By this we mean that $\lambda(\mathfrak{F}_1) = L^2(\mathfrak{F}_1, ds_1)/A(ds_1)$. We know from Theorems 1F and 2A that $ds_2 = |du + i*du|$ is an extremal metric for a subclass $\mathfrak{F}_2 \subset \mathfrak{F}$ with $\lambda(\mathfrak{F} - \mathfrak{F}_2) = \infty$.

Let $\mathfrak{F}' = \mathfrak{F}_1 \cap \mathfrak{F}_2$. Then $\lambda(\mathfrak{F}') = \lambda(\mathfrak{F})$ and ds_1, ds_2 are both extremal metrics for $\mathfrak{F}'$. We assume $\lambda(\mathfrak{F}') \neq \infty$. As we shall see, this requires in general that $ds_1 = k\, ds_2$ a.e. for some constant k.

Let $ds = \frac{1}{2}(ds_1 + ds_2)$. Since $L(\mathfrak{F}', ds_1) = L(\mathfrak{F}', ds_2) = 1$ we have $L(\mathfrak{F}', ds) = 1$ and $A(ds_1) = A(ds_2) = 1/\lambda(\mathfrak{F}')$. Note that

$$2A(ds) = A(ds_1) + A(ds_2) - \tfrac{1}{2}A(|ds_1 - ds_2|).$$

Consequently

$$\lambda(\mathfrak{F}') = \frac{2}{A(ds_1) + A(ds_2)} \leq \frac{1}{A(ds)}$$

and equality holds if and only if $ds_1 = ds_2$ a.e. But equality does hold since

$$\lambda(\mathfrak{F}') \geq \frac{L^2(\mathfrak{F}', ds)}{A(ds)} = \frac{1}{A(ds)}.$$

Therefore $|\operatorname{grad} u| = |\operatorname{grad} h|$, and this implies

$$dh = \cos\theta\, du - \sin\theta * du$$

for some constant θ. Let δ be a compact cycle which is homologous to α^0. Then $\int_\delta dh = \int_\delta du = 0$, $\int_\delta * du = ||du||^2 \neq 0$, and consequently $\theta = 0$.

The proof of the second statement is similar and will be omitted.

Remark. The above proof shows that the hypotheses of Theorem 3A may be weakened somewhat. Indeed, our reasoning remains valid if the integral inequalities of the theorem are replaced by

$$\int_c |dh+i*dh| \geq 1 \qquad \text{(almost all } c \in \mathfrak{F}),$$

$$\int_c |dh+i*dh| \geq ||dh||^2 \qquad \text{(almost all } c \in \mathfrak{F}^*).$$

This generalization will be useful later (3.1D).

3B. Monotone properties. Since the extremal length of a family increases as the family gets smaller, we can obtain (Marden-Rodin [2]) monotone properties of du using Theorem 2A.

Theorem. *The Dirichlet integral* $||du(\alpha^0,\alpha^1,\gamma^0,\gamma^1)||$ *does not decrease when components of* γ^0 *or* γ^1 *are placed in* α^0, α^1, *or* γ^1. *It does not increase when components of* α^0, α^1, *or* γ^1 *are placed in* γ^0 *or* γ^1.

3C. Properties of p. Let z be a local parameter near $\zeta \in R$ such that ζ corresponds to $z = 0$. Then

$$p(\zeta,\alpha^1,\gamma^0,\gamma^1) = \log|z| + k_z(\zeta,\alpha^1,\gamma^0,\gamma^1) + o(1).$$

For small $r>0$ set $\alpha^0 = \{|z| = r\}$. Let $\mathfrak{F}_r = \mathfrak{F}(\alpha^0,\alpha^1,\gamma^0,\gamma^1)$ and $\mathfrak{F}_r^* = \mathfrak{F}^*(\alpha^0,\alpha^1,\gamma^0,\gamma^1)$; the notation was explained in 1.1D. Let $\mathfrak{G}$ be the family of curves in the annulus $S = \{r' < |z| < r\}$ which go from one contour of S to the other. Then every curve in $\mathfrak{F}_{r'}$ contains a curve in $\mathfrak{G}$ and a curve in $\mathfrak{F}_r$. We shall show that this implies $\lambda(\mathfrak{F}_{r'}) \geq \lambda(\mathfrak{F}_r) + \lambda(\mathfrak{G})$.

Let ds_1, ds_2 be linear densities in S and $R_r = R - \{|z| < r\}$ respectively. Normalizing by a multiplicative constant we may assume $L(\mathfrak{G},ds_1) = A(S,ds_1)$ and $L(\mathfrak{F}_r,ds_2) = A(R_r,ds_2)$. Let ds be the density equal to ds_1 in S and to ds_2 in R_r. Then $L(\mathfrak{F}_{r'},ds) \geq L(\mathfrak{G},ds_1) + L(\mathfrak{F}_r,ds_2)$ and we find that

$$\lambda(\mathfrak{F}_{r'}) \geq \frac{L^2(\mathfrak{F}_{r'},ds)}{A(ds)} \geq L(\mathfrak{G},ds_1) + L(\mathfrak{F}_r,ds_2) \tag{10}$$

by our normalization. The right-hand member of (10) can be made arbitrarily close to $\lambda(\mathfrak{G}) + \lambda(\mathfrak{F}_r)$. By Theorem II.6.1H $\lambda(\mathfrak{G}) = (2\pi)^{-1}\log(r/r')$. Therefore

$$\lambda(\mathfrak{F}_{r'}) \geq \lambda(\mathfrak{F}_r) + \frac{1}{2\pi}\log\frac{r}{r'},$$

and it follows that $2\pi\lambda(\mathfrak{F}_r) + \log r$ increases as $r \to 0$. We claim that its limit is:

Theorem. $1 - k_z(\zeta,\alpha^1,\gamma^0,\gamma^1) = \lim_{r\to 0}(2\pi\lambda(\mathfrak{F}_r) + \log r)$.

3D. To prove Theorem 3C we shall replace z by *a* parameter obtained from $p = p(\zeta,\alpha^1,\gamma^0,\gamma^1)$. Recall the effect of a change of parameter at ζ: If w is a parameter which makes ζ correspond to $w = 0$ and $a = (dw/dz)_\zeta$ then

$$\log|w| = \log|z| + \log|a| + o(1)$$

and (see 1.3B)

$$k_w = k_z - \log|a|.$$

Suppose the disk $\{|w| < s\}$ lies inside $\{|z| < r\}$. Then $2\pi\lambda(\mathfrak{F}_{w=s}) \geq 2\pi\lambda(\mathfrak{F}_{z=r})$ where the notation is self-explanatory. Also $\log r = \log s - \log|a| + o(1)$, and we have

$$2\pi\lambda(\mathfrak{F}_{z=r}) + \log r \leq 2\pi\lambda(\mathfrak{F}_{w=s}) + \log s - \log|a| + o(1).$$

If the disk $\{|z| < r\}$ lies inside $\{|w| < s\}$ then the inequality is reversed. Hence

$$\log\left|\frac{dw}{dz}\right|_{z=0} + \lim_{r\to 0}[2\pi\lambda(\mathfrak{F}_{z=r}) + \log r] = \lim_{s\to 0}[2\pi\lambda(\mathfrak{F}_{w=s}) + \log s].$$

Now take w to be the single-valued function e^{p+ip^*} and note that $|dw/dz|_{z=0} = e^{k_z}$. If we choose $\alpha^0 = \{|w| = s\}$ then

$$u(\alpha^0,\alpha^1,\gamma^0,\gamma^1) = \frac{p - \log s}{1 - \log s}.$$

Consequently $\lambda(\mathfrak{F}_{w=s}) = \|du\|^{-2} = (1-\log s)^2\|dp\|^{-2}$ where the norm is taken over $R-\{|w|\leq s\}$. This norm for dp can be calculated by Green's formula and is $[2\pi(1-\log s)]^{1/2}$. Hence $2\pi\lambda(\mathfrak{F}_{w=s}) = 1-\log s$ and the theorem follows.

3E. Uniqueness of *dp*. Let h be harmonic on R except for a logarithmic singularity at ζ. If z is a local parameter near ζ which makes ζ correspond to $z=0$ then define the constant $k(h)$ by

$$h(z) = \log|z| + k(h) + o(1).$$

Let $\mathfrak{F}_r = \mathfrak{F}_r(\alpha^0,\alpha^1,\gamma^0,\gamma^1)$ where $\alpha^0 = \{|z| = r\}$. Let $B(h) = \lim_{\Omega\to R}\int_{\partial\Omega} h * dh$. Note that $B(p) = 2\pi$ for $p = p(\zeta,\alpha^1,\gamma^0,\gamma^1)$. We then have (Marden-Rodin [2]):

Theorem. *Let h be harmonic on R except for a logarithmic singularity at ζ, and suppose that*

$$\int_c dh \geq \int_c dp$$

for almost all $c\in\mathfrak{F}_r$. Then

$$B(h) - 2\pi k(h) \geq B(p) - 2\pi k(p)$$

and equality holds only if $h = p$.

The theorem is independent of the local parameter z; thus we may choose $z = e^{p+ip^*}$. We conclude from Theorem 1.1F that for almost all $c\in\mathfrak{F}_r$

$$\int_c dp = 1 - \log r.$$

Hence if we set $h_1 = h/(1-\log r)$ and let $\Delta\subset R$ correspond to $\{|z|<r\}$ then we may apply Theorem 3A to obtain $\|dh_1\| \geq \|du(\partial\Delta,\alpha^1,\gamma^0,\gamma^1)\|$. Thus

$$\|dh_1\|^{-2} \leq \lambda(\mathfrak{F}_r) \tag{11}$$

and equality holds only if $h_1 = u$. By Green's formula

$$||dh_1||^2 = \frac{B(h) - 2\pi k(h) - 2\pi \log r - \delta(r)}{(1 - \log r)^2}$$

where $\delta(r) \geq 0$ and $\lim_{r\to 0}\delta(r) = 0$. Thus (11) is equivalent to

$$\text{(12)} \qquad \frac{2\pi(1 - \log r)^2}{B(h) - 2\pi k(h) - 2\pi \log r - \delta(r)} + \log r \leq 2\pi\lambda(\mathfrak{F}_r) + \log r.$$

The right side tends to $1 + k(p)$ by Theorem 3C, since $k(p) = 0$ with respect to this parameter z. The left side is no less than

$$\text{(13)} \qquad \frac{2\pi + (B(h) - 2\pi k(h) - 4\pi) \log r}{B(h) - 2\pi k(h) - 2\pi \log r}$$

which is obtained by neglecting $\delta(r)$. The quantity (13) is non-increasing as $r \to 0$ and has as limit $k(h) + 2 - (2\pi)^{-1}B(h)$. Thus (12) leads to

$$\text{(14)} \qquad 2\pi k(h) - B(h) \leq 2\pi k(p) - B(p).$$

Furthermore, equality in (14) implies equality in (12) because the right side of (12) increases as $r \to 0$, as was remarked in 3C, and (13) decreases to its limit.

If equality holds in (14) then it holds in (11). Hence in such a case we would have $h_1 = u(\partial\Delta, \alpha^1, \gamma^0, \gamma^1)$ for all sufficiently small r. This implies that $h = p$, as may be seen by considering an exhaustion of R. The proof of Theorem 3E is therefore complete.

3F. Corresponding to the previous theorem is a uniqueness property (Marden-Rodin [2]) of p with respect to the family $\mathfrak{F}^*$.

Theorem. *In the notation of Theorem 3E, assume* $\int_c *dh \geq 2\pi$ *for almost all* $c \in \mathfrak{F}^*$. *Then*

$$\text{(15)} \qquad 2\pi k(h) - B(h) \leq 2\pi k(p) - B(p).$$

Equality holds in (15) *only if* $h = p$.

For the proof take $z = e^{p+ip^*}$ as a local parameter at ζ and let $\Delta \subset R$ correspond to $\{|z| < r\}$. Set $h_1 = 2\pi h \|dh\|_{R-\Delta}^{-2}$ and note that h_1 satisfies the hypotheses of Theorem 3A on $R - \Delta$. Hence

$$(16) \qquad 4\pi^2 \|dh\|_{R-\Delta}^{-2} \leq \left\| d\left(\frac{p - \log r}{1 - \log r}\right)\right\|_{R-\Delta}^2$$

and equality here implies that $h = p$. Computing the norms in (16) yields the equivalent inequality

$$(17) \qquad \frac{4\pi^2}{B(h) - 2\pi k(h) - 2\pi \log r - \delta(r)} \leq \frac{2\pi}{1 - \log r}$$

where $\delta(r) \geq 0$. On observing that $B(p) = 2\pi$ and that $k(p) = 0$ for this local parameter, we see that (17) contains the assertion of the theorem.

3G. Theorem 3C gives a geometric interpretation of the quantity $k_z(\zeta, \alpha^1, \gamma^0, \gamma^1)$. This allows us to discover immediately some of its monotonicity properties (Marden-Rodin [2]):

Theorem. *The quantity $k_z(\zeta, \alpha^1, \gamma^0, \gamma^1)$ does not decrease when components of γ^0 or γ^1 are placed in α^1 or γ^1. It does not increase when components of α^1 or γ^1 are placed in γ^0 or γ^1.*

Extremal length interpretations also yield the next result.

Theorem. *The condition $k(\zeta, \alpha^1, \gamma^0, \gamma^1) = -\infty$ is independent of the point ζ.*

4. Capacities

4A. Notation. For the remainder of this section we consider the special cases when one of γ^0 or γ^1 is the empty subset $\varnothing$ of the ideal boundary β of R. In such a situation we define

$$u_\alpha = u(\alpha^0, \alpha, \varnothing, \gamma^1), \qquad v_\alpha = u(\alpha^0, \alpha, \gamma^0, \varnothing)$$

where the partitions satisfy (A1), (A2) of 1.1A. Often α^0 will denote the boundary of a regularly embedded subregion of R.

In this case the notation must be understood according to the conventions of 1.1D. In particular, we note that the partition of β into (α,γ^1) or (α,γ^0) need only have α closed and nonempty in order to satisfy the required properties (A1), (A2).

Given a point $\zeta \in R$ define the functions

$$p_\alpha = p(\zeta,\alpha,\varnothing,\gamma^1), \qquad q_\alpha = p(\zeta,\alpha,\gamma^0,\varnothing).$$

In order that the partitions (α,γ^1), (α,γ^0) satisfy the requirements (A1), (A2) of 1.2A it is again only necessary that α be a nonempty closed subset of β.

4B. Capacities. We recall, in a slightly different form, the notion of capacity introduced in 1.2C. If $p = p(\zeta,\alpha^1,\gamma^0,\gamma^1)$ and z is a local parameter near ζ which makes ζ correspond to z_0 then p has a development

$$p(z) = \log|z-z_0|+k+o(1) \tag{18}$$

where k is a constant $\geq -\infty$. This constant depends on the parameter z. If we write $k = k_{z_0}$ and if w is another parameter near ζ with w_0 corresponding to ζ then we find that

$$k_{w_0} = k_{z_0} - \log\left|\frac{dw}{dz}\right|_{z=z_0}.$$

Thus $e^{k_z}|dz|$ is an invariant linear density. The quantity e^{k_z} was called the generalized capacity of α^1. If p in (18) is the function p_α or q_α then we shall refer to e^k as the *capacity of* α or the L_0-*capacity of* α respectively.

The next result is a special case of previous theorems and thus requires no proof.

Theorem. *Let α be a nonempty closed subset of the ideal boundary of R and let $\alpha^0 = \partial\Omega$ for a regularly embedded subregion $\Omega \subset R$. Let $\bar{R}$ be the Kerékjártó-Stoïlow compactification of R. The following conditions are equivalent.*

(a) $\|du_\alpha\| = 0$.

(b) *The family of compact cycles in R which separate α^0 and α has extremal length zero.*

(c) *The family of arcs in* $\bar{R}$ *which connect* α^0 *and* α *has infinite extremal length.*

(d) *The capacity of* α *is zero.*

If these conditions are satisfied then α is called *parabolic* or *weak*. Note that if α is parabolic and if α' is a nonempty closed subset of α then α' is also parabolic (see 3B, 3G).

4C. The previous theorem corresponds to the case $\gamma^0 = \varnothing$ in earlier results. For $\gamma^1 = \varnothing$ we have the related (Marden-Rodin [2])

Theorem. *Let* $\alpha \neq \varnothing$ *be a closed subset of* β *and let* $\alpha^0 = \partial\Omega$. *The following conditions are equivalent.*

(a) $\|dv_\alpha\| = 0$.

(b) *The family of curves in* R *which separate* α^0 *and* α *has extremal length zero.*

(c) *The family of arcs in* R *which connect* α^0 *and* α *has infinite extremal length.*

(d) *The* L_0*-capacity of* α *is zero.*

We shall say that α is L_0*-parabolic* if it satisfies the above conditions.

The reader should have no difficulty in showing, by extremal length considerations, that parabolicity or L_0-parabolicity of α depends only on the surface R in a neighborhood of α. This is of course stronger than asserting merely that it is a property of the ideal boundary of R.

It is easily seen, for example by comparing conditions (c) of the preceding two theorems, that parabolicity implies L_0-parabolicity. These two conditions are not equivalent, however, as the example of Figure 2 shows. In this figure $R - \Omega$ is an annulus less circular slits which accumulate to the outer boundary α in such a way that α^0 cannot be connected to α by a rectifiable arc contained in R. Therefore (c) of Theorem 4C is satisfied and consequently α is L_0-parabolic. However, any arc from α^0 to α, even if it passes through the slits, belongs to the family mentioned in (c) of Theorem 4B. We conclude that this family has finite extremal length and a fortiori α is not parabolic.

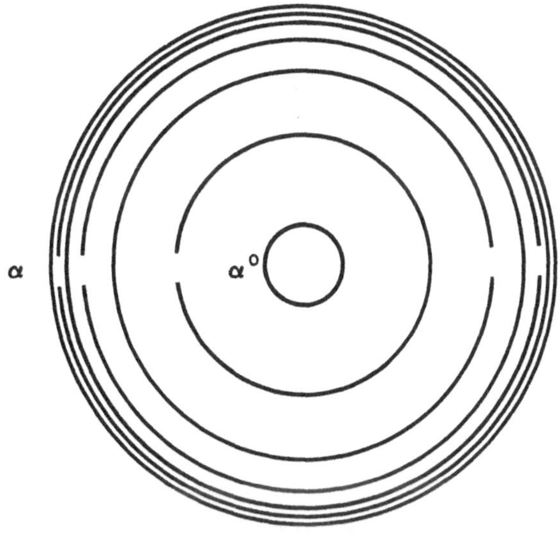

Figure 2

We remark in passing that α exemplifies an "unstable" boundary component. The fascinating phenomenon of such components will be discussed in 4.3.

4D. Extremal properties. Theorems 3E and 3F yield interesting properties of p_α and q_α. For example, consider Theorem 3F applied to p_α. The class $\mathfrak{F}^*$ consists of cycles on $R-\{\zeta\}$ which separate ζ from α. If h has zero flux across each ideal boundary component which is not in α then $\int_c *dh = 2\pi$ for every $c \in \mathfrak{F}^*$. Therefore h satisfies the hypotheses of that theorem. For better comparison of the resulting statement with classical theory we introduce a normalization on h and obtain (Sario [12]):

Theorem. *Let h be harmonic on R except for a logarithmic singularity at ζ, and suppose that the flux of h vanishes across each boundary component not contained in α. If h is normalized by an additive constant so that $k(h) = k(p_\alpha)$ then*

$$\int_\beta h * dh \geq 2\pi.$$

Equality holds if and only if $h = p_\alpha$.

4E. The next result follows from Theorem 3E applied to q_α.

Theorem. *Suppose h is harmonic on $R-\{\zeta\}$ and* $\lim_{z\to\zeta} h(z)-q_\alpha(z)=0$. *If*

$$\liminf_{z\to\alpha} h(z)>1$$

then $\int_\beta h*dh>2\pi$.

4F. Green's function. If α is the entire ideal boundary β and $p_\beta\not\equiv-\infty$, then $g(z,\zeta)=1-p_\beta(z)$ is *Green's function on R with pole at ζ*. By the construction of p_β we see that $g(z,\zeta)>0$. For a fixed ζ one has $\lim_{z\to\beta} g(z)=0$ in the following sense: for almost all curves c in R which tend to β

$$\lim_{\substack{z\to\beta\\ z\in c}} g(z)=0.$$

The assumption that β is not parabolic implies that such curves c exist. A special family of them will be considered in 4G.

The family $\mathfrak{F}^*(\zeta,\beta,\varnothing,\varnothing)$, relevant to the present special case, consists of all compact cycles on $R-\zeta$ which are homologous to $\partial\Delta$ where Δ is a disk in R containing ζ. If h_1 is harmonic on R except for a logarithmic singularity at ζ then $\int_c *dh_1=2\pi$ for all $c\in\mathfrak{F}^*$. If h_1 is normalized as in Theorem 4E then we conclude that $\int_\beta h_1*dh_1\geq 2\pi$. By Green's formula and the normalization on h_1

$$\int_\beta h_1*dp_\beta-p_\beta*dh_1=0.$$

Hence

$$2\pi=\int_\beta p_\beta*dh_1=\int_\beta h_1*dp_\beta\leq 2\pi\sup h_1$$

since $*dp_{\beta\Omega}>0$ on $\partial\Omega$ for an exhaustion $\{\Omega\}$ of R. We have proved that $\sup h_1\geq 1$ and $B(h_1)\geq 2\pi$. On applying this reasoning to $h_1=1-h$ we obtain the

Theorem. *If h is harmonic on $R-\{\zeta\}$ and satisfies* $\lim_{z\to\zeta} h(z)-g(z,\zeta)=0$ *then* $\inf h\leq 0$ *and* $\int_\beta h*dh\geq 0$.

4G. In 4F it was observed that $g(z,\zeta)\to 0$ as $z\to\beta$ along almost any curve. Consider the family of zero curves of $*dg$ for fixed ζ. Except for possibly a countable number which pass through a critical point of $g(z)$, these curves travel from ζ to β. They may be indexed in a natural way by $\theta \in [0,2\pi)$; namely, use $z = e^{-g-ig^*}$ as a local parameter at ζ and let c_θ be the zero curve $*dg = 0$ which coincides with $\arg z = \theta$ near ζ. Then we have (Brelot-Choquet [1]; cf. Nakai [1], Maeda [1]):

Theorem: *Let $g(z,\zeta)$ be the Green's function on R with pole at ζ. Then*

$$\lim_{\substack{z \to \beta \\ z \in c_\theta}} g(z,\zeta) = 0 \tag{19}$$

except for a set of θ of measure zero.

For the proof we need only remove a small disk Δ about ζ so that $g(z)$ has finite Dirichlet integral over $R-\Delta$. Let $\mathcal{G}$ be the family $\{c_\theta\}$ when each c_θ is restricted to $R-\Delta$. By Theorem II.6.2D, any subfamily of $\mathcal{G}$ with infinite extremal length corresponds to a set of θ's of measure 0. Since we know (19) is true for almost all curves c_θ the proof is complete.

§3. EXPONENTIAL MAPPINGS OF PLANE REGIONS

On a planar Riemann surface R the functions u and p can be used to construct biholomorphic functions. We begin by studying some general properties of these mappings for an arbitrary partition of the ideal boundary β of R. For detailed information on the geometric properties of the image regions we specialize to the cases in which γ^0 or γ^1 is empty. The image regions are then seen to be disks or annuli with circular slits or radial slits and "incisions". Finally, we discuss some methods for examining the image regions under an arbitrary partition of β.

The results obtained in this section are mainly due to Marden-Rodin [2]. Related works are Oikawa [7]; Oikawa-Suita [2]; Reich [1]; Reich-Warschawski [1], [2]; Strebel [2]; and Nickel [2]; among others.

1. Extremal slit annuli and disks

1A. The mappings U, P. Let R be a planar Riemann surface, $\zeta \in R$, and $(\alpha^1,\gamma^0,\gamma^1)$ an admissible partition of β in the sense of 1.2A with α^1 a single boundary component. For the remainder of this section we tacitly assume that R is not degenerate, i.e. that the functions u, p we consider are not $\equiv 0$ or $\equiv -\infty$.

Modulo an additive multiple of 2π, we can define a harmonic conjugate p^* for the function $p = p(\zeta,\alpha^1,\gamma^0,\gamma^1)$. Consequently $P = e^{p+ip^*}$ is a single-valued holomorphic function on R. If R is the interior of a compact bordered surface then the argument principle may be used to show that P maps R univalently onto a disk with circular or radial slits. More precisely, the contour of R corresponding to α^1 is mapped onto a circle about the origin, the contours in γ^1 are mapped onto circular arcs, and the contours in γ^0 onto radial segments.

If R is an arbitrary surface we may conclude that P is univalent on R. To study the image region we make the following definition:

A region R contained in $\{|z|<r\}$ and with ideal boundary partition $(\alpha^1,\gamma^0,\gamma^1)$ is called an *extremal slit disk* (of radius r and with respect to the partition $(\alpha^1,\gamma^0,\gamma^1)$) if

$$p(0,\alpha^1,\gamma^0,\gamma^1) = 1+\log\frac{|z|}{r}.$$

Similarly, if R is a planar surface with $(\alpha^0,\alpha^1,\gamma^0,\gamma^1)$ an admissible partition of β and with α^0 and α^1 single components of β then $u = u(\alpha^0,\alpha^1,\gamma^0,\gamma^1)$ has a harmonic conjugate u^* which is unique up to an additive multiple of $\int_c *du = ||du||^2$, where c is a cycle in $\mathfrak{F}^*$. Hence

$$U = \exp\frac{2\pi(u+iu^*)}{||du||^2}$$

is a single-valued holomorphic function on R. In the compact bordered case U maps R onto an annulus with circular slits corresponding to γ^1, radial slits to γ^0, and inner and outer contours to α^0 and α^1.

For an arbitrary planar R the function U must remain univalent. We shall say that a region R contained in $\{r_0<|z|<r_1\}$ is an *extremal slit annulus* of radii r_0, r_1 with respect to a partition $(\alpha^0,\alpha^1,\gamma^0,\gamma^1)$ if

$$u(\alpha^0,\alpha^1,\gamma^0,\gamma^1) = \frac{\log|z|-\log r_0}{\log r_1-\log r_0}.$$

1B. A reduction to annuli. The following result shows that in investigating the geometry of extremal slit disks and annuli we may limit our attention to the latter only.

Let R be an extremal slit disk contained in $\{|z|<r_1\}$ and corresponding to a partition $(\alpha^1,\gamma^0,\gamma^1)$. Let $\Delta\subset R$ be a closed disk centered at $z = 0$ and set $\alpha^0 = \partial\Delta$. Then $R-\Delta$ is an extremal slit annulus for the partition $(\alpha^0,\alpha^1,\gamma^0,\gamma^1)$. Conversely, if the inner contour of an extremal slit annulus is isolated from the other boundary components then an extremal slit disk is obtained by adding to the region the closed disk bounded by that contour.

The proof follows easily from the observation (cf. the last paragraph of 1.1C) that there is a regular exhaustion $\{\Omega_n\}$ of R such that $p = \lim p_n$ and $u = \lim u_n$ where p_n, u_n are defined on Ω_n and $\Omega_n-\Delta$. Thus u on $R-\Delta$ determines p on R and conversely.

1C. Area of slits. For an arbitrary R, planar or not, we know that $\|du\|_R^2 = \int_c *du$ for almost every $c\in\mathfrak{F}^*$ (see (9)). This formula contains the fact that the "slits" in an extremal slit annulus (or disk) have zero area:

Theorem. *If R is an extremal slit annulus then the 2-dimensional Lebesque measure of ∂R is zero.*

We may assume the annulus A is $\{1<|z|<r\}$. Then $u = (\log|z|)/\log r$, and a direct computation gives $\|du\|_A^2 = 2\pi/\log r$. On the other hand we may pick a compact $c\in\mathfrak{F}^*$ such that $\|du\|_R^2 = \int_c *du = (\log r)^{-1}\int_c d\arg z = 2\pi/\log r$. (The existence of such c follows from the topological model of open surfaces in 0.1.2C.) Then $\|du\|_R = \|du\|_A$ and therefore $A-R$ has measure zero.

1D. Characterization of *R*. The next theorem (Marden-Rodin [2]) refers to a plane region R contained in $\{1<|z|<r\}$ whose ideal boundary is given an admissible partition $(\alpha^0,\alpha^1,\gamma^0,\gamma^1)$.

Theorem. *Let* $ds = |dz|/(|z|\log r)$. *The following conditions are equivalent:*

(a) *R is an extremal slit annulus.*
(b) $\int_c ds\geq 1$ *and* $\int_d ds\geq 2\pi/\log r$ *for almost all* $c\in\mathfrak{F}$, $d\in\mathfrak{F}^*$.
(c) $\lambda(\mathfrak{F})\leq(\log r)/2\pi$ *and* $\int_c ds\geq 1$ *for almost all* $c\in\mathfrak{F}$.
(d) $\lambda(\mathfrak{F})\geq(\log r)/2\pi$ *and* $\int_c ds\geq 2\pi/\log r$ *for almost all* $c\in\mathfrak{F}^*$.

Remark. If $\gamma^0 = \emptyset$ then the condition $\int_c ds \geq 2\pi/\log r$ holds automatically for $c \in \mathfrak{F}^*$.

For the proof let $h = \log|z|/\log r$. Then $ds = |dh + i * dh|$ and $A(ds) = \|dh\|_R^2 \leq \|dh\|_{1<|z|<r}^2 = 2\pi/\log r$.

If (a) holds then $h = u$ and by Theorem 1C, $\|dh\|_R^2 = \|dh\|_{1<|z|<r}^2 = 2\pi/\log r = \|du\|_R^2$. Then (b), (c), (d) follow since they represent known properties of u.

Assume (c) holds. Since $\lambda(\mathfrak{F}) = \|du\|^{-2}$, the first inequality of (c) gives $\|du\|^{-2} \leq \|dh\|_R^{-2}$, and the remark to Theorem 2.3A yields (a). Similarly, (d) implies (a) by the same remark 2.3A.

Since (c) and (d) each imply (a) it follows that (b) implies (a).

The remark to Theorem 1D is clear. Indeed, $\mathfrak{F}^*$ then consists of compact cycles which wind once around $z = 0$. Thus

$$\int_c ds \geq \left| \frac{1}{\log r} \int_c \frac{dz}{z} \right| = \frac{2\pi}{\log r}.$$

2. Special cases

2A. Circular slit annulus. The geometry of ∂R, when R is an extremal slit annulus, is difficult to determine in the case of a general partition of β. We shall return to this problem in 3, after investigating the important special cases $\gamma^0 = \emptyset$ or $\gamma^1 = \emptyset$.

If R is an extremal slit annulus or disk for a partition with $\gamma^0 = \emptyset$ we shall refer to R as an *extremal circular slit annulus* or *disk*. If the partition has $\gamma^1 = \emptyset$ we say that R is an *extremal radial slit annulus* or *disk*. We shall prove (Reich-Warschawski [1], [2]):

Theorem. *If R is an extremal circular slit annulus of radii 1 and r, then ∂R consists of $\{|z| = 1\}$, $\{|z| = r\}$, points, and circular slits with centers at the origin. The set consisting of all real t such that $\{|z| = t\}$ contains a slit of positive length has measure zero.*

To prove the theorem consider a connected subset σ of some boundary component of R. Suppose σ is not a point or a circular arc. Then there exists a family $\{c_t\}_{t \in T}$ such that c_t is an arc on $\{|z| = e^t\}$ which intersects σ, and T is a proper interval. Each c_t may be selected so that, as an arc, the initial point $c_t(0)$ lies in R and the terminal point $c_t(1)$ lies in σ.

By Theorem 1.1F there is a constant C such that $\int_{c_t \cap R} du = C - u(c_t(0))$ for almost all c_t. Theorem II.6.2D shows that this formula applies to c_t for a.e. $t \in T$.

Furthermore $\int_{c_t \cap R} du = \int_{c_t} du$ for a.e. $t \in T$. Indeed, $c_t \cap (A - R)$, where $A = \{1 < |z| < r\}$ has linear measure zero (along c_t) for a.e. $t \in T$ since the areal measure of $A - R$ is zero. Therefore we can compute

$$\int_{c_t \cap R} du = \int_{c_t} \frac{d \log|z|}{\log r} = 0$$

for a.e. $t \in T$. Combining these observations we have $u(c_t(0)) = C$ for a set of t's of positive measure. But $u(c_t(0)) = t/\log r$ is not a constant, and we have a contradiction. Therefore σ is a circular arc.

The last statement of the theorem is an immediate consequence of the fact that the slits have area zero.

2B. Minimal circular slit annulus. As in 1.3A let α^0 be a fixed isolated component of β. For each component α in $\beta - \alpha^0$ form

$$du_\alpha = du(\alpha^0, \alpha, \varnothing, \beta - \alpha^0 - \alpha).$$

The corresponding mapping U_α sends R into an extremal circular slit annulus of radii 1 and $e^{2\pi/||du_\alpha||^2}$. We know from Theorem 1.3A that among all $\alpha \in \beta - \alpha^0$ there is one whose corresponding mapping U_α gives the thinnest extremal slit annulus. Let us refer to the image of this U_α as a *minimal circular slit annulus*. It enjoys the following property (Reich-Warschawski [1], [2]):

Theorem. *If R is a minimal circular slit annulus then each circular slit subtends an angle $< \pi$.*

For the proof let $\bar{\alpha}$ denote the outer boundary component of R. Thus $\bar{\alpha} = \{|z| = r\}$ where $\log r = 2\pi/||du_{\bar{\alpha}}||^2$. We have seen that $ds = |dz|/(|z| \log r)$ is the extremal metric for $\mathfrak{F}^*(\bar{\alpha}) = \mathfrak{F}^*(\alpha^0, \bar{\alpha}, \varnothing, \beta - \alpha^0 - \bar{\alpha})$ (more precisely, for almost all curves in $\mathfrak{F}^*(\bar{\alpha})$). Let α be any component of $\beta - \alpha^0 - \bar{\alpha}$. Then we have

$$\frac{L^2(\mathfrak{F}^*(\alpha), ds)}{A(ds)} < \lambda(\mathfrak{F}^*(\alpha)) \tag{20}$$

where $\mathfrak{F}^*(\alpha) = \mathfrak{F}^*(\alpha^0,\alpha,\varnothing,\beta-\alpha^0-\alpha)$ and the inequality is strict because the extremal metric for $\mathfrak{F}^*(\alpha)$ is essentially different from ds. Since $A(ds) = ||du_{\bar{\alpha}}||^2$ and $\lambda(\mathfrak{F}^*(\alpha)) = ||du_\alpha||^2$ we obtain from (20)

$$\inf \int_c ds < ||du_\alpha|| \cdot ||du_{\bar{\alpha}}|| \leq ||du_{\bar{\alpha}}||^2 = \frac{2\pi}{\log r}$$

for $c \in \mathfrak{F}^*(\alpha)$. Equivalently,

$$\inf \int_c \frac{|dz|}{|z|} < 2\pi \tag{20'}$$

for cycles c in R which separate the slit α from the inner circumference α^0. Geometrically, this means α cannot subtend an angle $\geq \pi$.

A similar theorem holds for "minimal circular slit disks."

2C. Extremal property of P_α. Consider the function $P_\alpha = e^{p_\alpha + ip_\alpha{}^*}$ where $p_\alpha = p(\zeta,\alpha,\varnothing,\beta-\alpha)$. It maps R onto an extremal circular slit disk. We shall examine the implications of Theorem 2.4D for this special case.

Assume F is a biholomorphic mapping of R with $F(\zeta) = 0$. Since F induces a homeomorphism of the compactifications $\bar{R}$ and $\overline{F(R)}$ it is meaningful to require that F makes the boundary component α correspond to the outer boundary of $F(R)$. These assumptions on F ensure that the function $h = \log|F|$ satisfies all the hypotheses in the first sentence of Theorem 2.4D. The conditions on $k(h)$, $k(p_\alpha)$ may then be interpreted in terms of derivatives at ζ. Indeed, if z is a local parameter at ζ with $z = 0$ corresponding to ζ then in this coordinate system $k(h) = \lim_{z\to 0} h(z) - \log|z| = \log|dF/dz|_{z=0}$.

We next obtain an interpretation of $\int_\beta h * dh$ which results in the following (Reich-Warschawski [1], [2]; Sario [15]):

Theorem. *If F is a univalent function on R with $F(\zeta) = 0$, $|F'(\zeta)| = |P_\alpha'(\zeta)|$, and $F(\alpha)$ corresponding to the outer boundary component of $F(R)$ then $\sup|F| \geq e$. Equality holds if and only if $F = P_\alpha$.*

To prove the theorem it is sufficient to show that $\int_\beta h*dh \leq 2\pi \sup \log|F|$. Let Ω be a regular subregion of R. The partition (α,γ^1) of β induces a partition $(\alpha_\Omega,\gamma_\Omega{}^1)$ of $\partial\Omega$ such that α_Ω consists of a single contour. Hence it suffices to prove that $\int_{\partial\Omega} h*dh \leq 2\pi \sup \log|F|$. Now $F(\alpha_\Omega)$ is the oriented boundary of a simply connected region containing $F(\Omega)$. For each contour σ in $\gamma_\Omega{}^1$ the cycle $F(\sigma)$ is the negatively oriented boundary of a region in the complement of $F(\Omega)$.

If Δ is a region in the w-plane with $0 \notin \Delta$ then

$$\int_{\partial\Delta} \log|w| d \arg w = \iint_\Delta \frac{d\xi d\eta}{|w|^2} \tag{21}$$

where $w = \xi + i\eta$. We set $w = F(z)$ and by (21) obtain

$$\int_\sigma h*dh \leq 0$$

for $\sigma \in \gamma_\Omega{}^1$. Since

$$\int_{\alpha_\Omega} h*dh \leq 2\pi \sup \log|F|,$$

$$\int_{\partial\Omega} h*dh = \int_{\alpha_\Omega} h*dh + \sum_\sigma \int_\sigma h*dh \leq 2\pi \sup \log|F|. \tag{22}$$

2D. Extremal radial slit annulus. We now turn our attention to properties of U, P for a partition with $\gamma^1 = \emptyset$, i.e. to the extremal radial slit disk and annulus mappings. Before investigating these mappings, however, we consider the following example.

Figure 3 suggests an annulus $\{1<|z|<r\}$ with a "radial incision" on the outer circumference and a countable number of radial slits γ^0. Let R denote the resulting region and suppose the radial slits accumulate to the incision in such a way that only the end point ζ of the incision is accessible from R. This means that a rectifiable arc from R to the incision must terminate at ζ. Since the extremal length of a family of arcs with a common end point

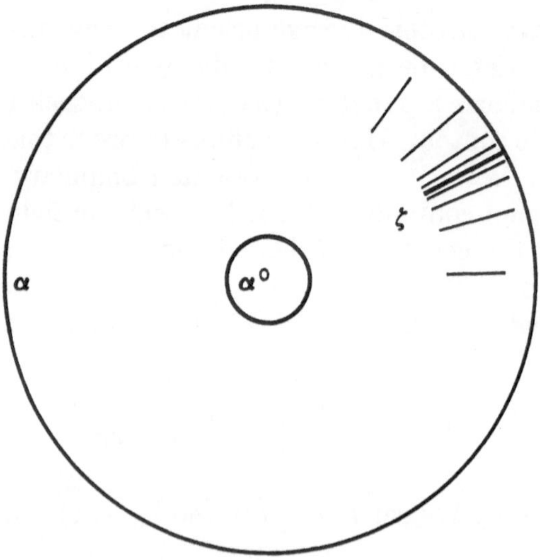

Figure 3

is infinite we see that almost every $c \in \mathfrak{F}$ is an arc from $\{|z| = 1\}$ to $\{|z| = r\}$. Therefore half of condition (c) of Theorem 1D is obviously satisfied. Let us show that $\lambda(\mathfrak{F}) \leq (\log r)/2\pi$ and thereby prove that R is an extremal radial slit annulus.

Let $c_\theta: [0,1] \to R$ be the arc $t \to [1+t(r-1)]e^{i\theta}$. Then $c_\theta \in \mathfrak{F}$ for almost all $\theta \in [0,2\pi)$. Using branches of $\log z$ as coordinate systems in R, consider any linear density $\rho(z)|d \log z|$. By the Schwarz inequality

$$L^2(\mathfrak{F},\rho) \leq \left[\int_{c_\theta} \rho\, d \log|z|\right]^2 \leq \log r \int_{c_\theta} \rho^2\, d \log|z|$$

for a.e. θ. Integration for $\theta \in [0,2\pi)$ gives $2\pi L^2(\mathfrak{F},\rho) \leq (\log r) A(\rho)$. Since ρ was arbitrary this means $\lambda(\mathfrak{F}) \leq (\log r)/2\pi$.

2E. If R is an extremal radial slit annulus of finite connectivity it is easy to see that R must be an annulus with a finite number of radial slits. The example above is typical of what may happen for arbitrary connectivity as the next result shows (Reich [1], Strebel [2]).

Theorem. *Let R be an extremal radial slit annulus of radii 1 and r. Then a component of ∂R is either a point, a radial slit,* *or one of the circumferences* $\{|z| = 1\}$, $\{|z| = r\}$ *with possible radial incisions. The set consisting of all* θ *such that* $\{\arg z = \theta\}$ *contains an incision or a radial slit (not a point) has measure zero.*

We first show that if σ is a component of γ^0 then σ is a radial slit or a point. Using an exhaustion $\{\Omega_n\}$ of R we may find a neighborhood S of σ on which $*du$ is exact. Let u^* be a harmonic conjugate of u on this neighborhood. We know that $u = \lim u_n$ where u_n is the u-function on Ω_n corresponding to the induced partition of $\partial\Omega_n$. Choose harmonic conjugates so that $u_n^* \to u^*$. An argument analogous to that of 1.1G shows that there exists a constant θ with

$$\int_\tau du^* = \theta - u^*(\tau(0))$$

for almost all arcs τ in $\bar{R}$, the compactification of R, which have initial point $\tau(0)$ in $S \subset R$ and end point $\tau(1)$ on σ. Now $u^*(z) = (\arg z)/\log r$. Reasoning as in 2A we may interpret the above conclusion as meaning that almost every arc in Cl R which ends on σ has its terminal point on a fixed radius. Therefore σ is a point or a radial slit.

2F. Now suppose σ is a component of ∂R which corresponds to α^0 or α^1. For definiteness let σ correspond to α^0. We shall show that for each $\varepsilon > 0$, $\sigma_\varepsilon = \sigma \cap \{1+\varepsilon < |z|\}$ is empty or else each component is a radial slit.

The family of all $c \in \mathfrak{F}$ with initial point not on $\{|z| = 1\}$ or terminal point not on $\{|z| = r\}$ has infinite extremal length because $\int du < 1$ along such c. To be precise we should first note that $c(0)$ is a point of $\bar{R}$, the compactification of R, rather than a point of Cl R. However, if $c: t \to c(t)$ does not oscillate then $c(0)$ may be defined by $\lim_{t\to 0} c(t)$ as a point of Cl R. Since the family of oscillating arcs has infinite extremal length, they may be discarded at the outset.

We conclude that if $\mathfrak{G}$ denotes the arcs τ in $\mathfrak{F}$ with $|\tau(0)| = 1$, $|\tau(1)| = r$ then $\lambda(\mathfrak{G}) = \lambda(\mathfrak{F})$. Consider the region $R_0 = R \cup \{1 < |z| < 1+\varepsilon\}$ with the partition $(\alpha_0^0, \alpha_0^1, \gamma_0^0, \varnothing)$ of ∂R_0 where α_0^0 corresponds to $\{|z| = 1\}$, α_0^1 corresponds to the component of ∂R_0 containing $\{|z| = r\}$, and γ_0^0 consists of the remainder of ∂R_0.

Let $\mathfrak{F}_0 = \mathfrak{F}(\alpha_0{}^0,\alpha_0{}^1,\gamma_0{}^0,\emptyset)$. Then $\mathfrak{F}_0 \supset \mathfrak{G}$ and consequently $\lambda(\mathfrak{F}_0) \leq \lambda(\mathfrak{F}) = (\log r)/2\pi$. A comparison of $\mathfrak{F}_0$ with the appropriate family for the full annulus gives $(\log r)/2\pi \leq \lambda(\mathfrak{F}_0)$. Therefore $\lambda(\mathfrak{F}_0) = \lambda(\mathfrak{F})$ and it follows by Theorem 1D that R_0 is an extremal radial slit disk. From this and from what we have already proved in 2E we infer that each component of σ_ϵ must be a radial slit.

The last statement of the theorem follows from Theorem 1C.

2G. Minimal radial slit annulus. The methods of 2B can also be applied to

$$dv_\alpha = du(\alpha^0,\alpha,\beta-\alpha^0-\alpha,\emptyset)$$

for a fixed isolated α^0.

Let $\bar{\alpha}$ denote the component of $\beta-\alpha^0$ which maximizes $\|dv_{\bar{\alpha}}\|$ and let R be the image under the associated radial slit annulus mapping, i.e. α^0, $\bar{\alpha}$ are the inner and outer boundary components of R. We shall refer to R as a *minimal radial slit annulus*. Then, as in 2B (20′) holds, i.e.

$$\inf \int_c \frac{|dz|}{|z|} < 2\pi,$$

where c now varies over $\mathfrak{F}^*(\alpha_0,\alpha,\beta-\alpha_0-\alpha,\emptyset)$ and α is any radial slit. However, it must be understood that curves of this class need lie only in $\bar{R}$. Consequently, the inequality does not give a bound for the logarithmic length of the slit α. In the special case of ∂R having only three components we can deduce the following:

Let R be a minimal radial slit annulus of connectivity three. If the radial slit extends from $\{|z| = a\}$ to $\{|z| = b\}$ then $b/a < e^\pi$.

3. Generalizations

3A. We return to the problem of the geometry of ∂R when R is an extremal slit annulus with respect to any admissible partition $(\alpha^0,\alpha^1,\gamma^0,\gamma^1)$. Often ∂R consists only of radial slits, circular slits with possible radial incisions, and the inner and outer circumferences with possible radial incisions. We shall briefly discuss a method of generalizing earlier results to this case.

3B. Removable boundary components. Let R be an extremal slit annulus and let δ be a subset of $\gamma^0 \cup \gamma^1$. If $R \cup \delta$ is an extremal slit annulus for the partition $(\alpha^0, \alpha^1, \gamma^0 - \delta, \gamma^1 - \delta)$ then δ is said to be *removable.* A sufficient condition for removability is provided by the (Marden-Rodin [2])

Theorem. *Let c be an analytic Jordan curve contained in an extremal slit annulus R. Let δ be the collection of boundary components inside of c and assume $\alpha_0 \notin \delta$. Then δ is removable.*

The proof depends on the fact that the function $u = u(\alpha^0, \alpha^1, \gamma^0, \gamma^1) = k \log|z|$ is actually harmonic on $R \cup \delta$. Let $v = u(\alpha^0, \alpha^1, \gamma^0 - \delta, \gamma^1 - \delta)$ be given on $R \cup \delta$, and let c be the oriented boundary of the simply connected region Δ encircled by it. Using an appropriate exhaustion one finds

$$||d(u-v)||_R^2 = -\int_c (u-v) * d(u-v) + ||d(u-v)||_\Delta^2 = 0.$$

3C. The previous result shows that isolated boundary components are removable. Before considering more general situations we introduce some definitions.

Let σ be a connected part of a component of ∂R where R is an extremal slit annulus. If there exists a $\delta \subset \beta$ such that δ is removable and $\gamma^0 - \delta$ has no cluster point on σ then σ is said to be *γ^0-isolated.* Replacing γ^0 by γ^1 we obtain the notion of *γ^1-isolated.*

It is possible to find conditions for σ to be γ^0- or γ^1-isolated, but we shall not pursue the matter here.

3D. The usefulness of the above notions lies in the fact that theorems such as 2A, 2E have analogues for a general partition if certain isolation conditions are added. For example, it can be shown that if $\sigma \in \gamma^1$ and σ is γ^1-isolated then σ is a point or circular slit with incisions (including the cases of a single radial or circular slit).

3E. A dual problem. Suppose the admissible partition conditions (A1), (A2) of 1.1A are replaced by (A1), (A2′) of that section. The question naturally arises: To what extent do previous results have analogues?

Let $\{\Omega_n\}$ be an exhaustion of R. Define an induced partition of

$\partial\Omega_n$ into $(\alpha_n^0, \alpha_n^1, \gamma_n^0, \gamma_n^1)$ as follows. In the notation of 1.1A the parts α_n^0, α_n^1 are as defined previously. A contour c on $\partial\Omega_n$ which does not belong to α_n^0 or α_n^1 shall belong to γ_n^1 if A^c contains only points of γ_n^1; otherwise c shall belong to γ_n^0.

The constructions in 1.1B, 1C, 1D can be repeated, and the classes $\mathfrak{F}$, $\mathfrak{F}^*$ defined. The proof of $\lambda(\mathfrak{F}) = \|du\|^{-2}$ is not difficult. On the other hand, the proof of $\lambda(\mathfrak{F}^*) = \|du\|^2$ requires a continuity method analogous to that of 2.1B.

§4. STABILITY

The problem of classifying boundary components according to their stability is far from settled at present. In this section we first present some results on conformal mapping which serve as an introduction to the interesting problems in this area. Next a method of classifying boundary components by capacities is discussed. Finally, the treatment of the mapping P_0+P_1 that was begun in II.2.2C is completed using extremal length methods developed earlier.

The stability results of this section are due to Sario [15]. The mapping property of P_0+P_1 is due partly to Grunsky [1], Schiffer [1], and Sario [16]. The complete proof was first given by Oikawa-Suita [1].

1. Parallel slit mappings

1A. The mappings P^θ. Recall the results of II.2.1. Let R be a planar Riemann surface and p_0^θ, p_1^θ the principal functions with singularity $\operatorname{Re}(e^{i\theta}/(z-\zeta))$ at a fixed point $\zeta \in R$. The meromorphic functions P_0^θ, P_1^θ with the p_i^θ as real parts map R onto a horizontal or vertical slit region in the plane. We shall be primarily interested in the function $P^\theta = e^{i\theta}P_0^{-\theta}$ for real values of θ. It has residue 1 at ζ and maps R onto a parallel slit region whose slits have inclination θ. From II.2.1A we have the important relation

$$(23) \qquad P^\theta = e^{i(\theta-\varphi)}[P^\varphi \cos(\theta-\varphi) - iP^{\varphi+\pi/2}\sin(\theta-\varphi)]$$

which holds for any real φ, θ. In the case $\theta = 0$ we write P_0, P_1 instead of P_0^0, P_1^0.

1B. It is clear that a univalent meromorphic function f on R

induces a homeomorphism of the compactifications of R and $f(R)$. Since $f(R)$ is embedded in the plane we may realize each of its ideal boundary points as a component of the topological boundary of $f(R)$. If α is an ideal boundary component of R let $f(\alpha)$ denote this realization of the corresponding ideal boundary point of $f(R)$.

In the case of an isolated ideal boundary component α it is easily seen that either $f(\alpha)$ is a point for all f or $f(\alpha)$ is a continuum (i.e. a compact connected set with more than one point) for all f. If α is not isolated such a dichotomy need not hold. However, if we consider only the functions P^θ we have (Sario [16]) the

Theorem. *If α is an ideal boundary point of R then one of the following alternatives occurs:*

(a) *$P^\theta(\alpha)$ is a point for all θ.*

(b) *$P^\theta(\alpha)$ is a continuum for all θ.*

(c) *There exists a φ such that $P^\varphi(\alpha)$ is a point and $P^\theta(\alpha)$ is a continuum for all $\theta \not\equiv \varphi$ (mod π).*

1C. The proof will make systematic use of (23) and the fact that a point w belongs to $P^\theta(\alpha)$ if and only if there is a sequence $\{z_n\}$ on R such that $z_n \to \alpha$ on $\bar{R}$ and $P^\theta(z_n) \to w$ on Cl $P^\theta(R)$.

Assume (a) and (b) are false. Then $P^\varphi(\alpha)$ is a point w_0 for some φ. This implies that $P^{\varphi+\pi/2}(\alpha)$ must be a proper slit. Indeed, for any θ we have

$$(24) \quad P^\theta(z) = e^{i(\theta-\varphi)}[P^\varphi(z)\cos(\theta-\varphi) - iP^{\varphi+\pi/2}(z)\sin(\theta-\varphi)]$$

for all $z \in R$. Let $z \to \alpha$ in (24) and set $P^{\varphi+\pi/2}(\alpha) = w_1$. Then $P^\theta(\alpha)$ is the single point

$$e^{i(\theta-\varphi)}[w_0\cos(\theta-\varphi) - iw_1\sin(\theta-\varphi)].$$

Hence $P^{\varphi+\pi/2}(\alpha)$ is a line segment containing two distinct points w_1, w_2 say. For $j = 1, 2$ let $\{z_n{}^j\}$ be a sequence on R with $P^{\varphi+\pi/2}(z_n{}^j) \to w_j$. An application of (24) to these sequences shows that $P^\theta(\alpha)$ contains the points

$$e^{i(\theta-\varphi)}[w_0\cos(\theta-\varphi) - iw_j\sin(\theta-\varphi)] \qquad (j = 1,2)$$

which will be distinct if and only if $\sin(\theta-\varphi) \neq 0$. Thus (c) must hold and the proof is complete.

1D. Extremal length properties. The parallel slit regions $P^\theta(R)$ are similar to the extremal slit annuli in some respects. Consider the image of P_0 in the w-plane for example. Denote it again by R.

Let S be a rectangle $\{0 < \text{Re}\, w < a,\ 0 < \text{Im}\, w < b\}$ which contains the complement of R. Set $S' = S \cap R$ and let $|S$ denote the set S' together with its boundary on the imaginary axis. Similarly let $\underline{S}$ consist of S' and its boundary on the real axis. We shall show that $e^{2\pi w/b}$ *maps* $|S$ *onto an extremal radial slit annulus and* $e^{2\pi i w/a}$ *maps* $\underline{S}$ *onto an extremal circular slit annulus.* Suppose these annuli lie in the W-plane.

The proof is an application of criterion (b) of Theorem 3.1D. We shall treat only the case of $|S$ since it is typical. It suffices to show that $\int_c ds \geq 1$ for all $c \in \mathfrak{F}$ and $\int_d ds \geq b/a$ for almost all $d \in \mathfrak{F}^*$. Here $\mathfrak{F}$ consists of all arcs in the image of $e^{2\pi w/b}$ which travel from $\{|W| = 1\}$ to $\{|W| = e^{2\pi a/b}\}$, and $\mathfrak{F}^*$ consists of curves which separate these circles. The metric ds is in this case $b|dw|/(2\pi a|w|)$.

The metric ds becomes $|dw|/a$ when transferred to S'. Consider the inverse image of an arc $c \in \mathfrak{F}$. Its projection onto the real axis travels from one vertical side of S to the other. Hence in the Euclidean metric $|dw|/a$ its length is ≥ 1. Thus $\int_c ds \geq 1$ for all $c \in \mathfrak{F}$.

We treat $\int_d ds$ for $d \in \mathfrak{F}^*$ by again transferring the integral to S'. The inverse image of d contains an arc τ, on the compactification of R, the end points of which lie on opposite horizontal sides of S. We shall see that $\int_\tau |d\, \text{Im}\, w| \geq b$ for almost all such τ. Note that $\text{Im}\, w = p_1$ where p_1 is the L_1-principal function whose singularity is the imaginary part of a simple pole at $\zeta = \infty$. Consider a countable exhaustion $\{\Omega_n\}$ of R and corresponding principal functions p_{1n} on Ω_n. We set $dp_{1n} \equiv 0$ on $R - \Omega_n$, and obtain by virtue of the L_1-behavior

$$\int_\tau dp_{1n} = p_{1n}(\tau(1)) - p_{1n}(\tau(0)). \tag{25}$$

An application of Lemma II.6.2B and (25) shows that

$$\int_\tau dp_1 = p_1(\tau(1)) - p_1(\tau(0)) = b$$

for almost all τ. (For example, see the proof of Theorem II.6.2G.) This completes the proof.

1E. Let $\mathcal{G}$ be the family of arcs on S' which travel from one of its vertical sides to the other. Let $\mathcal{G}^*$ be the family of arcs on the compactification of R with initial point on $\{\operatorname{Im} w = b\}$ and terminal point on the real axis. We shall show that $\lambda(\mathcal{G}) = 1/\lambda(\mathcal{G}^*) = a/b$.

Let A_1 be the extremal radial slit annulus onto which $|S$ is mapped by $e^{2\pi w/b}$. In A_1 consider the family $\mathfrak{F}_1$ of arcs joining the circumferences and the family $\mathfrak{F}_1^*$ of curves separating them. Denote by A_2 the extremal circular slit annulus onto which $e^{2\pi w/a}$ maps $\underline{S}$. In A_2 consider the family $\mathfrak{F}_2$ of arcs on the compactification of A_2 which join the circumferences and the family $\mathfrak{F}_2^*$ of arcs on A_2 which separate them.

By means of the mappings $e^{2\pi w/b}$, $e^{2\pi i w/a}$ we may compare $\mathcal{G}$, $\mathcal{G}^*$ with $\mathfrak{F}_1$, $\mathfrak{F}_1^*$ and $\mathfrak{F}_2$, $\mathfrak{F}_2^*$. Hence

$$a/b = \lambda(\mathfrak{F}_1) \leq \lambda(\mathcal{G}) \leq \lambda(\mathfrak{F}_2^*) = a/b,$$

$$b/a = \lambda(\mathfrak{F}_2) \leq \lambda(\mathcal{G}^*) \leq \lambda(\mathfrak{F}_1^*) = b/a,$$

and our assertion is proved.

It is clear that the reasoning beginning in 1D used only the boundary behavior of P_0 and did not depend on its residue at ζ. Therefore the above arguments hold for the mappings P_0^θ, and we have (Jenkins [1]) the

Theorem. *Let R be a parallel slit region and let S be a rectangle with sides of length a and b. Assume that S contains all the slits of R and that the sides of length a are parallel to the slits. Then*

$$\lambda(\mathcal{G}) = 1/\lambda(\mathcal{G}^*) = a/b$$

where $\mathcal{G}$ is the family of arcs in $S \cap R$ which join the sides of length b and $\mathcal{G}^*$ is the family of curves in $S \cap R$ which separate those sides.

2. The mapping $P_0 + P_1$

2A. A property of P^θ. Recall from II.2.2A that if R is the interior of a planar compact bordered surface, and P_0 and P_1 are the principal functions with a simple pole of residue 1 at ζ, then

P_0+P_1 maps R univalently onto a region whose complementary components are convex sets. We are now able to extend this result to any planar surface.

The proof will depend on the following property of the function P^θ:

Lemma. *Let w_0 be a boundary point of $P^\theta(R)$. In case w_0 lies on a proper slit assume it is an end point of that slit. For any $\varepsilon>0$ there is an open set U containing w_0 such that U is contained in $\{|w-w_0|<\varepsilon\}$ and $U\cap P^\theta(R)$ is connected.*

Consider the following assertion. *Given ζ_0, $\zeta_1 \in P_0^\theta(R)$ with* $\operatorname{Im}\zeta_0 = \operatorname{Im}\zeta_1$ *and given an open rectangle Q parallel to and containing the closed segment between ζ_0 and ζ_1, there is an arc in $Q\cap P_0^\theta(R)$ with end points ζ_0, ζ_1.*

We first show how this assertion implies the lemma. The mapping P^θ may be replaced by P_0^θ and hence w_0 may be assumed to be a point or to lie on a horizontal slit. There is a closed vertical line segment $\{\operatorname{Im}(w-w_0)\leq\delta,\ \operatorname{Re} w = \text{const.}\}$ contained in $P_0^\theta(R)\cap\{|w-w_0|<\varepsilon\}$ with $\delta>0$. Let U be defined by

$$U = \{w \mid |w-w_0|<\varepsilon, \operatorname{Im}(w-w_0)<\delta, w\in P_0^\theta(R)\}.$$

The assertion shows that U is arcwise connected since it implies that any point of U can be connected to the vertical segment by a path in U.

2B. We now turn to the proof of the above assertion. Let S be the rectangle $\{-L\leq \operatorname{Re} w, \operatorname{Im} w\leq L\}$ where L is so large that the interior of S contains ζ_0, ζ_1, all the slits of $P_0^\theta(R)$, and the closure of the rectangle Q. Assuming the assertion is false we can derive a contradiction to Theorem 1E by showing that the family $\mathcal{G}$ for S has $\lambda(\mathcal{G})>1$.

If ζ_0, ζ_1 cannot be joined within $Q\cap P_0^\theta(R)$, then there is an $\varepsilon_1>0$ such that any arc in $S\cap P_0^\theta(R)$ joining them has length $\geq|\zeta_0-\zeta_1|+\varepsilon_1$. Let ζ_0', ζ_1' be the points on ∂S with $\operatorname{Im}\zeta_0' = \operatorname{Im}\zeta_1' = \operatorname{Im}\zeta_0 = \operatorname{Im}\zeta_1$. There is an $\varepsilon_2>0$ such that any arc in $S\cap P_0^\theta(R)$ from ζ_0' to ζ_1' has length $\geq 2L+\varepsilon_2$. Therefore there is an $\varepsilon_3>0$ such that any arc in $S\cap P_0^\theta(R)$ from ζ_0' to the side of S containing ζ_1' has length $\geq 2L+\varepsilon_3$. We may choose ε_3 so small that $\varepsilon_3<L-|\operatorname{Im}\zeta_0'|$ and $\{|w-\zeta_0'|<\varepsilon_3\}\subset P_0^\theta(R)$. Note that any arc in $S \cap P_0^\theta(R)$ joining the semicircle $S \cap \{|w-\zeta_0'|<\varepsilon_3\}$ to the side of S containing ζ_1' has length $\geq 2L$.

Let ds be the linear density

$$ds = \begin{cases} 0 & \text{in} \quad S \cap \{|w-\zeta_0'|<\varepsilon_3\}, \\ |dw| & \text{in} \quad S - \{|w-\zeta_0'|<\varepsilon_3\}. \end{cases}$$

Then $L(\mathcal{G},ds) \geq 2L$ and $A(ds) < 4L^2$. Hence $\lambda(\mathcal{G}) > 1$, a contradiction.

2C. Convexity theorem. We are now ready to prove the convexity property (Oikawa-Suita [1]) mentioned earlier.

Theorem. *The meromorphic function P_0+P_1 maps R univalently onto a region R^* with the property that every component of the complement of R^* is convex.*

Actually we shall prove more than this. Let $f = P_0+P_1$ and consider $f(R) = R^*$ as a region in the extended w-plane. Then f is univalent since it is the limit, uniformly on compacta of R, of the corresponding univalent functions $P_{0\Omega}+P_{1\Omega}$ on exhausting regular subregions Ω of R. If α is an ideal boundary component of R we have denoted the realization of α in the w-plane by $f(\alpha)$. Thus $f(\alpha)$ is a component of ∂R^*. Let $f^*(\alpha)$ denote the component of the complement of R^* which contains $f(\alpha)$.

Given α we consider $P^\theta(\alpha)$ and $P^{\theta+\pi/2}(\alpha)$. The set

$$Q^\theta(\alpha) = \{w+w' | w \in P^\theta(\alpha),\, w' \in P^{\theta+\pi/2}(\alpha)\}$$

is a rectangle, possibly degenerate, with inclination θ and with sides congruent to $P^\theta(\alpha)$ and $P^{\theta+\pi/2}(\alpha)$. It will be shown that corresponding to the cases (a), (b), (c) of Theorem 1B the following conditions hold:

(a′) *$f^*(\alpha)$ is a point and coincides with $Q^\theta(\alpha)$ for all θ.*

(b′) *$f^*(\alpha)$ is a convex set which is not a line segment, and for any θ, $Q^\theta(\alpha)$ is the smallest rectangle of inclination θ which contains $f^*(\alpha)$.*

(c′) *$f^*(\alpha)$ is a proper line segment and coincides with $Q^\varphi(\alpha)$ for some φ. If $\theta \not\equiv \varphi$* (mod π) *then $f^*(\alpha)$ is a diagonal of the nondegenerate rectangle $Q^\theta(\alpha)$.*

2D. Proof of (b′). Suppose α satisfies (b) of Theorem 1B. First we verify the second part of (b′) which implies that $f^*(\alpha)$ is not a line segment.

Let $\zeta_0 \in f(\alpha)$. There is a sequence $\{z_n\}$ in R with $\lim z_n = \alpha$ on $\bar{R}$, and $\lim f(z_n) = \zeta_0$ on $\mathrm{Cl}\, R^*$. By extracting a subsequence we may assume $\lim P^\theta(z_n) = w_1$ and $\lim P^{\theta+\pi/2}(z_n) = w_2$ exist. Because of

$$f(z) = P^\theta(z) + P^{\theta+\pi/2}(z), \tag{26}$$

a consequence of (24), we obtain $\zeta_0 = w_1 + w_2$. This shows that $\zeta_0 \in Q^\theta(\alpha)$. Hence $f(\alpha) \subset Q^\theta(\alpha)$ and $f^*(\alpha) \subset Q^\theta(\alpha)$. To see that $Q^\theta(\alpha)$ is the smallest such rectangle containing $f^*(\alpha)$ it suffices to prove that each side of $Q^\theta(\alpha)$ has a nonempty intersection with $f(\alpha)$. For the remainder of the proof we make constant use of (26). Now a side of $Q^\theta(\alpha)$ is of the form $\{w_1 + w\}$ where w_1 is an end point of one of the slits $P^\theta(\alpha)$, $P^{\theta+\pi/2}(\alpha)$ and w varies over the other slit. Assume for definiteness that $w_1 \in P^\theta(\alpha)$. Choose a sequence z_n on R such that $z_n \to \alpha$, $P^\theta(z_n) \to w_1$, and moreover $P^{\theta+\pi/2}(z_n)$ tends to a limit point w_2 on $P^{\theta+\pi/2}(\alpha)$. Then $f(z_n)$ converges to a point on $f(\alpha)$ which must be $w_1 + w_2$. This is a point on the side of $Q^\theta(\alpha)$ originally selected. Thus the second assertion of (b′) is proved.

Now suppose $f^*(\alpha)$ is not convex. An elementary argument, which we leave to the reader, shows that there is a line of support l of $f^*(\alpha)$, distinct points ζ_0 and ζ_0' on $l \cap f(\alpha)$ and a point $\zeta^* \notin f^*(\alpha)$ such that ζ^* is on the line segment with end points ζ_0 and ζ_0'. Let l have direction θ.

Let $\{z_n\}$, $\{z_n'\}$ be sequences on R such that

$$z_n \to \alpha, \qquad z_n' \to \alpha,$$

$$f(z_n) \to \zeta_0, \qquad f(z_n') \to \zeta_0'.$$

Then $P^{\theta+\pi/2}(z_n)$ and $P^{\theta+\pi/2}(z_n')$ converge to an end point w_1 of $P^{\theta+\pi/2}(\alpha)$. Therefore $P^\theta(z_n)$, $P^\theta(z_n')$ converge to points w_0, w_0' on $P^\theta(\alpha)$. Furthermore, $w_0 \neq w_0'$.

Let $w^* \in P^\theta(\alpha)$ be the point defined by $w_1 + w^* = \zeta^*$. The line l^* through w^* orthogonal to $P^\theta(\alpha)$ clearly separates w_0, w_0'.

Choose a sequence $\varepsilon_n \searrow 0$ such that for each n $|P^{\theta+\pi/2}(z_n) - w_1| < \varepsilon_n$ and $|P^{\theta+\pi/2}(z_n') - w_1| < \varepsilon_n$. According to Lemma 2A we may assume, by passing to subsequences of $\{z_n\}$ and $\{z_n'\}$, that the points $P^{\theta+\pi/2}(z_n)$, $P^{\theta+\pi/2}(z_n')$ can be connected by an arc in $P^{\theta+\pi/2}(R)$ which remains within a distance ε_n from w_1. Let γ_n denote the inverse image on R of this arc under $P^{\theta+\pi/2}$.

Since l^* separates w_0, w_0' there are points $z_n^* \in \gamma_n$ such that $P^\theta(z_n^*) \in l^*$. By virtue of $z_n^* \to \alpha$ we have $P^\theta(z_n^*) \to w^*$. From $P^{\theta+\pi/2}(z_n^*) \to w_1$ it follows that $f(z_n^*) \to \zeta^*$. This contradicts $\zeta^* \notin f^*(\alpha)$.

The proof of (b′) is now complete. The proofs of (a′), (c′) are quite similar and will be left to the reader.

3. Stability

3A. Strong, weak, and unstable components. Let R be a planar surface and f a univalent meromorphic function on R. As in 4.1B we let $f(\alpha)$ be the component of $\partial f(R)$ which corresponds to α. If $f(\alpha)$ is a point for all such f then α is said to be *weak*. If $f(\alpha)$ is a continuum for all f then α is called *strong*. If α is neither weak nor strong it is described as being *unstable*. By the *stability* of α we mean its nature with respect to this classification of weak, strong, or unstable.

Theorem 2C, and its proof, give information about stability. It was shown that when for $f = P_0 + P_1$, $f(\alpha)$ is a proper slit only if $P^\varphi(\alpha)$ is a point for some φ, i.e. if α is strong then $f(\alpha)$ is a continuum which is not a slit. Since $f^*(\alpha)$, the component of the complement of $f(R)$ to which $f(\alpha)$ belongs, is convex and not a slit it contains an interior point. The plane cannot contain an uncountable number of disjoint open sets, and we have the

Theorem. *A planar surface has at most a countable number of strong boundary components.*

Clearly there exist regions whose topological boundary consists of an uncountable number of continua. We conclude:

Corollary. *There exist unstable boundary components.*

3B. A condition for weakness. We now give a necessary and sufficient condition for a boundary component to be weak. Because this condition is intrinsic it shows that the notion of weakness can be extended to nonplanar surfaces. Whether or not this extension is possible for the other classes of stability is unknown.

In 2.4A we considered the function p_α and the corresponding notion of capacity of α. Now suppose R is a planar surface and α is a boundary component of capacity zero. Recall that this condition is equivalent to $p_\alpha \equiv -\infty$, and is independent of the pole ζ and the local parameter used to define capacity. We shall prove the (Sario [15])

Theorem. *A necessary and sufficient condition for an ideal boundary component of a planar surface to be weak is that its capacity be zero.*

First suppose that the boundary component α has capacity zero. If it were not weak then the surface could be realized as a subregion of $\{|z| < 1\}$ with α corresponding to the circumference. We may assume that for some $0 < r < 1$, $\Delta = \{|z| \leq r\}$ is contained in the subregion. Remove Δ, set $\alpha^0 = \{|z| = r\}$, and denote the resulting subregion by R. If Theorem 2.4B is applied to R with the partition $(\alpha^0, \alpha, \emptyset, \beta - \alpha^0 - \alpha)$ we find that the family $\mathfrak{F}^*$ of cycles in R which separate α^0, α has extremal length zero. This is clearly impossible. Indeed, consider the annulus with boundary α^0, α and let $\mathfrak{G}^*$ be the family of cycles in it which separate α^0, α. Then $\mathfrak{F}^* \subset \mathfrak{G}^*$ and hence $\lambda(\mathfrak{F}^*) \geq \lambda(\mathfrak{G}^*) > 0$.

Conversely, suppose α is weak. If the capacity were positive then $P_\alpha = e^{p_\alpha + ip_\alpha^*}$ would be a univalent mapping of the surface into a circular slit disk with α corresponding to the circumference $\{|z| = e\}$. This is again a contradiction. For explicit criteria on stability see Oikawa [5] and Akaza-Oikawa [1].

3C. A condition for strength. From the results of 2.4C we know that the capacity of α cannot be less than the L_0-capacity of α. It is natural to ask whether having positive L_0-capacity for an ideal boundary component is equivalent to being strong. In one direction the implication is true (Sario [15]):

Theorem. *If α has positive L_0-capacity then it is strong.*

In fact, if α could be realized as a point then, given any disk Δ contained in the region, the family of arcs from $\alpha^0 = \partial\Delta$ to α

would have infinite extremal length. This contradicts the equivalence of (c) and (d) in Theorem 2.4C.

Whether or not the converse of the above theorem holds is an open question.

3D. Extendability. Let R now denote a surface of finite genus. A *continuation* of R to a Riemann surface S is a pair (f,S) where f is a biholomorphic mapping of R into S. If S is closed the continuation is said to be *compact.* If $S-f(R)$ is a nonempty set with an interior point then the continuation is said to be *essential.*

The surface R is said to be *uniquely continuable* if whenever (f_1,S_1), (f_2,S_2) are compact continuations of R there is a homeomorphism h of S_1 onto S_2 such that $f_2 = h\circ f_1$. We shall prove the (Jurchescu [1])

Theorem. *Let R be a Riemann surface of finite genus. The following conditions are equivalent:*

(a) *Every ideal boundary component of R has capacity zero.*
(b) *R is uniquely continuable.*
(c) *There are no essential continuations of R.*

We have observed that vanishing capacity of a boundary component is a local property (2.4C) and that for planar neighborhoods it implies weakness. Therefore, if (f,S) is a compact continuation of R then each component of $S-f(R)$ is a point. Thus S is homeomorphic to $\bar{R}$, the compactification of R. Since homeomorphisms of surfaces extend to homeomorphisms of their compactifications we conclude that (a) implies (b).

By considering an exhaustion of the range surface it is easy to see that if there is an essential continuation of R then there is a compact essential continuation, say (f_1,S_1). By attaching a handle to $S_1-f_1(R)$ we obtain a surface S_2 and a compact continuation (f_2,S_2) of R. Since S_1, S_2 have different genus R is not uniquely continuable. This shows that (b) implies (c).

Let α be a boundary component of R of positive capacity. Let Δ be a regular planar boundary neighborhood of α and let α^0 be the relative boundary of Δ. Since α has positive capacity, $\Delta\cup\alpha^0$ is conformally equivalent to an extremal circular slit annulus with inner contour α^0, and outer boundary corresponding to α. Let R_1 be this region together with the unbounded component of the plane determined by α (including α and ∞). If R_1 is welded to $R-\Delta$ along α^0 the obvious identification of part of the resulting surface with R provides an essential continuation of R. Thus (c) implies (a) and the proof is complete.

4. Vanishing capacity

4A. Nonuniqueness. Let R be an arbitrary Riemann surface with ideal boundary β. The function p_β of 2.4F degenerates to the constant $-\infty$ if β has capacity zero. We may nevertheless try to find a function which is harmonic on R except for a negative logarithmic pole at a point ζ and, by analogy with p_β, has constant boundary values on β in some sense.

Let $\{\Omega_n\}$ be an exhaustion of R with $\zeta \in \Omega_1$. Consider the functions $s_n = p_n - k_n$ where p_n is the capacity function of $\partial\Omega_n$ and e^{k_n} is the capacity of $\partial\Omega_n$. We shall refer to s_n as the normalized capacity function for Ω_n. Then $\lim_{z\to\zeta} s_n(z) - \log|z-\zeta| = 0$ when z is the local parameter used in defining k_n and the function s_n has constant value $1-k_n$ on $\partial\Omega_n$. Furthermore $\{k_n\}$ is a decreasing sequence. It is possible to show that the family $\{s_n\}$ is normal and that any limit function s tends to ∞ along almost every curve tending to β. We shall not give the proof here because the method employed does not furnish a unique limit function. However, we consider an example of Accola which illustrates the nonuniqueness.

Let R be the Riemann sphere punctured at $z = 1, \infty$. The origin $z = 0$ will play the role of ζ. Let Ω_n be the compact region bounded by

$$C_n = \{|z| = n\}$$

and

$$C_n' = \{|z-1| = r(n)\},$$

where $r(n)$ will be determined later.

Let ω_n be the harmonic function in Ω_n with boundary values $\log|z|$ for $z \in C_n \cup C_n'$. Then

$$s_n(z) = \log|z| - \omega_n(z) + \omega_n(0)$$

is the normalized capacity function for Ω_n.

Let Δ be a small disk centered at $z = 0$. For fixed n observe that $\|\omega_n - \log n\|_\Delta \to 0$ as the radius of C_n' shrinks to zero. Now determine $r(n)$ by the requirement

$$\|\omega_n - \log n\|_\Delta < \frac{1}{n}$$

for $n = 1, 2, \cdots$.

We have

$$\|s_n(z) - \log|z|\,\|_\Delta = \|\omega_n(z) - \omega_n(0)\|_\Delta < \frac{2}{n}.$$

Hence any convergent subsequence of $\{s_n\}$ must tend to $\log|z|$.

We have shown that for a certain exhaustion the limit function s has a removable singularity at $z = 1$. By the symmetry of R, a twice punctured sphere, there is an exhaustion for which the limit function has ∞ as a removable singularity. Thus the normalized capacity function is not unique.

4B. Evans potential. Let R be an open Riemann surface with ideal boundary β. An *Evans potential* s on R with pole $\zeta \in R$ is a function s on R, harmonic except for a negative logarithmic pole $\log|z-\zeta|$ at ζ, and such that s tends uniformly to ∞ near β in the sense that

$$\lim_{\Omega \to R} \inf_{z \in R - \Omega} s(z) = \infty \tag{27}$$

for any exhaustion $\{\Omega\}$ of R.

Suppose there exists an Evans potential s on R with pole $\zeta \in R$. Set $\Omega_n = \{z \in R | s(z) < n\}$. The capacity function p_n of $\partial\Omega_n$ is given by

$$p_n = s|\Omega_n - (n-1).$$

Therefore $p_\beta = \lim_n p_n = -\infty$ is the capacity function of β, i.e. β has capacity zero.

It can be shown that the converse is also valid (Evans [1], Selberg [1], Noshiro [1], Kuramochi [1], Nakai [2]). To this end let β have capacity zero and let Ω_0 be an open disk in R. Then there exists a harmonic function h on $R-\Omega_0$ such that $h|\partial\Omega_0 = 0$,

$$\int_{\partial\Omega_0} *dh = 2\pi, \tag{28}$$

and

$$\lim_{\Omega\to R}\ \inf_{z\in R-\Omega} h(z) = \infty \tag{29}$$

for any exhaustion $\{\Omega\}$ of R with $\Omega\supset\Omega_0$. For the proof we refer to Nakai [2] or Sario-Noshiro [1, p. 114].

Take a point $\zeta\in\Omega_0$ and a bordered disk Δ about ζ with $\Delta\subset\Omega_0$. For the bordered boundary neighborhood $A = (R-\Omega_0)\cup(\Delta-\{\zeta\})$ of the ideal boundary of $R-\{\zeta\}$ let L be the normal operator which is the direct sum of the Dirichlet operator in $\Delta-\{\zeta\}$ and the operator L_0 in $R-\Omega_0$. Define the harmonic function σ on A such that $\sigma(z)|\Delta-\{\zeta\} = \log|z-\zeta|$ and $\sigma|R-\Omega_0 = h$.

In view of (28), $\int_{\partial A} *d\sigma = 0$. Therefore the Main Existence Theorem I.1.2B applies and yields an L_0-principal function s with respect to σ. The function s is harmonic on $R-\zeta$ with (27) and has a negative logarithmic pole at ζ, i.e. s is an Evans potential on R.

CHAPTER IV
CLASSIFICATION THEORY

Classification theory is the investigation of families of Riemann surfaces which share some important property. As such, the subject has no bounds. In this chapter we shall confine ourselves to those results of classification theory on which the theory of principal functions has a direct bearing. In §1 we introduce the fundamental classes and prove the inclusion relations among them. In §2 we treat other properties and characterizations of these classes.

In keeping with the selective nature of this chapter, only occasional references will be given to the voluminous literature. For a comprehensive discussion of the topic we refer to the forthcoming monograph Nakai-Sario [11] "Classification theory."

§1. INCLUSION RELATIONS

To prepare for the proofs of the inclusion relations we first collect results from earlier chapters in a convenient form. Next we define the O-classes and demonstrate the inclusions. Finally we prove some alternative characterizations of these classes.

1. Properties of principal functions

1A. Reproducing differentials. Let p_0, p_I, and p_Q denote the principal functions on a Riemann surface R corresponding to the operators L_0, $(I)L_1$ and $(Q)L_1$ respectively. Unless mention is made to the contrary, these functions are assumed to have a single singularity s of the form $\operatorname{Re}(z-\zeta)^{-1}$ for a fixed $\zeta \in R$.

We shall make use of the following results from II.3 concerning reproducing differentials:

(1) $$(dh, dp_0 - dp_I) = 2\pi \left. \frac{\partial h}{\partial x} \right|_{z=\zeta} \quad \text{for all} \quad dh \in \Gamma_{he},$$

(2) $$(dh, dp_0 - dp_Q) = 2\pi \left. \frac{\partial h}{\partial x} \right|_{z=\zeta} \quad \text{for all} \quad dh \in \Gamma_{he} \cap \Gamma^*_{hse},$$

(3) $$(dh, dp_Q - dp_I) = 2\pi \left. \frac{\partial h}{\partial x} \right|_{z=\zeta} \quad \text{for all} \quad dh \in \Gamma_{hm},$$

(4) $$\Gamma_{he} = \Gamma_{hm} \oplus \Gamma_{he} \cap \Gamma^*_{hse}.$$

If the singularity s is of the form $\log(|z-\zeta_2|/|z-\zeta_1|)$ we also have

(5) $$(dh, dp_Q - dp_0) = 2\pi \int_{\zeta_1}^{\zeta_2} dh \quad \text{for all} \quad dh \in \Gamma_{he} \cap \Gamma^*_{hse}.$$

1B. Spans. The *I-span* of R at ζ is defined as $\partial(p_0 - p_I)/\partial x|_{z=\zeta}$, where $z = x + iy$ is a local parameter near ζ. By (1) we see that

(6) $$I\text{-span} = \frac{1}{2\pi} ||dp_0 - dp_I||^2.$$

Similarly, the Q-span is $\partial(p_0 - p_Q)/\partial x|_{z=\zeta}$, and from (2) we have

(7) $$Q\text{-span} = \frac{1}{2\pi} ||dp_0 - dp_Q||^2.$$

1C. Univalent functions. Let P_0, P_1 be the horizontal and vertical parallel slit mappings of a planar surface R and $\zeta \in R$ the common pole of P_0, P_1. By Theorem II.2.2D the function $\frac{1}{2}(P_0 + P_1)$ is univalent on R and maps R onto a region whose complementary area is $\frac{1}{2}\pi\partial(p_0 - p_Q)/\partial x|_{z=\zeta}$. Furthermore, this function maximizes the complementary image area among all univalent functions f on R with a simple pole of residue 1 at ζ. If

A_f denotes the complementary area of $f(R)$ for such univalent f we may record these results in the form

$$\text{(8)} \qquad Q\text{-span vanishes} \Leftrightarrow A_f = 0 \quad \text{for all} \quad f.$$

Recall that in any case $A_{P_0} = A_{P_1} = 0$.

1D. Capacities and extremal length. We now briefly recall some of the results of Chapter III. Let R denote a Riemann surface, bordered or not, and let $(\alpha^0,\alpha^1,\gamma^0,\gamma^1)$ be a suitable partition of the ideal boundary of R. In Chapter III we constructed a harmonic function $u(\alpha^0,\alpha^1,\gamma^0,\gamma^1)$ having boundary behavior (in the sense of an exhaustion) L_0 on γ^0, $(Q)L_1$ on γ^1, $u = 0$ on α^0, and $u = 1$ on α^1. The families $\mathfrak{F}(\alpha^0,\alpha^1,\gamma^0,\gamma^1)$, $\mathfrak{F}^*(\alpha^0,\alpha^1,\gamma^0,\gamma^1)$ can be described roughly as consisting of all arcs on $R\cup\alpha^0\cup\alpha^1\cup\gamma^1$ which go from α^0 to α^1, and of all curves on $R\cup\gamma^0$ which separate α^0 and α^1. We shall make frequent use of our earlier result

$$\text{(9)} \qquad ||du||^2 = \lambda(\mathfrak{F}^*) = \frac{1}{\lambda(\mathfrak{F})}.$$

For a suitable partition $(\alpha,\gamma^0,\gamma^1)$ of β we defined the capacity c_α and showed that if $c_\alpha > 0$ then there exists a function $p(\zeta,\alpha,\gamma^0,\gamma^1)$ which is harmonic on R except for a negative logarithmic pole at $\zeta\in R$, has value 1 on α, and has L_0-behavior on γ^0 and $(Q)L_1$-behavior on γ^1. If $c_\alpha = 0$ then $p_\alpha \equiv -\infty$. If α is a single component, R is planar, and $c_\alpha > 0$ then $P_\alpha = e^{p_\alpha + ip_\alpha^*}$ is a univalent mapping of R into a disk such that the image has outer contour corresponding to α.

Let Ω_0 be a regular subregion of R and form

$$u_\alpha = u(\partial\Omega_0,\alpha,\varnothing,\beta-\alpha)$$

on $R-\Omega_0$. The capacity c_α relative to R, $\zeta\in R$, and the partition $(\alpha,\varnothing,\beta-\alpha)$ of β will vanish if and only if $u_\alpha\equiv 0$. In the special case $\alpha = \beta$ the function u_β is called the *harmonic measure* of β relative to Ω_0.

2. Classes of Riemann surfaces

2A. Notation. For each Riemann surface R we consider the following classes of functions on R:

H = the class of real-valued harmonic functions;

K = the class of functions h in H such that $*dh$ has period zero along every dividing cycle on R;

A = the class of complex analytic functions;

S = the class of univalent ("schlicht") complex analytic functions.

A function f belonging to one of the above classes might possess one or more of the following properties:

B: $|f|$ is bounded on R.

D: the Dirichlet integral of f over R is finite.

BD: f has properties B and D.

P: $f>0$ on R.

E: the complement of $f(R)$ with respect to the complex plane has positive area.

For the remainder of this chapter, X will stand for one of the symbols H, K, A, or S, and Y will stand for B, D, BD, P, or E. Then O_{XY} is defined as the class of Riemann surfaces R such that the set of functions on R which belong to X and have property Y either is empty or consists only of the constant functions.

In addition to the above definitions we let S_I denote the class of Riemann surfaces on which the I-span vanishes at every point. Similarly S_Q consists of surfaces on which the Q-span vanishes everywhere.

For the sake of orientation we remark that for any X and Y the class O_{XY} contains all closed Riemann surfaces. The complex plane also belongs to every O_{XY}. In contrast, the open unit disk belongs to none of these classes.

2B. Classes O_{HY}. We first consider the classes which are defined with respect to harmonic functions. The following relations are valid (Virtanen [1], Sario [9]):

$$O_{HP} \subset O_{HB} \subset O_{HBD} = O_{HD} = S_I. \tag{10}$$

The first inclusion is trivial since a nonconstant HB-function can be used to construct a nonconstant HP-function. The second inclusion also follows immediately from the definitions.

To prove $O_{HBD} \subset O_{HD}$ we suppose $R \notin O_{HD}$. Then there exists a $dh \in \Gamma_{he}$ such that $\partial h/\partial x \neq 0$ at some point ζ. By (1) the corresponding function $p_0 - p_I$ cannot be constant. Since it is an HBD-function we have $R \notin O_{HBD}$. The opposite inclusion $O_{HD} \subset O_{HBD}$ is trivial, and the first equality of (10) is proved.

Since the I-span at ζ fails to vanish if and only if $p_0 - p_I$ is a nonconstant HD-function (see (6)) we have $O_{HD} = S_I$.

2C. Classes O_{AY} for planar surfaces. We begin the study of O_{AY} classes by considering planar surfaces only. We then have

$$(11) \qquad O_{AE} = O_{AB} \subset O_{ABD} = O_{AD} = S_Q \qquad \text{(planar surfaces)}.$$

The relation $O_{AE} \subset O_{AB}$ is trivial. For the opposite inclusion we make use of the following observation: *If $R \in O_{AB}$ is a plane region then the complement of R has zero area.* To see this we may assume $\infty \in R$, since linear transformations preserve sets of positive area. We compare the identity mapping z with the slit mapping P_0 which has a simple pole at ∞. If the complement of R has positive area then $P_0 \neq z$, since P_0 always leaves a complement of zero area. Since $P_0 - z$ is bounded it provides a nonconstant AB-function.

We now prove $O_{AB} \subset O_{AE}$. Suppose $R \in O_{AB}$ and let f be any analytic function on R. Then $f(R) \in O_{AB}$, for if g were a nonconstant AB-function on $f(R)$ then $g \circ f$ would be a nonconstant AB-function on R. Therefore $f(R)$ must have a complement of area zero, and $f \notin AE$. Thus $R \in O_{AE}$ and the proof of $O_{AE} = O_{AB}$ is complete.

The inclusions $O_{AB} \subset O_{ABD}$ and $O_{AD} \subset O_{ABD}$ are trivial. For the proof of $O_{ABD} \subset O_{AD}$ assume $R \notin O_{AD}$ and let $f = u + iv$ be a nonconstant AD-function. Then $du \in \Gamma_{he} \cap \Gamma^*_{hse}$ and $\partial u/\partial x \neq 0$ at some $\zeta \in R$. Hence $p_0 - p_Q$ is nonconstant by (2), and consequently the corresponding analytic function $P_0 - P_1$ is a nonconstant ABD-function. Thus $R \notin O_{ABD}$, and we have the desired relation $O_{ABD} \subset O_{AD}$. This proof also shows that $R \notin O_{AD}$ implies $R \notin S_Q$. Conversely $R \notin S_Q$ implies that $P_0 - P_1$ is a nonconstant AD-function for some ζ. Therefore $O_{AD} = S_Q$.

It is interesting to note that if the Q-span vanishes at one point ζ, it vanishes at all points and hence $R \in S_Q$. Indeed, assume that the Q-span vanishes at $\infty \in R$. Then by (8) every univalent mapping which leaves ∞ fixed maps R onto a region of complementary area zero. It follows that every univalent mapping of R has vanishing complementary image area. Indeed, such a mapping can be followed by a linear transformation to obtain a univalent mapping with ∞ fixed, and linear transformations preserve sets of measure zero. Hence $A_f = 0$ for univalent mappings f with an arbitrary pole. Using (8) again we see that the Q-span vanishes at every $\zeta \in R$.

Let S_A denote the class of planar surfaces on which the Q-span vanishes for at least one point. We may then state the above result as

(12) $$S_Q = S_A \qquad \text{(planar surfaces)}.$$

2D. Capacities on planar surfaces. Let C denote the class of Riemann surfaces R such that $c_\alpha = 0$ for every ideal boundary component α of R. We shall show that

(13) $$O_{SD} = O_{SB} = C \qquad \text{(planar surfaces)}.$$

The proof of (13) is quite easy using the results of 1D. The relation $O_{SD} \subset O_{SB}$ is obvious. It is also clear that $O_{SB} \subset C$ since $c_\alpha > 0$ for some α implies that P_α exists and is an SB-function (see 1D). Finally, to prove $C \subset O_{SD}$ suppose the origin 0 is in $R \notin O_{SD}$ and let f be an SD-function on R with $f(0) = 0$. Let $\alpha^0 = \{z \| f(z)| = r\}$ and note that $L(\mathfrak{F}^*, |df|) > 0$, $A(|df|) < \infty$. Hence $\lambda(\mathfrak{F}^*) > 0$, $c_\alpha > 0$, and $R \notin C$.

2E. Classes O_{AY} for nonplanar surfaces. For planar surfaces we made essential use of the fact that p_0, p_Q have single-valued harmonic conjugates. This is not true for nonplanar surfaces, but nevertheless our methods yield

(14) $$O_{AE} = O_{AB} \subset O_{AD}$$

for arbitrary Riemann surfaces.

The proof of $O_{AE} = O_{AB}$ is the same as in 2C. To show that $O_{AB} \subset O_{AD}$ assume $R \notin O_{AD}$ and let f be a nonconstant AD-function on R. Then f represents R as a covering surface of the plane which has finite area. Hence $f(R)$ has finite area and $f(R) \notin O_{AE} = O_{AB}$. Therefore there is a nonconstant AB-function g on $f(R)$, and $g \circ f$ is a nonconstant AB-function on R.

2F. Parabolic surfaces. Green's function $g(z,\zeta)$ on R with pole at ζ is defined as $g(z,\zeta) = 1 - p_\beta$ provided $p_\beta \not\equiv -\infty$ (III.2.4F). The class O_G of parabolic surfaces is, by definition, the class of Riemann surfaces which do not possess a Green's function. The next result shows that this definition does not depend on the choice of the pole ζ.

On a given surface R let $HB(\beta)$ and $HD(\beta)$ denote the classes of HB- and HD-functions respectively which are defined on a neighborhood of β, the neighborhood depending on the function.

Theorem. *For an open Riemann surface R the following conditions are equivalent:*

(a) *R carries no nonconstant negative subharmonic functions.*
(b) $\int_\beta * dh = 0$ *for every* $h \in HB(\beta)$.
(c) *The harmonic measure* u_β *vanishes identically.*
(d) $\int_\beta * dh = 0$ *for every* $h \in HD(\beta)$.
(e) *The capacity* c_β *vanishes.*
(f) $R \in O_G$.

(a)$\Rightarrow$(b). If (b) is false we can find an $HB(\beta)$-function s_1 defined in a bordered boundary neighborhood A and such that $\int_\beta * ds_1 = 2\pi$. Choose a point $\zeta \notin A$ and a disk Δ containing ζ and disjoint from A. Let $s_2 = \log|z-\zeta|$ in Δ, where z is a local parameter in Δ. The singularity function s equal to s_1 in A and to s_2 in Δ satisfies the flux condition of the Main Existence Theorem. Applying this theorem to L_1 say, we obtain a subharmonic function which is bounded from above. Thus (a) cannot hold.

(b)$\Rightarrow$(c). Use of an exhaustion shows that $\|du_\beta\|^2 = \int_\beta * du_\beta$. Since $0 \leq u_\beta \leq 1$ we can infer (c) from (b) at once.

(c)$\Rightarrow$(d). Suppose (d) failed for an HD-function h on $R - \Omega_0$, where Ω_0 is a regular subregion of R. Then $\int_c |*dh| \geq |\int_\beta * dh| = M > 0$ for every c in $\mathfrak{F}^*(\partial\Omega_0, \beta, \emptyset, \emptyset)$. Using $|dh + i * dh|$ as a linear density we find $\lambda(\mathfrak{F}^*) > 0$. By (9) the harmonic measure does not vanish.

(d)⇒(e). We have observed that $\|du_\beta\|^2 = \int_\beta *du_\beta$. Since $u_\beta \in HD(R-\Omega_0)$ we see that (d) implies (c) which in turn implies (e) as in 1.1D.

(e)⇒(f). This follows at once from 1D and the definition of Green's function.

(f)⇒(a). Assume (f) holds. From 1D we know that c_β and hence u_β must vanish. Let h be a negative subharmonic function on R and set $M = \max_{\partial\Omega_0} h$. Consider a regular region $\Omega \supset \Omega_0$ and form the harmonic measure $u_{\beta\Omega}$ of $\partial\Omega$ with respect to Ω_0. Then $h \leq M(1-u_{\beta\Omega})$ on $\partial(\Omega - \mathrm{Cl}\,\Omega_0)$ and hence also on $(\Omega - \mathrm{Cl}\,\Omega_0)$. Letting $\Omega \to R$ we obtain $h \leq M$ on R. Thus h attains its maximum and must be a constant.

Remark 1. Condition (c) and 1D show that parabolicity may also be defined in terms of extremal length. Thus

$$\lambda(\mathfrak{F}^*(\partial\Omega_0,\beta,\varnothing,\varnothing)) = 0,$$

$$\lambda(\mathfrak{F}(\partial\Omega_0,\beta,\varnothing,\varnothing)) = \infty$$

are each equivalent to $R \in O_G$.

Remark 2. As an illustration of Theorem 2F we note the following Picard type theorem: *A nonconstant meromorphic function on a parabolic Riemann surface cannot omit a set of values of positive capacity.* The proof is clear, for if the image region carried a nonconstant negative subharmonic function, it could be lifted back to the surface using the meromorphic function.

2G. Summary. Some of the preceding results, and others that may be proved similarly, can be combined to give the following diagram of inclusion relations.

$$\begin{array}{ccccccccccc}
 & & & & & & O_{ABD} & & & & \\
 & & & & & & \cup \quad \cup & & & & \\
 & & & & O_{AE} & = & O_{AB} \subset O_{AD} & & & & \\
 & & & & \cup & & \quad\quad \cup & & & & \\
C & & O_{KP} & \subset & O_{KB} & \subset & O_{KBD} & = & O_{KD} & = & S_Q \\
\cup & & \cup & & \cup & & \cup & & \cup & & \cup \\
O_G & \subset & O_{HP} & \subset & O_{HB} & \subset & O_{HBD} & = & O_{HD} & = & S_I
\end{array}$$

In this brief survey of classification theory we shall not treat the important problem of investigating which of the above inclusions are strict. Typical among them are $O_G < O_{HP} < O_{HB} < O_{HD}$, $O_G < O_{AB} < O_{AD}$ (see Ahlfors-Sario [1, pp. 251–264]).

§2. OTHER PROPERTIES OF THE *O*-CLASSES

In this section we first relate the O-classes to properties of normal operators. Then we discuss implications for null sets in the plane and, in general, for dependence of O-classes on the ideal boundary. We also obtain relations with extremal distance problems.

1. Normal operators and ideal boundary properties

1A. Degeneracies of normal operators. Let A be a bordered boundary neighborhood of R. We consider the problem of finding conditions under which the various normal operators L on A may coincide.

Theorem. *The following conditions are equivalent:*

(a) $L_0 = (I)L_1$.

(b) $R \in O_{HD}$.

(c) *If* $L(f)$ *is an HD-function on* A *for all* $f \in H(\partial A)$ *then* $L = L_0$.

(a)$\Rightarrow$(b). This follows at once from (1).

(b)$\Rightarrow$(c). Using the Main Existence Theorem it is easy to see that (b) implies the following property for R: If s is an HD-function on A, $\int_\beta * ds = 0$, and $s|\partial A = 0$ then $s \equiv 0$. Applying this property to $s = Lf - L_0 f$ for $f \in C(\partial A)$ we conclude that (c) holds.

(c)$\Rightarrow$(a) is obvious.

1B. The above theorem has an analogue for the canonical partition Q and the class of KD-functions:

Theorem. *The following conditions are equivalent:*

(a) $L_0 = (Q)L_1$.

(b) $R \in O_{KD}$.

(c) *If* $L(f)$ *is a KD-function on* A *for all* $f \in H(\partial A)$ *then* $L = L_0$.

The proof is similar to 1A and will be omitted.

1C. If we compare $(I)L_1$ and $(Q)L_1$ we obtain the

Theorem. *The following conditions are equivalent:*

(a) $(I)L_1 = (Q)L_1$.
(b) $\Gamma_{hm} = 0$.
(c) *Every HD-function on R is a KD-function.*
(d) *There do not exist two disjoint closed subsets of ideal boundary components with positive capacity.*

(a)⇒(b) is a consequence of (3), and (b)⇒(c) follows from (4).

To prove (c)⇒(d) we suppose (d) is false. Let α^0, α^1 be disjoint closed subsets of β with positive capacity. Then p_{α^0}, p_{α^1} exist and $h = p_{\alpha^0} - p_{\alpha^1}$ has a removable singularity at ζ. It is easily seen that h violates condition (c).

Now consider (d)⇒(a). If (a) is false, the Main Existence Theorem yields an HD-function h on R such that $|\int_c *dh| = M > 0$ for some dividing cycle c. This cycle determines two disjoint closed subsets α^0, α^1 of β. The family $\mathfrak{F}^* = \mathfrak{F}^*(\alpha^0, \alpha^1, \varnothing, \varnothing)$ of cycles which separate α^0, α^1 has positive extremal length as may be seen by using $ds = |dh + i*dh|$ as a linear density. It follows that both α^0 and α^1 have positive capacity.

Remark. The class of surfaces which satisfy the conditions of Theorem 1C is the only class thus far considered which contains the unit disk.

1D. Removable singularities. Let E be a compact subset of the extended plane P. The set E is said to be a *removable singularity for XY*-functions or an *XY-null set* if every XY-function which is defined on a deleted neighborhood of E can be continued to an XY-function on the full neighborhood.

Let $R = \mathsf{P} - E$. If E is an XY-null set then $R \in O_{XY}$ since the classes XY on P contain at most the constant functions. The converse is also true for many of the classes XY. For example, suppose $R \in O_{AD}$. Let f be an AD-function on $A = (\mathrm{Cl}\, G) - E$, where G is an open neighborhood of E with analytic boundary curves. Let L be the Dirichlet operator for G and apply the Main Existence Theorem to R, L, and the singularity function $u = \mathrm{Re}\, f$. Note that the flux condition $\int_\beta *du = 0$ is satisfied. We obtain a harmonic function u_1 on R such that $u_1 = u+L(u_1-u)$ on A. Thus u_1 is HD on A and hence also on R. Since R is planar $u_1 = \operatorname{Re} f_1$ for a single-valued AD-function f_1 on R. By virtue of $R \in O_{AD}$, f_1 and consequently u_1 must be constant. This means $u = Lu$, and therefore u is extendable to E.

1E. The same reasoning works for AB. For HD and HB the proof is somewhat more difficult because the flux condition needed to apply the Main Existence Theorem is not so evident. However, if we assume $R \in O_G$ the required flux condition is given immediately by (b) and (d) of Theorem 1.2F, i.e. the proof in 1D applies *mutatis mutandis* to give the

Theorem. *Let R be a plane region with complement E. The following conditions are equivalent:*

(a) $R \in O_G$.
(b) *E is an HB-null set.*
(c) *E is an HD-null set.*

1F. To complete the discussion of removability and classification we show that $O_G = O_{HB} = O_{HD}$ for planar surfaces, and thus restore symmetry to Theorem 1E.

Theorem. *For planar surfaces $O_G = O_{HP} = O_{HB} = O_{HD}$.*

The diagram in 1.2G shows that it will be sufficient to prove $O_{HD} = O_G$. Let C' denote the class of surfaces which satisfy the conditions of Theorem 1C.

Lemma. *For arbitrary Riemann surfaces $C \cap C' = O_G$.*

To prove the lemma we first note that the inclusion $O_G \subset C \cap C'$ follows easily from previous results. For the opposite inclusion we assume $R \in C' - O_G$ and prove $R \notin C$.

Let u be the harmonic measure of β with respect to Ω_0, a regular subregion, and consider a canonical subregion $\Omega \supset \Omega_0$. Let the contours of $\partial\Omega$ be $\beta_1, \cdots, \beta_n$ and for $i = 1, \cdots, n$ let A_i be the component of $R - \Omega$ bordered by β_i. We show that $\int_{\beta_i} {*du} \neq 0$ for exactly one value of i. If not, suppose β_i, β_j are the two exceptional contours. Then there are nonzero constants a, b such that the singularity function

$$s = \begin{cases} au & \text{in } A_i, \\ bu & \text{in } A_j, \\ 0 & \text{in } A_k \qquad (k \neq i,j), \end{cases}$$

satisfies $\int_\beta *ds = 0$. The Main Existence Theorem applied to L_0 yields an HD-function h on R with nonzero flux across β_i and β_j. This contradicts the hypothesis $R \in C'$.

It follows that the flux of u is concentrated at a single ideal boundary point α. Therefore, if $ds = |du + i*du|$ is used as a linear density for estimating the extremal length of $\mathfrak{F}^*(\partial\Omega_0, \alpha, \emptyset, \beta - \alpha)$ it follows that this length is >0. Hence $c_\alpha > 0$ and $R \notin C$. This completes the proof of the lemma.

1G. If $R \notin C'$ then $u = u(\alpha^0, \alpha^1, \emptyset, \beta - \alpha^0 - \alpha^1)$ is a nonconstant HD-function on R. Thus $O_{HD} \subset C'$ and from the above lemma we immediately obtain:

Corollary. *For arbitrary Riemann surfaces* $C \cap O_{HD} = O_G$.

For planar surfaces $O_{HD} \subset C$. Indeed, if $c_\alpha > 0$ for some ideal boundary point α then $\operatorname{Re} P_\alpha$ is a nonconstant HD-function. Therefore in the case of planar surfaces this corollary reduces to $O_{HD} = O_G$. This completes the proof of Theorem 1F.

1H. Properties of the ideal boundary. A class O_{XY} is said to represent a property of the ideal boundary if whenever two Riemann surfaces have conformally equivalent bordered boundary neighborhoods, either both surfaces belong to O_{XY} or neither one does. The normal operator method may be used to show that the classes O_G, O_{HP}, O_{HB}, O_{HD}, O_{KD}, C represent properties of the ideal boundary. We give the typical proof for O_{HP}.

Let R, S be Riemann surfaces. Take bordered boundary neighborhoods A, B in R, S and consider a conformal homeomorphism f of A onto B. We assume f is defined on ∂A and maps it onto ∂B.

Suppose $S \notin O_{HP}$ and let h be a nonconstant HP-function on S. Then $s = h \circ f$ is a nonconstant HP-function on A. It qualifies as a singularity function for the Main Existence Theorem since $\int_{\partial A} *ds = \int_{\partial B} *dh = 0$. We obtain a harmonic function p on R

with $p = s+L_1(p-s)$ in A. Therefore p is bounded from below on A and hence also on R. If p were constant then we could have $s = L_1 s$. This means s takes its maximum on ∂A. Hence h takes its maximum on ∂B, contradicting the fact that h is nonconstant. It follows that for a suitable constant M the function $p+M$ is positive and nonconstant. Thus $R \notin O_{HP}$. By the symmetry of R and S we conclude that O_{HP} represents an ideal boundary property.

11. An example. We present an example which shows that the classes O_{AB}, O_{AD} do not represent properties of the ideal boundary (Ahlfors [3], Mori [1], Myrberg [1], Royden [5]).

Consider the domain D obtained by removing the points 0, 2 from $\{|z|<3\}$. Let D_1, D_2 be copies of D with two identical sequences of slits, one tending to 0 and the other to 2. Let R be the Riemann surface formed by attaching D_1 and D_2 along opposite edges of corresponding slits. There is a natural mapping $\pi: R \to D$. The branch points are at the ends of the slits, and we shall assume that they do not project by π into $\{|z| = 1\}$. Obviously $R \notin O_{AB}$ since π is analytic.

We construct a surface $S \in O_{AB}$ with the same ideal boundary as R. Cut R along one of the curves which projects on $\{|z| = 1\}$; denote it by c. For each θ, identify the point on the outside of c which projects to $e^{i\theta}$ with the point on the inside of c which projects to $e^{i(\theta+\alpha)}$, where α is a fixed irrational multiple of 2π. Denote the resulting surface by S. The function π remains analytic on $S-c$.

Suppose f is any AB-function on S. For each $z \in D$ with $|z| \neq 1$ let $\pi^{-1}(z) = \{z_1, z_2\}$. We shall show that $f(z_1) = f(z_2)$. It will then follow that f takes the same value on a dense subset of c and hence must be a constant. Since R and S obviously have conformally equivalent bordered boundary neighborhoods our example will be complete.

For $z \in D$ and $|z| \neq 1$ the function $g(z) = (f(z_1)-f(z_2))^2$ belongs to the class AB in a deleted neighborhood of 0 and 2, and therefore can be continued to an analytic function on all of $\{|z|<1\}$ and $\{1<|z|<3\}$. Since $g(z) = 0$ if z is the projection of a branch point and since there are sequences of such points tending to 0 and 2 we conclude that $g \equiv 0$. Thus $f(z_1) = f(z_2)$ as asserted.

2. Extremal distances

2A. Let α^0,α^1 be disjoint compact subsets of an arbitrary Riemann surface R. Let $\bar{\mathcal{G}}$ be the family of arcs in $\bar{R}$, the compactification of R, with initial point in α^0 and terminal point in α^1. Let $\mathcal{G}$ be the subfamily of $\bar{\mathcal{G}}$ consisting of arcs in R. We define two notions of *extremal distance between* α^0, α^1 by

$$\lambda(\alpha^0,\alpha^1) = \lambda(\mathcal{G}), \quad \bar{\lambda}(\alpha^0,\alpha^1) = \lambda(\bar{\mathcal{G}}),$$

and then have the following (Rodin [4]):

Theorem. *A necessary and sufficient condition that* $\lambda(\alpha^0,\alpha^1) = \bar{\lambda}(\alpha^0,\alpha^1)$ *for all pairs of compact subsets* α_0, α_1 *of* R *is that* $R \in O_{KD}$.

To begin the proof, suppose $R \in O_{KD}$ and let α^0, α^1 be compact subsets of R which may be assumed to be disjoint. Let R_1 be a component of $R-(\alpha^0 \cup \alpha^1)$. Then the ideal boundary of R_1 may be partitioned into $(\alpha_1{}^0,\alpha_1{}^1,\beta_1)$, the parts determined by α^0, α^1, and the ideal boundary β of R. For our purposes we may assume that $\alpha_1{}^0$, $\alpha_1{}^1$ are nonempty and then prove that $\lambda(\alpha_1{}^0,\alpha_1{}^1) = \bar{\lambda}(\alpha_1{}^0,\alpha_1{}^1)$ on R_1. Let A be a bordered boundary neighborhood of β_1 in R_1 such that the closure of A in $\bar{R}_1$ does not contain points of $\alpha_1{}^0$, $\alpha_1{}^1$. Since $R \in O_{KD}$, it is easily seen from Theorem 1B that the operators L_0, $(Q)L_1$ defined for A must coincide.

On R_1 let $u_0 = u(\alpha_1{}^0,\alpha_1{}^1,\beta_1,\varnothing)$, $u_1 = u(\alpha_1{}^0,\alpha_1{}^1,\varnothing,\beta_1)$. It follows from the definition of these functions, and the fact that $\alpha_1{}^0$, $\alpha_1{}^1$ are separated from β_1, that they have L_0- and $(Q)L_1$-behavior respectively near β_1. Hence $u_0 = u_1$. By 1.1D the family of open arcs on $R_1 \cup \beta_1$ which tend to $\alpha_1{}^0$ in one direction and to $\alpha_1{}^1$ in the other direction has the same extremal length as the subfamily of arcs on R_1 with the same property.

It now follows that $\lambda(\alpha_1{}^0,\alpha_1{}^1) = \bar{\lambda}(\alpha_1{}^0,\alpha_1{}^1)$. Indeed, the families $\mathcal{G}$, $\bar{\mathcal{G}}$ differ from the families considered above only in that the latter contain arcs which might oscillate near $\alpha_1{}^0$, $\alpha_1{}^1$. But a family of oscillating curves has infinite extremal length and thus may be ignored. This proves the sufficiency of the condition in the theorem.

2B. To prove the necessity assume that extremal distances λ, $\bar{\lambda}$ on R are always equal. Let ζ_1, ζ_2 be arbitrary points on R and let $p_0(\zeta_1,\zeta_2)$, $p_1(\zeta_1,\zeta_2)$ be the L_0- and $(Q)L_1$-principal functions with negative and positive logarithmic poles at ζ_1 and ζ_2 respectively. We shall show that these functions are essentially equal and hence the reproducing differential $\psi = d(p_0(\zeta_1,\zeta_2) - p_1(\zeta_1,\zeta_2))$ vanishes. Since

$$(dh,\psi) = 2\pi(h(\zeta_2) - h(\zeta_1))$$

for all $dh \in \Gamma_{he} \cap \Gamma^*_{hse}$ (see (5)) it will follow that $R \in O_{KD}$.

We normalize the principal functions p so that for a fixed parameter w near ζ_1, $p(w) - \log|w - \zeta_1| \to 0$ as $w \to \zeta_1$. Let G be a subregion of R such that $\zeta_1 \in G$ and $R -$ Cl G is a topological disk about ζ_2 with analytic boundary ∂G. Denote by p_{0G} the harmonic function on $G - \{\zeta_1\}$ with a normalized negative logarithmic pole at ζ_1, L_0-behavior near the boundary of R, and with constant value, say $l(G)$, on ∂G. The function p_{1G} is defined similarly except that it has $(Q)L_1$-behavior near the ideal boundary β of R.

We have

$$p_{0G} = p(\zeta_1,\partial G,\beta,\varnothing) + a_0,$$

$$p_{1G} = p(\zeta_1,\partial G,\varnothing,\beta) + a_1,$$

where a_0, a_1 are constants.

Consider disks $\{|w - \zeta_1| < r\} = W_r$ about ζ_1 and the corresponding families $\mathfrak{F}(\partial W_r,\partial G,\beta,\varnothing)$ and $\mathfrak{F}(\partial W_r,\partial G,\varnothing,\beta)$. The extremal lengths of these families are equal, since they are the same as $\lambda(\partial W_r,\partial G)$ and $\bar{\lambda}(\partial W_r,\partial G)$, which coincide by hypothesis. From Theorem III.2.3C we see that the constants k_w associated with $p(\zeta_1,\partial G,\beta,\varnothing)$ and $p(\zeta_1,\partial G,\varnothing,\beta)$ are equal. It follows that these functions are identical and hence p_{0G}, p_{1G} differ by a constant. By our normalization this implies $p_{0G} = p_{1G}$.

It remains only to infer from $p_{0G} = p_{1G}$ that $p_0(\zeta_1,\zeta_2) = p_1(\zeta_1,\zeta_2)$.

2C. In the sequel let p_G, p denote either of the pairs p_{0G}, $p_0(\zeta_1,\zeta_2)$ or p_{1G}, $p_1(\zeta_1,\zeta_2)$. By Theorem I.3.2B integrals of the following form must vanish $(G' \supset G)$:

$$\int_\beta p_G * dp_G, \qquad \int_\beta p_G * dp_{G'}, \quad \text{etc.}$$

We make tacit use of these facts in the remaining computation.

Recall that $l(G)$ denotes the constant value of p_G on ∂G. If $G \subset G'$ then $l(G) \leq l(G')$. Indeed, we have

$$(15) \qquad 0 \leq D_G(p_{G'} - p_G) = \int_{\partial G} (p_{G'} - p_G) * d(p_{G'} - p_G).$$

On noting that

$$\int_{\partial G} p_G * dp_{G'} - p_{G'} * dp_G = 0,$$

$$\int_{\partial G} p_G * dp_G = \int_{\partial G} p_G * dp_{G'} = 2\pi l(G),$$

$$\int_{\partial G} p_{G'} * dp_{G'} \leq \int_{\partial G'} p_{G'} * dp_{G'} = 2\pi l(G'),$$

one computes from (15) that $l(G) \leq l(G')$.

Let z be a local parameter for a neighborhood of ζ_2 such that ζ_2 corresponds to $z = 0$. For sufficiently small r set $G_r = R - \{|z| \leq r\}$.

The function $p - p_{G_r}$ is harmonic on Cl G_r. Because of its normal operator behavior near β it is bounded by its maximum value on ∂G_r. In order to estimate this maximum we introduce

$$m_r = \min_{|z|=r} p(z), \qquad M_r = \max_{|z|=r} p(z),$$

$$G' = \{\zeta \in R | -\infty \leq p(\zeta) < m_r\}, \qquad G'' = \{\zeta \in R | -\infty \leq p(\zeta) < M_r\}.$$

Since $G' \subset G_r \subset G''$ we have $l(G') \leq l(G_r) \leq l(G'')$. It is obvious that $p_{G'} = p|G'$ and $p_{G''} = p|G''$. Hence $l(G') = m_r$ and $l(G'') = M_r$. Therefore $m_r \leq l(G_r) \leq M_r$ and

$$|p - p_{G_r}| \leq M_r - m_r.$$

From $p(z) = -\log|z| + h(z)$ with h regular at ζ_2 we have

$\lim_{r\to 0}(M_r - m_r) = 0$. Consequently

$$\lim_{r\to 0} p_{0G_r} = p_0(\zeta_1,\zeta_2), \qquad \lim_{r\to 0} p_{1G_r} = p_1(\zeta_1,\zeta_2).$$

In 1G we showed that $p_{0G_r} = p_{1G_r}$, and thus have the desired conclusion $p_0(\zeta_1,\zeta_2) = p_1(\zeta_1,\zeta_2)$.

It is interesting to compare Theorem 2A with Remark 1 in 1.2F which characterizes O_G in terms of extremal length.

2D. Plane regions. If R is a plane region we can replace $\bar{\lambda}(\alpha^0,\alpha^1)$ of the previous theorem by the extremal distance between α^0, α^1 taken on the entire sphere. That is, we shall show that if the removal of a point set E from the extended plane does not change the extremal distance between compact sets α^0,α^1 then the complement R of E is in O_{KD}.

Our assumption means that the function $p(\alpha^0,\alpha^1,E,\emptyset)$ on R is equal to the function $p(\alpha^0,\alpha^1,\emptyset,\emptyset)$ for the extended plane, i.e. for any α^0, α^1 the set E is a removable singularity for $p(\alpha^0,\alpha^1,E,\emptyset)$. It follows immediately that E is a removable singularity for every $p_0(\zeta_0,\zeta_1)$ where this principal function was defined in 2B above. We intend to differentiate $p_0(\zeta_0,\zeta_1)$ as a function of ζ_1. The following symmetry property will be useful in justifying the differentiation.

Let $p_0(z;\zeta_0,\zeta_1)$ denote the value of $p_0(\zeta_0,\zeta_1)$ at z.

Lemma. *On an arbitrary Riemann surface* $p_0(z;\zeta_0,\zeta_1) = p_0(\zeta_1;\zeta_0,z)$.

For the proof let α^0, α^1, δ be the oriented boundaries of disjoint disks about ζ_0, ζ_1, z respectively. We shall choose these disks so that the integrals in Green's formula

$$(16) \qquad \int_c p_0(\zeta_0,\zeta_1) * dp_0(\zeta_0,z) = \int_c p_0(\zeta_0,z) * dp_0(\zeta_0,\zeta_1),$$

$$c = \beta - \alpha^0 - \alpha^1 - \delta,$$

will be easy to compute. We may assume that $p_0(\zeta_0,\zeta_1)$ and $p_0(\zeta_0,z)$ are constant along α^1 and δ respectively, merely by taking α^1, δ to be appropriate level curves. Let α^0 be a circle of radius r

with respect to the fixed local coordinate at ζ_0 which was used to normalize these functions.

Consider the integral on the left side of (16). Along β it is zero as was remarked in 2C. It also vanishes along α^1. Along δ it gives $-2\pi p_0(z;\zeta_0,\zeta_1)$. Making use of the normalization at ζ_0 the integral is seen to give $2\pi \log r + o(1)$ along α^0, where $o(1) \to 0$ as $r \to 0$. The right side of (16) may be evaluated similarly and we find that

$$2\pi p_0(z;\zeta_0,\zeta_1) + o(1) = 2\pi p_0(\zeta_1;\zeta_0,z) + o(1).$$

Letting $r \to 0$ we obtain the assertion of the lemma.

2E. We have seen that $p_0(z;\zeta_0,\zeta_1)$ is a harmonic function of ζ_1 for each fixed z, ζ_0. Since $R \subset \mathbf{C}$ we may differentiate with respect to Re ζ_1 and obtain another function $p_0'(z;\zeta_0,\zeta_1)$. This harmonic function of z is easily seen to be regular at ζ_0 and to have a singularity $\text{Re}(z-\zeta_1)^{-1}$ at ζ_1. If R were the interior of a compact bordered surface, so that the normal derivative of $p_0(\zeta_0,\zeta_1)$ vanished along the border, then the same would be true of $p_0'(z;\zeta_0,\zeta_1)$, i.e. this function would be the L_0-principal function with singularity $\text{Re}(z-\zeta_1)^{-1}$. By considering an exhaustion of R we obtain the same result in general.

We now know that for $\zeta \in R$ the L_0-principal function with singularity $\text{Re}(z-\zeta)^{-1}$ can be continued harmonically over E. Therefore E is a removable singularity for the horizontal slit mapping P_0 with pole at ζ of residue 1. It follows that E is a removable singularity for the horizontal slit mapping P_0^θ with residue $e^{i\theta}$. Using II.2.1A we conclude that $P_0 = P_1$. By (11) this implies that $R \in O_{AD}$ since the span vanishes.

Recall that $O_{AD} = O_{KD}$ for planar surfaces. We have just proved one half of the following theorem (Ahlfors-Beurling [1]), the other half of which is an immediate consequence of Theorem 2A.

Theorem. *A plane point set E is an AD-null set if and only if the removal of E does not change extremal distances.*

CHAPTER V
ANALYTIC MAPPINGS

The main theorems on the distribution of values of meromorphic functions in the plane have counterparts in the more general setting of analytic mappings between Riemann surfaces. This chapter is devoted to the development of such a general value distribution theory. Principal functions, in particular the estimates I.1.2G to the Main Existence Theorem, provide an essential tool for this work.

The classical theory of meromorphic functions in a disk or plane is not needed for the present development. Indeed, the celebrated Main Theorems for these cases will be treated in 3.2 as a special case.

We shall establish the First Main Theorem for mappings into arbitrary Riemann surfaces, and the Second Main Theorem for those into closed Riemann surfaces. A comprehensive treatment of value distribution theory for mappings between arbitrary Riemann surfaces was given in Sario-Noshiro [1].

§1. THE PROXIMITY FUNCTION

We shall study analytic mappings $f:R\to S$ between Riemann surfaces by means of a suitable metric on S. This metric is constructed from the Laplacian of a function $s(\cdot,a)$ on S which is bounded from below and tends to ∞ near a. It is called a proximity function for S and is constructed from certain principal functions. In this section we derive the properties of s (Sario [21], [25], [26]) and its associated metric which will be needed later.

1. Use of principal functions

1A. Construction of $p(\zeta,\zeta_1)$. Let S denote a Riemann surface, open or closed. A function p harmonic in a punctured neighborhood of a point $a\in S$ is said to have a logarithmic pole of order n at a if, in terms of a local parameter z in this neighborhood, $p(z) - n \log|z - z_0|$ has a removable singularity at z_0, the value of z corresponding to a. This notion is independent of the particular choice of parameter. Indeed, if w is another parameter which makes a correspond to w_0 then $w = \varphi(z)$ for a biholomorphic function φ. Hence $w - w_0 = (z - z_0)\tilde{\varphi}(z)$ where $\tilde{\varphi}(z)$ does not vanish at z_0. Thus $n \log|w - w_0|$ and $n \log|z - z_0|$ differ by a function $\log|\tilde{\varphi}(z)|$ which is harmonic at z_0. Since $p(z)$ tends to $+\infty$ or $-\infty$ according as n is negative or positive, we refer to these singularities as positive or negative logarithmic poles of order $|n|$ respectively.

Let ζ_0 and ζ_1 be distinct points of S. The Main Existence Theorem can be used to construct a function p with a positive logarithmic pole of order 2 at ζ_0, a negative logarithmic pole of order 2 at ζ_1, and with the property $p = L_0 p$ in a bordered boundary neighborhood of S which does not contain ζ_0 or ζ_1. If S is closed, i.e. has empty ideal boundary, we consider the condition concerning L_0 to be vacuously satisfied.

If $p = L_0 p$ in some boundary neighborhood, then it satisfies a similar equation in any smaller boundary neighborhood, as may be seen from Theorem I.3.2A. Therefore these conditions determine p up to an additive constant.

We now recall the construction of p in detail, in order to prepare for a closer study of its properties.

1B. Let D_0 and D_1 be disjoint closed parametric disks about ζ_0 and ζ_1. For $i = 0, 1$ let $z^{(i)}$ denote a parameter in D_i which makes ζ_i correspond to 0. Furthermore, in case S is open, let A be a bordered boundary neighborhood of S which is disjoint from D_0 and D_1.

Define a function σ in $D_0 \cup D_1 \cup A$ as follows: $\sigma(z^{(0)}) = -2 \log|z^{(0)}|$ in D_0, $\sigma(z^{(1)}) = 2 \log|z^{(1)}|$ in D_1, $\sigma \equiv 0$ in A. Set $W = S - \{\zeta_0, \zeta_1\}$ and $W' = W \cap (D_0 \cup D_1 \cup A)$. Let L_0 be the normal operator for A, and for $i = 0, 1$ let K_i be the Dirichlet operator for $D_i - \{\zeta_i\}$ (see I.1.2D). Set $L = L_0 \oplus K_0 \oplus K_1$, a normal operator for W'. Since $\int * d\sigma = 0$ along the ideal boundary of W the Main Existence Theorem I.1.2B yields a function p harmonic on W and satisfying $p - \sigma = L(p - \sigma)$ in W'. Then $p = L_0 p$ in A and p has the required logarithmic poles at ζ_0 and ζ_1.

1C. The function p constructed above is determined only up to an additive constant. To achieve uniqueness it will be convenient to adopt the following normalization:

In terms of the fixed local parameter $z^{(0)}$ at ζ_0 the function $p(z^{(0)})+2\log|z^{(0)}|$ has a removable singularity at the origin, $z^{(0)}=0$, which corresponds to ζ_0. Thus p can be adjusted by adding a suitable constant so that

$$\lim_{z^{(0)}\to 0} p(z^{(0)})+2\log|z^{(0)}| = 0. \tag{1}$$

The resulting normalized function of ζ is denoted by $p(\zeta,\zeta_1)$.

To summarize, $p(\zeta,\zeta_1)$ is the harmonic function of ζ on S with a positive logarithmic pole of order 2 at ζ_0, a negative logarithmic pole of order 2 at ζ_1, L_0-behavior near the ideal boundary of S, and with property (1) for the fixed parameter $z^{(0)}$ near ζ_0.

1D. The proximity function $s(\zeta,a)$. For a point a on S, distinct from ζ_0 and ζ_1, we define the proximity function for a as

$$s(\zeta,a) = \log[(1+e^{p(\zeta,\zeta_1)})(1+e^{p(a,\zeta_1)})]-p(\zeta,a). \tag{2}$$

As a function of ζ, $s(\zeta,a)$ has the following properties. It can be extended continuously to ζ_0,ζ_1 and thereby becomes infinitely differentiable for all $\zeta\neq a$. Near a it satisfies, in terms of a local parameter z, $s(z,a) = -2\log|z-a|+h(z)$ where $h(z)$ is continuous at a. Furthermore, it is symmetric and bounded from below by a constant independent of a (Sario [21], [26]):

Theorem. *For points a and b of S distinct from ζ_0 and ζ_1,*

$$s(a,b) = s(b,a).$$

1E. Proof of symmetry. As we see from (2), the theorem is equivalent to the statement $p(a,b) = p(b,a)$. This was essentially Lemma IV.2.2D, but for convenience we shall repeat the proof here.

To begin, we select disjoint Jordan curves α_0, α_1, α_2 about ζ_0, a, b respectively. We may assume $p(\zeta,a)$ is constant for $\zeta\in\alpha_1$.

Indeed, α_1 may be chosen as a circle for the parameter

$$\exp[\tfrac{1}{2}(p(\zeta,a)+ip(\zeta,a)^*)]$$

where $p(\zeta,a)^*$ is a local harmonic conjugate for the function $p(\zeta,a)$ on S. Similarly assume $p(\zeta,b)$ is constant for $\zeta\in\alpha_2$. Let α_0 be the circle $|z^{(0)}| = r$ where $z^{(0)}$ is the distinguished local parameter at ζ_0. Orient α_0, α_1, α_2 so that they bound three disjoint disks.

Let $\{\Omega\}$ be an exhaustion of S. For any surface S write $p(\zeta,a) = p_S(\zeta,a)$ to emphasize that this function was defined for S. Applying Green's formula to the functions $p_\Omega(\zeta,a)$ we obtain

$$(3)\quad \int_{\partial\Omega-\alpha_0-\alpha_1-\alpha_2} p_\Omega(\zeta,a) *dp_\Omega(\zeta,b)$$

$$= \int_{\partial\Omega-\alpha_0-\alpha_1-\alpha_2} p_\Omega(\zeta,b) *dp_\Omega(\zeta,a).$$

Consider the integral on the left side of (3). It is easily seen to vanish over $\partial\Omega$ and α_1. Over α_2 it yields $4\pi p_\Omega(b,a)$ and over α_0 it gives $8\pi\log r+o(1)$ where $o(1)\to0$ as $r\to0$. The right side of (3) can be evaluated similarly, and we obtain

$$(4)\qquad p_\Omega(b,a) = p_\Omega(a,b)$$

on letting $r\to0$. By Theorem I.3.2A, $p_\Omega(b,a)\to p_S(b,a)$ and $p_\Omega(a,b)\to p_S(a,b)$ as $\Omega\to S$. Thus (4) implies the symmetry of $p(a,b)$ as desired.

1F. Boundedness from below. We shall now prove the following result, due to Sario [26]:

Theorem. *There is a constant M such that*

$$s(\zeta,a)>M$$

for all $\zeta\in S$ and all $a\in S-\{\zeta_0,\zeta_1\}$.

From (2) we obtain two lower estimates for the proximity

function:

(5) $$s(\zeta,a) \geq -p(\zeta,a),$$

(6) $$s(\zeta,a) \geq p(\zeta,\zeta_1) - p(\zeta,a).$$

Thus the proof of the theorem rests on finding bounds for the principal function p.

1G. Bounds for $p(\zeta,a)$. Recall that D_0 is a fixed parametric disk about ζ_0 which does not contain ζ_1, and $z^{(0)}$ is a distinguished local parameter defined there. We now introduce some smaller open disks $G_3 \subset G_2 \subset G_1 \subset G_0$ corresponding respectively to $\{|z^{(0)}| < r_i\}$ with $0 < r_3 < r_2 < r_1 < r_0$ and such that $D_0 = \mathrm{Cl}\, G_0$. The estimates on p needed for the theorem are given by the

Lemma. *On ∂G_1 the function $|p(\cdot,a)|$ is bounded uniformly for all $a \in \mathrm{Cl}\, G_2$ and on ∂G_3 this function is bounded uniformly for all $a \notin G_2$. Furthermore, $|p(\cdot,\zeta_1)|$ is bounded on $\partial G_1 \cup \partial G_3$.*

We assume the lemma temporarily and turn to the proof of Theorem 1F. It will suffice to give a lower bound for $s(\zeta,a)$ in each of the two cases $\zeta \in G_0$, $\zeta \notin G_0$.

If $\zeta \notin G_0$ then according to (5) we need only show that $p(\zeta,a)$ is bounded above uniformly in a. If $a \in \mathrm{Cl}\, G_2$ then $p(\cdot,a)$ is uniformly bounded on ∂G_1 by the lemma. Since it is harmonic on $S - G_1$ and has L_0-behavior there, this bound is valid on all of $S - G_1$. If $a \notin G_2$ the lemma shows that $p(\cdot,a)$ is uniformly bounded on ∂G_3. It is harmonic on $S - G_3$ except for a negative logarithmic pole at a, is bounded on ∂G_3, and has L_0-behavior near the ideal boundary. Therefore the bound on ∂G_3 is a valid upper bound on all of $S - G_3$.

If $\zeta \in G_0$ we use (6). Estimates for this case need bear no relationship to the preceding ones. Therefore by expanding the system of disks we may assume ζ is actually in the smallest one G_3. According to the lemma there are uniform bounds for $p(\cdot,\zeta_1) - p(\cdot,a)$ which are valid on either ∂G_1 or ∂G_3, depending on the location of a. The singularity at ζ_0 is removable and therefore this difference is harmonic in G_1 except for a possible positive logarithmic pole at a. Thus it is bounded from below in G_3 independently of a.

1H. Proof of lemma. For the proof of Lemma 1G essential use will be made of Theorem I.1.2G.

Consider the case $a \in \mathrm{Cl}\, G_2$. To apply the estimates of the Main Existence Theorem we make explicit a construction of $p(\zeta,a)$. Define a singularity function σ as $\log 2|(z^{(0)}-a)/z^{(0)}|$ in D_0, and $\equiv 0$ in the bordered boundary neighborhood A. Let Ω' be a regular subregion of S which contains $\mathrm{Cl}(S-A)$ and set $\Omega = \Omega' - \mathrm{Cl}\, G_1$. By Theorem I.1.2G there is a principal function $q(\cdot,a)$, differing from $p(\cdot,a)$ by a constant, which satisfies

$$||q(\cdot,a)||_{\partial\Omega} \leq k||\sigma - K_0\sigma||_{\partial G_1} \leq k(||\sigma||_{\partial G_1} + ||\sigma||_{\partial D_0})$$

where k depends only on the geometry of S and Ω. Clearly $||\sigma||_{\partial G_1} + ||\sigma||_{\partial D_0}$ can be bounded independently of $a \in \mathrm{Cl}\, G_2$. Thus $q(\cdot,a)$ is uniformly bounded on ∂G_1 for $a \in \mathrm{Cl}\, G_2$.

Now $p(\cdot,a) - q(\cdot,a)$ is a constant $l(a)$ which can be shown to be bounded uniformly for $a \in \mathrm{Cl}\, G_2$. In G_0 we have

$$|l(a)| \leq |p(z^{(0)},a) - \sigma(z^{(0)})| + |q(z^{(0)},a) - \sigma(z^{(0)})|.$$

The normalization of $p(\cdot,a)$ at ζ_0 makes the first term on the right tend to zero as $z^{(0)} \to 0$. The second term is uniformly bounded on ∂G_1 if $a \in \mathrm{Cl}\, G_2$ as we remarked above. This proves the first assertion of the lemma.

Suppose now that $a \notin G_2$. In order to prove that $||p(\cdot,a)||_{\partial G_3}$ is uniformly bounded it suffices to prove, according to the symmetry of p established in 1E, that $||p(\cdot,a)||_{S-G_2}$ is bounded for $a \in \partial G_3$. By the normal operator behavior of p near the ideal boundary of S, $||p(\cdot,a)||_{S-G_2}$ is bounded by $||p(\cdot,a)||_{\partial G_2}$ for $a \in \partial G_3$.

Thus in essence we must prove that if a is constrained to lie in a disk about ζ_0, then $p(\cdot,a)$ is uniformly bounded on the circumference of a slightly larger disk. But this is precisely what the first assertion of the lemma shows.

The last statement of the lemma is in substance a special case of the second statement. Thus the proof is complete.

2. The conformal metric

2A. The proximity function $s = s(\cdot,a)$ will be used to form a metric on S. We begin by specifying as area element the second-order differential $d\omega = d*ds$. In terms of local coordinates

$d\omega = \Delta s\, dx \wedge dy$. The Laplacian Δs can be computed from (2) and is

$$\Delta s = \frac{e^p}{(1+e^p)^2} |\text{grad } p|^2 \tag{7}$$

where p stands for $p(\cdot, \zeta_1)$. An invariant form of (7) is

$$d\omega = \frac{e^p}{(1+e^p)^2} dp \wedge *dp.$$

An important property of $d\omega = d*ds$ is its independence of the point a used to define s.

Examining the local behavior of p near ζ_0, ζ_1 we find from (7) that Δs is singularity free and different from zero at these points. Since it is nonnegative we may form the conformal metric $\lambda(z)|dz|$ with $\lambda = \sqrt{\Delta s}$. This metric has zeros exactly at the zeros of grad p. At any other point the function $p+ip^*$ serves as a local parameter. In terms of this parameter we have the following useful form for the conformal metric:

$$\lambda = \frac{e^{p/2}}{1+e^p}. \tag{8}$$

2B. Area of S. Let c_t denote the level curve $p^{-1}(t)$, oriented so that $*dp$ increases along the positive direction of c_t. The area of S with respect to the area element $d\omega$ is

$$\iint_S d\omega = \int_{-\infty}^{\infty} \int_{c_t} \frac{e^t}{(1+e^t)^2} *dpdt. \tag{9}$$

We shall show that $\int_{c_t} *dp = 4\pi$ for almost all t. Then (9) will yield the following result (Sario [29]):

Theorem. *The surface S has area 4π in the metric $\lambda(z)|dz|$.*

The curve c_t is weakly homologous to a compact cycle surrounding ζ_1. Indeed, for large negative values t' the level curve $c_{t'}$ is such a cycle and $c_t - c_{t'}$ bounds the open set $\{\zeta | t' < p(\zeta) < t\}$.

By Theorem II.6.2G, $\int_{c_t - c_{t'}} *dp = 0$ except for a set of values T such that the family $\{c_t - c_{t'}\}_{t \in T}$ has infinite extremal length. Equivalently, $\int_{c_t} *dp = 4\pi$ except for t in a set T with $\lambda(\{c_t\}_{t \in T}) = \infty$. By Theorem II.6.2D, T has measure zero. This proves the theorem.

2C. Curvature. The Gaussian curvature of a positive differentiable metric $\rho(z)|dz|$ is defined as $\kappa = -(\Delta \log \rho)/\rho^2$ where ρ and the Laplacian Δ are both computed in terms of the same local parameter. The curvature κ is then a function on the surface, independent of the choice of local parameter.

For the metric $\lambda(z)|dz|$ the curvature κ is thus defined everywhere on S except at the zeros of grad p. Using (8) we find that $\kappa \equiv 1$ whenever it is defined.

§2. ANALYTIC MAPPINGS

We begin by proving the First Main Theorem for analytic mappings between arbitrary Riemann surfaces. We then give the Second Main Theorem for the case in which the image surface is closed. As corollaries we obtain defect and ramification relations and a generalization of Picard's celebrated theorem.

1. First Main Theorem

1A. Notation. Let $f: R \to S$ be an analytic mapping of Riemann surfaces. On R choose a parametric disk R_0 with border β_0. A subregion Ω of $R - \mathrm{Cl}\, R_0$ will be called a *regular adjacent subregion* if $\Omega \cup \mathrm{Cl}\, R_0$ is a regular subregion of R. Given such an Ω, form the harmonic function u which vanishes on β_0, is constant on $\beta_\Omega = \partial\Omega - \beta_0$, and satisfies $\int_{\beta_0} *du = 1$. Let k denote the constant value of u on β_Ω. By analogy with the case in which Ω is an annulus, we may refer to $e^{2\pi k}$ as the *modulus* of Ω. For $h \in [0,k]$ let β_h be the level line $u^{-1}(h)$ and let Ω_h be the adjacent region $u^{-1}((0,h))$. Also define $R_h = R_0 \cup \Omega_h$.

On S we have the function $s(\cdot, a)$ and its associated metric and area element. We consider three mapping properties of f: the area of the Riemannian image $f(R_h)$ of R_h, the number of times $f(R_h)$ covers a given point $a \in S$, and the average distance between $f(\beta_h)$ and a.

1B. The fundamental functions. For $a \in S - \{\zeta_0, \zeta_1\}$ let $\nu(h,a)$ denote the number of a-points of f in R_h, i.e., the number of times, counting multiplicities, that $f|R_h$ takes the value a. The *counting function* for a-points is defined as

$$A(h,a) = 4\pi \int_0^h \nu(t,a)\,dt.$$

The *mean proximity function*, which measures how close $f(\beta_h)$ is to a, is defined as

$$B(h,a) = \int_{\beta_h} (s \circ f - M) * du$$

where $M = \inf\{s(\zeta,a) | a,\zeta \in S\}$. Note that M depends only on S, ζ_0, and ζ_1. Furthermore, $B(h,a)$ is nonnegative. The fact that M is finite is the content of Theorem 1.1F.

Finally, the *characteristic function* of f is

$$C(h) = \int_0^h \left[\iint_{f(R_t)} d\omega \right] dt.$$

Thus $C'(h)$ is the area of the Riemannian (multiple sheeted) image of R_h and is independent of a (see 1.2A). We then have (Sario [21], [26]):

First Main Theorem. *For a complex analytic mapping $f: R \to S$ between arbitrary Riemann surfaces and for any point $a \in S$*

$$A(k,a) + B(k,a) = C(k) + D(a) \tag{10}$$

where $D(a)$ is a nonnegative bounded function of a.

1C. To prove (10) take disjoint disks Δ_j about the distinct a-points z_j of f in Ω_h. Set $\alpha_j = \partial\Delta_j$ and apply Green's formula to $v(z) = h - u(z)$ and $s \circ f(z) = s(f(z),a)$:

$$\int_{\beta_h - \beta_0 - \Sigma\alpha_j} v * d(s \circ f) - (s \circ f) * dv = \iint_{\Omega_h - \Sigma\Delta_j} v\, d * d(s \circ f). \tag{11}$$

We intend to let the disks Δ_j shrink to their centers z_j. During this limit process $\int_{\alpha_j}(s\circ f)*dv\to 0$ since v is harmonic in Δ_j. For $p_a=p(\cdot,a)$ we also have $\lim\int_{-\alpha_j}*d(s\circ f)=\lim\int_{\alpha_j}*d(p_a\circ f)$, which is 4π times the multiplicity $n(z_j)$ of the a-point z_j. Thus the integral on the left in (11), when evaluated over $-\Sigma\,\alpha_j$, has the limit $4\pi\Sigma_j n(z_j)v(z_j)$. This quantity can be expressed as the integral $4\pi\int_0^h(h-x)d\nu_a(x)$ where $\nu_a(x)=\nu(x,a)$. After partial integration we have

$$\int_{-\Sigma\alpha_j} v*d(s\circ f)-(s\circ f)*dv\to 4\pi\int_0^h \nu_a(x)dx-4\pi h\nu_a(0)$$

and (11) gives

$$(12)\quad 4\pi\int_0^h \nu_a(x)dx-4\pi h\nu_a(0)+\int_{\beta_h-\beta_0}(s\circ f)*du$$
$$-h\int_{\beta_0}*d(s\circ f)=\iint_{\Omega_h} v\,d*d(s\circ f).$$

Putting disjoint disks Δ_j^0 about the a-points of f in R_0, with $\partial\Delta_j^0=\alpha_j^0$, we have

$$\int_{\beta_0-\Sigma\alpha_j^0}*d(s\circ f)=\iint_{R_0-\Sigma\Delta_j^0} d*d(s\circ f).$$

As before, $\int_{-\Sigma\alpha_j^0}*d(s\circ f)\to 4\pi\nu_a(0)$ and we obtain

$$(13)\qquad \int_{\beta_0}*d(s\circ f)+4\pi\nu_a(0)=\iint_{R_0} d*d(s\circ f).$$

Combining (12) and (13) we get

$$(14)\quad 4\pi\int_0^h \nu_a(x)dx+\int_{\beta_h-\beta_0}(s\circ f)*du$$
$$=h\iint_{R_0} d\omega\circ f+\iint_{\Omega_h} v\,d\omega\circ f$$

where $d\omega\circ f=d*d(s\circ f)$.

Set $D(a) = B(0,a)$. Then $D(a)$ is bounded and nonnegative. Indeed, s is uniformly bounded from below and bounded from above away from its pole. The reader may verify that $D(a)$ is finite and varies continuously as a passes through $f(\beta_0)$. Thus the assertion follows.

Comparing (14) with statement (10) of the First Main Theorem we see that the proof will be complete once it is shown that the second term on the right side of (14) is $\int_0^h [\iint_{\Omega_t} d\omega \circ f] dt$. To do this it will suffice to prove that the h-derivative of that term is $\iint_{\Omega_h} d\omega \circ f$.

1D. The proof of the First Main Theorem now rests on establishing

$$\iint_{\Omega_h} d\omega \circ f = \frac{d}{dh} \iint_{\Omega_h} (h-u)\, d\omega \circ f. \tag{15}$$

For right-handed derivatives, the right side of (15) becomes

$$\lim_{\Delta h \to 0+} \frac{1}{\Delta h} \left[\iint_{\Omega_{h+\Delta h}} (h+\Delta h - u) d\omega \circ f - \iint_{\Omega_h} (h-u) d\omega \circ f \right] \tag{16}$$

$$= \lim_{\Delta h \to 0+} \iint_{\Omega_{h+\Delta h} - \Omega_h} \frac{h+\Delta h - u}{\Delta h} d\omega \circ f + \iint_{\Omega_h} d\omega \circ f.$$

On $\Omega_{h+\Delta h} - \Omega_h$ we have $0 < h + \Delta h - u < \Delta h$, and hence the first member on the right side of (16) vanishes. A similar calculation holds for left-handed derivatives. The proof of (15), and hence also of the First Main Theorem, is now complete.

2. Second Main Theorem

2A. If f omits points $a_i \in S$ then $\sum_{i=1}^q A(h,a_i) = 0$ for all h. We shall obtain an upper bound for the number q of such Picard points by writing, according to the First Main Theorem, $\sum_1^q A(h,a_i) = qC(h) + \sum_1^q D(a_i) - \sum_1^q B(h,a_i)$ and by obtaining an upper bound for $\sum B(h,a_i)$.

2B. For technical reasons it will be convenient to work not with $B(k,a_i)$ directly, but rather with the integral of its integral.

We introduce the following notation: if $\Phi(h)$ is a function of $h \in [0,k]$ set $\Phi_0 = \Phi$ and $\Phi_j(h) = \int_0^h \Phi_{j-1}(t)\,dt$ for $j \geq 1$.

With this convention we derive from (10)

$$A_j(k,a) + B_j(k,a) = C_j(k) + D_j(a) \qquad (j \geq 0). \tag{17}$$

Suppose we are given distinct points $a_1, \cdots, a_q$ on S. Construct $s = s(\cdot, a)$ as in 1.1D by choosing ζ_0, ζ_1 different from $a_1, \cdots, a_q$. Let the zeros of $d\omega = d*ds$ be $a_{q+1}, \cdots, a_{q+l}$. They will be finite in number if S is compact, and henceforth we assume this to be the case.

First we give an interpretation of the number l. The zeros of $d*ds$ are those of dp, as may be seen from (7). Thus l is the number of zeros of grad p, or, equivalently, of the meromorphic differential $dp + i*dp$. This differential has a simple pole at ζ_0 and at ζ_1. As a consequence of the Riemann-Roch theorem we found in II.5.1E that every meromorphic differential on S has degree $2g-2$ where g is the genus of S. Hence the number of zeros l of dp is precisely $2g$.

We may obtain further notational simplification by setting

$$\Phi(h) = \sum_{i=1}^{q+l} \Phi(h, a_i)$$

for any function of the form $\Phi(h,a)$. We also write $O(\varphi(h))$ for any function of the form $\varphi(h)\psi(h)$ with $\psi(h)$ bounded. Noting that $D_2(a) = O(h^2)$ we have from (17)

$$A_2(h) + B_2(h) = (q+2g)C_2(h) + O(h^2). \tag{18}$$

Our goal is to find an upper bound for $B_2(h)$.

2C. At points of Ω_h where grad $u \neq 0$ we can define a function μ by

$$\mu^2 du \wedge *du = d*d(s \circ f). \tag{19}$$

On S we also introduce a new area element $dm = \sigma\, d\omega$ for which the density σ becomes infinite at the points a_i. Specifically, take

$$\sigma(\zeta) = \exp[s(\zeta) - 2\log(s(\zeta) - (q+l)M)]$$

where by convention $s(\zeta) = \sum_1^{q+l} s(\zeta, a_i)$.

We shall need the fact that the dm-area of S is finite. As a consequence of Theorem 1.2B, we have only to check that $\int\int dm$ is finite over a neighborhood of each a_i. This is a straightforward computation with local coordinates, and we leave it to the reader.

From the definition of $B(h, a_i)$ we have

$$B(h) = \int_{\beta_h} \log \sigma \circ f * du + 2 \int_{\beta_h} \log(s \circ f + O(1)) * du + O(1).$$

Making use of the estimates $B(h) \leq (q+l)C(h) + O(1)$, an obvious consequence of (10), and of the convexity of the logarithm

$$\int_0^t \log \Psi \, dh \leq t \log \left\{ \frac{1}{t} \int_0^t \Psi \, dh \right\}$$

one obtains

$$B(h) < \int_{\beta_h} \log \sigma \circ f * du + 2 \log C(h) + O(1). \tag{20}$$

Recall the definition (19) of μ. We shall obtain a bound for the integral in (20) by expressing it as $F + G$, where

$$F(h) = \int_{\beta_h} \log(\mu^2 \cdot \sigma \circ f) * du,$$

$$G(h) = -2 \int_{\beta_h} \log \mu * du,$$

and then estimating F and G separately. Thus far we have

$$B_2(h) < F_2(h) + G_2(h) + 2[\log C(h)]_2 + O(h^2). \tag{21}$$

2D. Estimate of $F_2(h)$. Set $H(h) = \int_{\beta_h} \mu^2 \cdot \sigma \circ f * du$. Then $F(h) \leq \log H(h)$. From this one obtains for sufficiently large Ω, by integration and by convexity of the logarithm,

$$F_1(h) \le h \log \left\{ \frac{1}{h} H_1(h) \right\} = h \log H_1(h) - h \log h,$$

$$F_2(h) < h^2 \log H_2(h) + O(h^2 \log h).$$

Here and elsewhere we assume $h > 1$, a condition which can be assured for sufficiently large Ω by taking R_0 small enough. Recall that $dm \circ f = \mu^2 \cdot \sigma \circ f \, du \wedge *du$. Thus

$$H_1(h) = \int_0^h \int_{\beta_t} dm \circ f = \iint_{f(\Omega_h)} dm$$

$$= \iint_{a \in S} [\nu(h,a) - \nu(0,a)] dm \le \iint_{a \in S} \nu(h,a) dm.$$

Another integration yields $H_2(h) \le (4\pi)^{-1} \int\int_S A(h,a) dm$. According to the First Main Theorem $A(h,a) \le C(h) + O(1)$. Since $C(h)$ is independent of a and since $\int\int_S dm < \infty$ we conclude that $H_2(h) \le C(h) O(1) + O(1)$. Collecting these results we have

$$F_2(h) < h^2 \log C(h) + O(h^2 \log h). \tag{22}$$

2E. Evaluation of G_2. The function $G(h) = -2\int_{\beta_h} \log \mu * du$ has derivative $G'(h) = -2\int_{\beta_h} *d \log \mu$, as may be seen exactly as in 1D. Let $\{\Gamma_j\}$ be small disks about the singularities of $\log \mu$, and set $\gamma_j = \partial \Gamma_j$. Then

$$G'(h) = 2\int_{-\Sigma\gamma_j - \beta_0} *d \log \mu - 2 \iint_{\Omega_h - \cup \Gamma_j} d*d \log \mu. \tag{23}$$

The last integrand can be written

$$\frac{\Delta \log \mu}{\mu^2} \cdot \mu^2 du \wedge *du,$$

and as such can be easily transferred to S using f, at least where f and $u + iu^*$ are locally 1–1. Indeed, $\mu^2 du \wedge *du$ was defined in

(19) as $d\omega$ pulled back to R. The curvature $\kappa = -(\Delta \log \lambda)/\lambda^2$ on S is independent of the local parameter. Hence locally $-(\Delta \log \mu)/\mu^2 = \kappa \circ f \equiv 1$, and

$$-\iint_{\Omega_h - \cup \Gamma_j} d * d \log \mu \to \iint_{f(\Omega_h)} d\omega$$

as the Γ_j shrink to points. We conclude that the last term in (23) tends to $2C'(h)$.

In terms of the parameters $z = x + iy$ on S and $\zeta = \xi + i\eta$ on Ω_h we have

$$\mu^2 |\operatorname{grad} u|^2 d\xi \wedge d\eta = (\lambda^2 \circ z \circ f)|(z \circ f)'|^2 d\xi \wedge d\eta.$$

Thus the singularities of $\log \mu$ are negative poles due to the zeros $a_{q+1}, \cdots, a_{q+l}$ of λ and the ramification points of f, and positive poles due to the zeros of grad u. Therefore

$$2\int_{-\Sigma\gamma_j} * d \log \mu = 4\pi[-\nu(h,\lambda) - \nu(h,f') + \nu(h, \operatorname{grad} u)]$$

where $\nu(h,\varphi)$ denotes the number of zeros of φ in Ω_h. Although $\nu(h,a)$ refers to the zeros of $f - a$, no confusion should arise from this formal inconsistency.

As in 2B, we can interpret the number of zeros of grad u as $\frac{1}{2}(2\tilde{p} - 2)$ where $\tilde{p}$ is the genus of the double of Ω_h. Indeed, $du + i * du$ extends analytically to the double with symmetrically placed zeros. If Ω_h has connectivity c and genus p then the double $\tilde{\Omega}_h$ has genus $2p + c - 1$. Hence $\nu(h, \operatorname{grad} u) = 2p + c - 2$.

In (23) we may put $\int_{\beta_0} * d \log \mu = O(1)$. We have shown that

$$G'(h) = 4\pi[2p + c - 2 - \nu(h, f') - \sum_{i=q+1}^{l} \nu(h, a_i)] + 2C'(h) + O(1). \tag{24}$$

The Euler characteristic $e(W)$ of a surface W of genus p with c contours is defined as $e(W) = 2p + c - 2$. We set

$$E(h) = 4\pi \int_0^h e(\Omega_t)\, dt$$

and integrate (24) three times. This gives

$$(25)\quad G_2(h) = -\sum_{q+1}^{l} A_2(h,a_i) - A_2(h, f') + E_2(h) + 2C_2(h) + O(h^3)$$

where

$$A(h,f') = 4\pi \int_0^h \nu(t,f')\,dt.$$

2F. The estimates of F_2 and G_2 in (22) and (25) may now be substituted into (21). The term $2[\log C(h)]_2$ of (21) is bounded from above by $2h^2 \log C(h)$. The convexity of $C(h)$, easily established, shows that the last term in (22) is subsumed in $O(h^2 \log C(h))$. Therefore we have proved:

$$B_2(h) < E_2(h) - A_2(h, f')$$

$$- \sum_{q+1}^{l} A_2(h,a_i) + 2C_2(h) + O(h^3 + h^2 \log C(h)).$$

This is the upper bound we sought for use in (18). We conclude (Sario [21]):

Second Main Theorem. *Let S be a closed Riemann surface, $f: R \to S$ an analytic mapping of an arbitrary Riemann surface R, R_0 a parametric disk of R, and Ω a regular adjacent subregion in R of modulus* $\log k$. *Then for any subset $\{a_1, \cdots, a_q\} \subset S$*

$$(26)\quad \sum_{1}^{q} (C_2(k) - A_2(k,a_i)) + e(S)C_2(k) + A_2(k, f')$$

$$< E_2(k) + O(k^3 + k^2 \log C(k)).$$

3. Defects and ramifications

3A. Admissible functions. Interesting consequences can be drawn from the Second Main Theorem for functions f whose characteristic function $C(k)$ grows so rapidly that

$$(27)\qquad \lim_{R_k \to R} \frac{k^3 + k^2 \log C(k)}{C_2(k)} = 0$$

where $R_k = R_0 \cup \Omega$. Such functions shall be called *admissible*. We shall not analyze this property here except to note the following necessary condition for a function to be admissible (Sario-Noshiro [1]):

Theorem. *If $f: R \to S$ is admissible then $f(R-K)$ is dense in S for any compact subset K of R.*

Indeed, if $f|R-K$ omitted a neighborhood of $a \in S$ then $s \circ f$ would be bounded on $R-K$. Hence $B(k,a) = \int_{\beta_k} s \circ f * du$ would be bounded for all $k = k(\Omega)$, $\Omega \subset R$. Also, $A(k,a) = 4\pi \int_0^k \nu(t,a)\,dt$ would be $O(k)$. By the First Main Theorem $C(k)$ would also be of the form $O(k)$. It is easy to see that under these conditions (27) cannot be satisfied.

3B. Defect-ramification relation. For an analytic mapping f of R into S we define the *defect* $\delta(a)$ of f at $a \in S$ as

$$\delta(a) = 1 - \limsup_{R_k \to R} \frac{A_2(k,a)}{C_2(k)}. \tag{28}$$

If a is a Picard point of f then obviously $\delta(a) = 1$. We shall see that such points are scarce if f is admissible.

The *ramification* index of f is

$$\theta = \liminf_{R_k \to R} \frac{A_2(k, f')}{C_2(k)}. \tag{29}$$

It measures the density of branch points of $f(R)$ over S. Finally, we compare the Euler characteristic of a subregion of R with its image area under f to obtain the *Euler index*

$$\eta = \liminf_{R_k \to R} \frac{E_2(k)}{C_2(k)}. \tag{30}$$

Note that $\eta = 0$ if f is admissible and R has finite Euler characteristic.

We use the Second Main Theorem and divide both sides of (26) by $C_2(h)$. This gives the following defect and ramification relation (Sario [21]):

Theorem. *For any admissible mapping of R into a closed Riemann surface S*

$$\sum_{a \in S} \delta(a) + \theta \leq \eta - e(S). \tag{31}$$

3C. Consequences. Let P denote the number of points of S omitted by f. We have remarked that $\delta(a) = 1$ for such Picard points, and hence we deduce from (31): *The number P of Picard points of an admissible map $R \to S$ satisfies*

$$P \leq \eta - e(S). \tag{32}$$

An interesting special case of (32) is found by taking S to be a torus and R to be any surface with finite Euler characteristic. Call such an R a finite Riemann surface. Since $\eta = 0$ and $e(S) = 0$ we find that *any admissible mapping of a finite surface into a torus is surjective.*

Take S to have genus greater than 1. Then $e(S) > 0$, and if $\eta = 0$ then (31) is impossible since the left side is nonnegative. We infer that *there is no admissible mapping of a finite surface into a closed surface of genus* >1.

We also take note of the ramification relation

$$\theta \leq \eta - e(S).$$

§3. MEROMORPHIC FUNCTIONS

We specialize S to be the Riemann sphere. The conformal metric and other quantities on S can then be given a more explicit form. Consequences of specializing R are also examined. In particular, if R is a disk or plane we obtain the two Main Theorems of the Nevanlinna theory.

1. The classical case

1A. Proximity function. We wish to examine the proximity function $s(\zeta, a)$ when S is the Riemann sphere. For simplicity take $\zeta_0 = 0$, $\zeta_1 = \infty$. Let us use ζ as a global parameter on S which may take the value ∞.

The harmonic function $p(\zeta,a)$ is characterized up to an additive constant as being harmonic on $S-\{0,a\}$ and having a positive logarithmic pole at 0 of order 2 and a negative one at a of the same order. Thus

$$(33) \qquad p(\zeta,a) = \begin{cases} 2\log \dfrac{|\zeta-a|}{|a\zeta|} & \text{if} \quad a \neq \infty, \\ \\ 2\log \dfrac{1}{|\zeta|} & \text{if} \quad a = \infty, \end{cases}$$

except possibly for an additive constant. Recall that the additive constant was determined so that $p(\zeta,a)+2\log|\zeta|\to 0$ as $\zeta\to\zeta_0$. Therefore (33) is exact.

The proximity function $s(\zeta,a)$ was defined in (2). Thus we have $s(\zeta,a) = \log[(1+|\zeta|^2)(1+|a|^2)/|\zeta-a|^2]$. When the extended plane S is realized by means of stereographic projection as a sphere of diameter 1, the chordal distance between the points ζ and a is denoted by $[\zeta,a]$. It is readily verified that

$$s(\zeta,a) = 2\log\frac{1}{[\zeta,a]}.$$

The area element $d\omega = d*ds$ is easily calculated:

$$d\omega = \frac{4d\xi\, d\eta}{(1+|\zeta|^2)^2}$$

where $\zeta = \xi+i\eta$.

1B. Thus if S is the Riemann sphere, so that the analytic mappings under consideration are merely meromorphic functions on R, we can give a more concrete form to the A, B, C, D functions. In particular we have the following special case of (32): The number P of Picard points for an admissible meromorphic function on R satisfies

$$P \leq \eta + 2.$$

1C. Specializing *R*. In the classical theory of value distribution the surface R is a disk or plane $\{|z| < e^{2\pi h_0} \leq \infty\}$. Let us first identify the function u for an $\Omega \subset R$.

For simplicity take $R_0 = \{|z| < e^{2\pi h_1} < e^{2\pi h_0}\}$ and $\Omega = \{e^{2\pi h_1} < |z| < e^{2\pi(h_1+h)}\}$. Then $u = u_\Omega = (2\pi)^{-1} \log |z| - h_1$. The A, B, C functions take the following form:

$$A(h,a) = 4\pi \int_0^h \nu(t,a)\,dt,$$

$$B(h,a) = \frac{1}{\pi} \int_{|z|=e^{2\pi(h_1+h)}} \log \frac{1}{[f(z),a]}\,d\theta,$$

$$C'(h) = 4 \iint_{|z|<e^{2\pi(h_1+h)}} \frac{|f'(z)|^2}{(1+|f(z)|^2)^2}\,dxdy.$$

We have $\eta = 0$ and $e(S) = -2$. The corresponding Main Theorems are essentially the classical ones of Nevanlinna except that the extra integrations have resulted in a simpler form.

1D. Classical defect-ramification relations. A simplification of the defect and ramification index can be obtained by exploiting the fact that the A, B, C functions depend only on the variable h and not on the region Ω.

We replace equations (28) and (29) by

$$\tilde{\delta}(a) = 1 - \limsup_{h \to h_0} \frac{A(h,a)}{C(h)},$$

$$\tilde{\theta} = \liminf_{h \to h_0} \frac{A(h,f')}{C(h)}.$$

By l'Hospital's rule $\tilde{\delta} \leq \delta$, $\tilde{\theta} \leq \theta$ and the defect and ramification relation (31) becomes

$$\sum \tilde{\delta}(a) + \tilde{\theta} \leq 2.$$

A corollary of this result is the Picard theorem: an admissible meromorphic function can omit no more than two points.

1E. Admissible functions. To complete the discussion of the classical case we must describe the admissible functions. If R is the plane, $h_0 = \infty$, then we find that a function f is admissible if

$$\lim_{h\to\infty} \frac{h}{C(h)} = 0.$$

This condition holds if the image of f has infinite spherical area.

If R is a disk, $h_0 < \infty$, the condition becomes

$$\limsup_{h\to h_0} C(h)(h_0 - h) = \infty,$$

which implies that the spherical area of the image of the subdisk R_h grows more rapidly than $1/(h_0 - h)$.

2. R_p-surfaces

2A. In the classical case in which R is a disk or the plane there exists the standard exhaustion $\{\Omega\}$ of R by concentric disks. The functions u_Ω are then simply restrictions to Ω of a globally defined function. This fact alone was used above to simplify the defect and ramification relations.

Such a device can be exploited in more general situations if R has an "Evans potential" p_β with the following properties: Except for a single negative logarithmic pole p_β is harmonic on R and tends uniformly to a constant $k \leq \infty$ near β in the sense that the sets $\Omega_h = p_\beta^{-1}([-\infty, h])$ are relatively compact for $h < k$ and $\cup\{\Omega_h | h < k\} = R$. Such a surface is said to be of class R_p.

It is fairly clear that if R is the interior of a compact bordered surface then the capacity function will serve as an Evans potential. However, if such an R is modified by removing one point then the resulting surface is not of class R_p. Indeed, the puncture must be a removable singularity of p_β and therefore some sets Ω_h will fail to be relatively compact on the modified surface. However, if R is parabolic then an Evans potential always exists for any choice of pole (see III.4.4B).

CHAPTER VI

PRINCIPAL FORMS AND FIELDS ON RIEMANNIAN SPACES

Having discussed various aspects of the normal operator method on Riemann surfaces, and consequently on 2-dimensional Riemannian spaces, we now turn to their counterparts on higher dimensional Riemannian spaces. As far as functions, i.e. 0-forms, are concerned, it is plausible that all results thus far discussed, save those involving complex analyticity, can be reproduced for higher dimensions. That this is actually the case is shown in §1.

A new aspect is brought in for higher dimensions by the problem of constructing harmonic p-forms with given behavior at the ideal boundary. The solutions will be called principal forms with respect to given singularities. We shall show in §2 that, on some parallel locally flat spaces, the analogues of L_0- and L_1-operators, as well as the corresponding principal forms, can be constructed in a natural manner.

On arbitrary Riemannian spaces the solution of our original problem remains open at this writing. However, by abandoning the maximum principle in the definition of the normal operator we are led to a related problem of equal significance. In §3 we give a complete solution to this problem both for principal forms and principal fields. The latter are the counterparts of principal harmonic differentials on Riemann surfaces (Ahlfors-Sario [1, Ch. V]).

The three sections of the chapter can be read rather independently of one another.

§1. PRINCIPAL FUNCTIONS ON RIEMANNIAN SPACES

After a brief exposition of the fundamentals of Riemannian spaces, the Main Existence Theorem of principal functions is established. Applications are then made to the theory of the span and to classification theory, the interpolation problem, and some special topics in locally flat spaces.

The main results in this section were obtained in Sario-Schiffer-Glasner [1], Sario-Weill [1], Ow [1], Sario-Fukuda [1], Sario [28], and Nakai-Sario [6].

1. Fundamentals of Riemannian spaces

1A. Riemannian spaces. By a *Riemannian n-space R*, we mean a connected countable oriented C^∞ manifold of dimension n, with a C^∞ metric tensor g_{ij} yielding a positive definite symmetric form $g_{ij}\xi^i\xi^j$. Here and hereafter we follow the Einstein summation convention on repeated indices. We shall use the letter x to denote both the generic point in R and also the local parametric system $x = (x^1, \cdots, x^n)$ at the generic point.

If we can always choose $x = (x^1, \cdots, x^n)$ so that $g_{ij}(x) \equiv \delta_{ij}$ (Kronecker's delta) then R is called *locally flat*, or *locally Euclidean.*

We denote by (g^{ij}) the inverse matrix and by g the determinant of (g_{ij}). The metric tensor g_{ij} determines the *arc element ds* by

$$ds^2 = g_{ij}dx^i dx^j \tag{1}$$

and the *volume element*

$$dV = \sqrt{g}\, dx = \sqrt{g}\, dx^1 \wedge \cdots \wedge dx^n \tag{2}$$

in terms of the local parameter $x = (x^1, \cdots, x^n)$.

An open set G of R will be called *smooth* if the relative boundary ∂G of G has the following property: For every $x \in \partial G$ there exists a parametric ball V at x such that $V \cap \partial G$ is an open subset of a hyperplane in terms of some local parameters. Moreover if G is relative compact in R, then G will be called *regular.* A regular region G is *canonical* if each component of ∂G is the entire border of a component of $V - G$.

By means of the arc element, R is provided with the natural distance function

$$d(x_1, x_2) = \inf \int_\gamma ds \tag{3}$$

where the infimum is taken with respect to the arcs γ which join x_1 and x_2. As a metric space R satisfies the second countability axiom, and also admits an *exhaustion* $\{R_m\}_1^\infty$, i.e. a sequence of regular subregions R_m of R with

$$\bar{R}_m \subset R_{m+1}, \qquad R = \bigcup_1^\infty R_m.$$

In the particular case $n = 2$ the Lichtenstein [1] - Korn [1] theorem shows that there always exists an isothermal parametric system characterized by $g_{11} = g_{22}$ and $g_{12} = g_{21} = 0$ over any entire parametric disk; thus R becomes a Riemann surface. Conversely, for a Riemann surface R there always exists a positive density ρ such that $\rho|dz|$ is invariant. This gives isothermal coordinates with $g_{ij} = \rho^2\delta_{ij}$ for the parameter $z = x^1 + ix^2$, and R becomes a Riemannian space.

1B. Differential forms. We denote by $\mathfrak{A}^p$ $(0 \leq p \leq n)$ the space of *differential p-forms*, or simply p-forms,

$$\varphi = \varphi_{\alpha_1 \cdots \alpha_p} dx^{\alpha_1} \wedge \cdots \wedge dx^{\alpha_p}. \tag{4}$$

Here and in the sequel the summation is with respect to $1 \leq \alpha_1 < \cdots < \alpha_p \leq n$ and the $\varphi_\alpha = \varphi_{\alpha_1 \cdots \alpha_p}$ are covariant tensors of rank p. Here 0-forms are merely functions.

The *Hodge star operator* $*$ is the operator $\varphi \rightarrow *\varphi$ from $\mathfrak{A}^p$ into $\mathfrak{A}^{n-p}$ given by

$$*\varphi = (*\varphi)_{\beta_1 \cdots \beta_{n-p}} dx^{\beta_1} \wedge \cdots \wedge dx^{\beta_{n-p}} \tag{5}$$

with covariant tensors

$$(*\varphi)_{\beta_1 \cdots \beta_{n-p}} = \operatorname{sgn} \begin{pmatrix} 1 \cdots p & p+1 \cdots n \\ \gamma_1 \cdots \gamma_p & \beta_1 \cdots \beta_{n-p} \end{pmatrix} \sqrt{g}\, \varphi_{\alpha_1 \cdots \alpha_p} g^{\gamma_1 \alpha_1} \cdots g^{\gamma_p \alpha_p} \tag{6}$$

where $\alpha = (\alpha_1, \cdots, \alpha_p)$, $\beta = (\beta_1, \cdots, \beta_{n-p})$, and $\gamma = (\gamma_1, \cdots, \gamma_p)$ have the entries in increasing order.

The *exterior derivative* $d\varphi$ of the p-form φ in (4) with differentiable $\varphi_{\alpha_1\cdots\alpha_p}$ is the $(p+1)$-form given by

$$(7)\quad d\varphi = d\varphi_{\alpha_1\cdots\alpha_p}\wedge dx^{\alpha_1}\wedge\cdots\wedge dx^{\alpha_p},\quad d\varphi_{\alpha_1\cdots\alpha_p} = \frac{\partial}{\partial x^i}\varphi_{\alpha_1\cdots\alpha_p}dx^i.$$

The *coderivative* $\delta\varphi$ of the differentiable p-form φ is defined as

$$(8)\qquad \delta\varphi = (-1)^{np+n+1}*d*\varphi.$$

Thus $\delta\varphi$ is a $(p-1)$-form for $p\geq 1$ and $\equiv 0$ for $p=0$.

The Laplace-Beltrami operator Δ is defined as

$$(9)\qquad \Delta = \delta d + d\delta.$$

Hence $\Delta\varphi$ is a p-form along with φ. We call φ a *harmonic p-form* if the $\varphi_{\alpha_1\cdots\alpha_p}$ are of class C^∞ and $\Delta\varphi = 0$.

For a detailed description of these subjects we refer the reader to the monographs of de Rham [2] and Hodge [4].

1C. 0-forms. In this section we are primarily interested in 0-forms, i.e. functions. For a sufficiently smooth function u formula (7) implies $du = (\partial u/\partial x^i)dx^i$, so that

$$(10)\quad *du = \sum_{j=1}^{n}(-1)^{j-1}\sqrt{g}\,g^{ij}\frac{\partial u}{\partial x^i}\,dx^1\wedge\cdots\wedge dx^{j-1}\wedge dx^{j+1}\wedge\cdots\wedge dx^n,$$

and thus

$$(11)\qquad \Delta u = -\frac{1}{\sqrt{g}}\frac{\partial}{\partial x^i}\left(\sqrt{g}\,g^{ij}\frac{\partial u}{\partial x^j}\right).$$

For a regular region G and a smooth $(n-1)$-form φ *Stokes' formula* reads

$$(12)\qquad \int_G d\varphi = \int_{\partial G}\varphi.$$

Given smooth functions u and v on an open set G, their *Dirichlet inner product* $D_G(u,v)$ is

$$D_G(u,v) = \int_G du \wedge *dv = \int_G g^{ij} \frac{\partial u}{\partial x^i} \frac{\partial v}{\partial x^j} dV \tag{13}$$

whenever the integral is meaningful. The Dirichlet integral $D_G(u)$ of u is

$$D_G(u) = D_G(u,u) = \int_G du \wedge *du = \int_G g^{ij} \frac{\partial u}{\partial x^i} \frac{\partial u}{\partial x^j} dV. \tag{14}$$

From (12) it follows for smooth functions u and v on the closure of a regular region G that

$$D_G(u,v) = \int_{\partial G} u *dv + \int_G u \Delta v \, dV. \tag{15}$$

Taking coordinates $x = (x^1, \cdots, x^n)$ such that $x^n \equiv 0$ for $x \in \partial G$, we obtain a Riemannian structure for each component of ∂G. The surface element dS on ∂G is, by definition, the volume element of ∂G considered as an $(n-1)$-dimensional Riemannian space in the above sense. If v is smooth on $\bar{G}$ then the normal derivative $\partial v/\partial n$ of v is given by

$$\frac{\partial v}{\partial n} dS = *dv \tag{16}$$

on ∂G. In particular, if $v|\partial G$ is constant then by (10)

$$\frac{\partial v}{\partial n} = (-1)^{n-1} \sqrt{g}\, g^{nn} \frac{\partial v}{\partial x^n}. \tag{17}$$

We can write (13) as

$$D_G(u,v) = \int_{\partial G} u \frac{\partial v}{\partial n} dS + \int_G u \Delta v \, dV \tag{18}$$

and (14) as

$$(19)\quad D_G(u) = \int_G |\operatorname{grad} u|^2 dV = \int_{\partial G} u \frac{\partial u}{\partial n} dS + \int_G u \Delta u \, dV$$

where we set

$$(20)\qquad |\operatorname{grad} u|^2 = g^{ij} \frac{\partial u}{\partial x^i} \frac{\partial u}{\partial x^j}.$$

1D. Green's functions. A *harmonic function* u is a harmonic 0-form and thus a C^2 function with $\Delta u = 0$. Therefore it is a solution of the second order strongly elliptic partial differential equation

$$(21)\qquad \frac{\partial}{\partial x^i}\left(\sqrt{g}\, g^{ij} \frac{\partial u}{\partial x^j}\right) = 0.$$

We shall denote by $H(E)$, with $E \subset R$, the totality of harmonic functions whose domains contain E. As in the case of Riemann surfaces we continue to denote by $HP(R)$, $HB(R)$, and $HD(R)$ the classes of functions in $H(R)$ which are positive, bounded, and Dirichlet finite, respectively. We also set

$$HBD(R) = HB(R) \cap HD(R).$$

The existence of a fundamental solution for (21) leads to the existence of *Green's function* $q_V(\cdot, a)$, for each parametric ball V. The function $q_V(\cdot, a)$, with pole $a \in V$, is the unique function on $\bar{V} - a$ with the following properties:

(D1) $q_V(\cdot, a) \in HP(V - a) \cap C(\bar{V} - a)$,

(D2) $q_V(\cdot, a) | \partial V = 0$,

(D3) $q_V(x, a) = O(d(x, a)^{2-n}) \qquad (n > 2)$,

(D3)′ $q_V(x, a) = O(-\log d(x, a)) \qquad (n = 2)$,

(D4) $\int_{\partial V} * dq_V(\cdot, a) = -1$.

The normalization in (D4) is immaterial. For Riemann surfaces we took the multiplicative constant to be -2π, but in this chapter we always take it to be unity.

Given $u\in H(V)\cap C(\bar{V})$, we obtain by (12)

$$u(a) = -\int_{\partial V} u*dq_V(\cdot,a). \tag{22}$$

Actually for $u\in C(\bar{V})$ formula (22) is a characterizing property for $u\in H(V)$.

A superharmonic function v on R is a lower semicontinuous function on R such that $-\infty<v\leq\infty$, $v\not\equiv\infty$, and

$$v(a)\geq -\int_{\partial V} v*dq_V(\cdot,a) \tag{23}$$

for every parametric ball V. If $-v$ is superharmonic, then v is called subharmonic.

1E. Harmonic functions. The existence of a local Green's function for (21) yields the following results:

(E1) *Harnack's inequality,* which states that *Harnack's function*

$$k_R(x,y) = \inf\{c|c^{-1}u(x)\leq u(y)\leq cu(x) \quad \text{for all} \quad u\in HP(R)\} \tag{24}$$

is finite and $\log k(x,y)$ is a pseudometric on R.

(E2) *Maximum principle:* $u\in H(R)$ does not take its maximum on R unless it is a constant.

(E3) *Completeness:* If $\{u_m\}\subset H(R)$ is bounded on each compact subset and converges to a function u then $u\in H(R)$.

(E4) *Monotone compactness:* If $\{u_m\}\subset H(R)$ is an increasing sequence bounded at a point of R then it converges to a $u\in H(R)$.

(E5) *Bounded compactness:* $HB(R)$ is sequentially compact.

In view of these properties of harmonic functions Perron's method is applicable for the present equation (21) (cf. 0.2.1C), and thus the *Dirichlet problem* is solvable for regular regions and continuous boundary functions. The mixed boundary value problem is also solvable for regular regions and admissible continuous boundary functions.

Finally, a sequence $\{u_m D\} \subset H(R)$ which converges at a point and converges in Dirichlet norm $D_R(\cdot)^{1/2}$, can be shown to converge uniformly on compacta of R (see 3D below).

For a discussion of the equation (21) we refer the reader to Feller [1] and to the monographs of Miranda [1], Duff [5], and Hörmander [1].

2. The main theorem

2A. Normal operators. Let A be the complement of a regular subregion of a Riemannian space R. Let α denote the border of A, $C(\alpha)$ the space of continuous real-valued functions on α, and $H_1(A)$ the space of real-valued functions which are continuous on A and harmonic on its interior. In a natural way we may consider $C(\alpha)$ and $H_1(A)$ as real vector spaces.

Definition. *A normal operator L for A is a linear transformation of $C(\alpha)$ into $H_1(A)$ such that for all $f \in C(\alpha)$*

(A1) $(Lf)|\alpha = f$,

(A2) $\min_\alpha f \leq Lf \leq \max_\alpha f$,

(A3) $\int_\beta *dLf = 0$.

In condition (A3) β is any cycle in the interior of A homologous to α. By Green's formula, a consequence of Stokes' formula, the integral is well defined; it is called the flux of Lf across the ideal boundary of A.

2B. The main theorem. As the counterpart of Theorem I.1.2B, we have the following (Sario-Schiffer-Glasner [1]; cf. Sario [28], Sario-Weill [1]):

Theorem. *Let L be a normal operator for A and let $s \in H_1(A)$. A necessary and sufficient condition for the existence of a $p \in H(R)$ which satisfies*

$$p - s = L(p - s) \quad in \quad A \tag{25}$$

is $\int_\beta *ds = 0$. Here p is uniquely determined up to an additive constant.

To be precise (25) should be written $p|A - s = L(p|\alpha - s|\alpha)$, but we shall often omit these references to restriction operators when it is safe to do so. The function p is called the L-principal function corresponding to the singularity s; in symbols $p = p[s,L]$.

In view of (E1) to (E5) the proof in I.1.2C to 2F could be applied verbatim to the present case. Although in essence there is no distinction we shall now give a direct proof.

2C. The necessity of $\int_\beta * ds = 0$ is obvious. Conversely, suppose this condition is satisfied. Without loss of generality we assume that $s|\alpha = 0$. Let A_0 be the closure of a regular subregion of R which contains the closure of $R-A$ in its interior. Set $\alpha_0 = \partial A_0$. Let K be the "Dirichlet operator" which associates with each continuous function f on α_0 the solution of the Dirichlet problem in A_0 with boundary values f. It suffices to find $p|\alpha_0$, for then

$$p|A_0 = Kp, \qquad p|A = s+Lp.$$

We set $T = LK$ and obtain $(I-T)p = s$ on α_0, with I the identity operator. The formal solution

$$p = \sum_0^\infty T^m s \tag{26}$$

is the desired $p|\alpha_0$ provided the series converges uniformly.

For the convergence one shows as in I.1.2E that the assumption $\int_{\alpha_0} * ds = 0$ implies sign $T^m s|\alpha \neq$ const. Here we borrow from the theory of partial differential equations the fact that the vanishing of a harmonic function and all its first partial derivatives on a hyperplane implies it vanishes identically (uniqueness of the solution of the Cauchy problem for (21)).

By (E1) there exists a constant $q \in (0,1)$ such that

$$q \min_{\alpha_0} u \leq u|\alpha \leq q \max_{\alpha_0} u \tag{27}$$

for $u \in H_1(A_0)$ with sign $u|\alpha \neq$ const. Therefore

$$q^m M_0 \leq T^m s|\alpha_0 \leq q^m M_1 \tag{28}$$

where

$$M_0 = \min_{\alpha_0} s, \qquad M_1 = \max_{\alpha_0} s. \tag{29}$$

This gives the uniform convergence of (26).

2D. The above reasoning provides us with the following important by-product (Sario-Schiffer-Glasner [1]): Let

$$Q = \frac{1}{1-q}$$

with q in (27) and set

$$||\varphi||_E = \sup_E |\varphi|$$

for functions φ on a set E.

Theorem. *The principal function* $p = p[s,L]$ *with* $s|\alpha = 0$ *constructed in* 2C *satisfies*

$$M_0Q \leq p|A_0 \leq M_1Q, \tag{30}$$

$$M_0Q \leq p - s \leq M_1Q. \tag{31}$$

The mapping $s \to p - s$ *from* $\{s|s \in H_1(A), s|\alpha = 0, \int_\beta *ds = 0\}$ *into* $H_1(A)$ *is linear and bounded:*

$$||p-s||_A \leq Q||s||_{\alpha_0}. \tag{32}$$

In fact (28) gives (30). Hence by the maximum principle for $p|A_0$ and for $p|\alpha = (p-s)|\alpha$, (A2) and (25) give (31). The rest is trivial.

2E. Operators L_0 and L_1. Let $\Omega \subset R$ be a regular subregion with disconnected border $\partial\Omega$. Partition $\partial\Omega$ into sets α and β of its components. For $f \in C(\alpha)$ let u_0, u_1 be in $H_1(\Omega)$ with $u_i|\alpha = f$, $\int_\beta *du_i = 0$, and

$$\frac{\partial u_0}{\partial n} = 0 \quad \text{on} \quad \beta, \tag{33}$$

$$u_1|\beta = c \text{ (const)}. \tag{34}$$

Clearly the operators $L_i, i = 0, 1$, defined by $u_i = L_i f$ are normal. Moreover the operator $(P)L_1$ for a partition P of β may be given exactly as in I.2.1C.

Transition to the general A of operators L_0 and $(P)L_1$ may be achieved in the same manner as in I.2.2.

3. Functions with singularities

3A. Construction on regular subregions. We shall now construct principal functions with singularities. First we consider the case of a regular subregion.

Let Ω be a regular region with its border β partitioned into sets β_j, $j = 1, \cdots, j_\Omega$. Take parametric balls V_a, V_b, centered at a, b, with disjoint closures in Ω. Given real constants μ, λ consider the class $P_{\mu+\lambda}$ of functions $p \in H(\bar{\Omega}-a-b)$ with the properties

$$(35) \qquad p|\bar{V}_a = (\mu+\lambda)q_{V_a}(\cdot,a)+e, \qquad e \in H_1(\bar{V}_a),$$

$$(36) \qquad p|\bar{V}_b = -(\mu+\lambda)q_{V_b}(\cdot,b)+f, \qquad f \in H_1(\bar{V}_b),$$

$$(37) \qquad \int_{\beta_j} *dp = 0, \qquad j = 1, \cdots, j_\Omega.$$

We choose the normalization $f(b) = 0$.

Take a regular region Ω_0 in R with $\bar{V}_a \cup \bar{V}_b \subset \Omega_0 \subset \bar{\Omega}_0 \subset \Omega$. Set $A = (\Omega-\Omega_0) \cup (\bar{V}_a - a) \cup (\bar{V}_b - b)$, $\alpha = \partial V_a + \partial V_b - \partial\Omega_0$. Define $s \in H_1(A)$ by $s|\bar{V}_a - a = q_{V_a}(\cdot,a)$, $s|\bar{V}_b - b = -q_{V_b}(\cdot,b)$, $s|\Omega-\Omega_0 = 0$. Since $\int_\alpha *ds = 0$, Theorem 2B applies, and we obtain $\bar{p}_i = p[s,L_i]$, $i = 0,1$, where $L_1 = (P)L_1$ with respect to the partition of β. For $i = 0,1$ let $\bar{h}_i = \bar{p}_i - q_{V_a}(\cdot,a)$ in $\bar{V}_a$, $\bar{k}_i = \bar{p}_i + q_{V_b}(\cdot,b)$ in $\bar{V}_b$, and take $p_i = \bar{p}_i - \bar{k}_i(b)$. Then $h_i = \bar{h}_i - \bar{k}_i(b)$ and $k_i = \bar{k}_i - \bar{k}_i(b)$ are the functions e and f corresponding to $p_i \in P_1$, and

$$(38) \qquad p_{\mu\lambda} = \mu p_0 + \lambda p_1 \in P_{\mu+\lambda}.$$

3B. Extremal property of $p_{\mu\lambda}$. The function $p_{\mu\lambda}$ has the following minimum property (Sario-Schiffer-Glasner [1]):

$$(39) \qquad \int_\beta p*dp + (\lambda-\mu)e(a) = \lambda^2 h_1(a) - \mu^2 h_0(a) + D_\Omega(p-p_{\mu\lambda})$$

for every $p \in P_{\mu+\lambda}$.

In fact, by Green's formula we have

$$D(p-p_{\mu\lambda}) = \int_\beta p*dp+p_{\mu\lambda}*dp_{\mu\lambda}-p*dp_{\mu\lambda}-p_{\mu\lambda}*dp.$$

Here

$$\int_\beta p_{\mu\lambda}*dp_{\mu\lambda} = \mu\lambda\int_\beta p_0*dp_1 = \mu\lambda\int_\beta p_0*dp_1-p_1*dp_0$$

$$= \mu\lambda\int_{\partial V_a} h_0*d(q_{V_a}(\cdot,a)+h_1)-h_1*d(q_{V_a}(\cdot,a)+h_0)$$

$$+\mu\lambda\int_{\partial V_b} k_0*d(-q_{V_b}(\cdot,b)+k_1)-k_1*d(-q_{V_b}(\cdot,b)+k_0).$$

By (22) we now conclude that

$$\int_\beta p_{\mu\lambda}*dp_{\mu\lambda} = \mu\lambda(h_1(a)-h_0(a)).$$

Similarly we can show that

$$\int_\beta p*dp_{\mu\lambda} = \lambda((\mu+\lambda)h_1(a)-e(a)),$$

$$\int_\beta p_{\mu\lambda}*dp = \mu(e(a)-(\mu+\lambda)h_0(a)),$$

and (39) follows.

3C. The span of Ω. For $\mu = 1$, $\lambda = -1$ formula (39) takes the form

(40) $$D(u)-2u(a) = h_1(a)-h_0(a)+D(u-p_0+p_1)$$

for every u in the class P_0 of regular harmonic functions in Ω with $u(b) = 0$ and $\int_{\beta_j}*du = 0$, $j = 1,\cdots,j_\Omega$.

For a regular Ω we introduce the *span* $S = S(\Omega, \{\beta_j\}_1^{j_\Omega};a,b)$ as

(41) $$S = h_0(a)-h_1(a).$$

Taking $u \equiv 0$ in (40) we obtain

$$S = D(p_0 - p_1). \tag{42}$$

3D. A convergence theorem. As an application of the span we pause here to prove the following

Theorem. *Suppose that with each regular region Ω of a Riemannian space R there is associated a unique function $u_\Omega \in H(\Omega)$. If for $\Omega \subset \Omega' \subset R$*

$$\lim_{\Omega, \Omega' \to R} D_\Omega(u_\Omega - u_{\Omega'}) = 0 \tag{43}$$

then for a fixed $x_0 \in R$ the function $u_\Omega - u_\Omega(x_0)$ tends to a harmonic function on R uniformly on every compact of R.

In the 2-dimensional case the theorem can be easily proved by taking a local analytic function f_Ω with $u_\Omega = \operatorname{Re} f_\Omega$ (cf. 0.2.3F). If R is locally flat, then the harmonicity of $\partial u_\Omega / \partial x^i$ may be exploited in the proof (see e.g. the monograph of Brelot [1, p. 11], or Sario [28]). In the general case the problem is not as simple. One of the standard proofs makes use of the reproducing kernel (see e.g. Duff [5, p. 165] or Courant-Hilbert [1]).

The following proof, due to Sario-Schiffer-Glasner [1], is perhaps the most direct. It starts with the inequality

$$(u(a) - u(b))^2 \leq S \cdot D(u) \tag{44}$$

for every $u \in H(\bar{\Omega})$ with $\int_{\beta_j} {*du} = 0$, $j = 1, \cdots, j_\Omega$. In fact, if u is normalized by $u(b) = 0$ then by (40)

$$t^2 D(u) - 2t\, u(a) + D(p_0 - p_1) = D(tu - p_0 + p_1) \geq 0$$

for every t, and hence $u(a)^2 \leq SD(u)$. For an arbitrary u, $u - u(b)$ can be used to obtain (44).

From (22) and Green's formula it follows that

$$q_V(b,a) = q_V(a,b) \tag{45}$$

for a parametric ball V and $a, b \in V$. This with (E.1) implies that

$$\|q_V(\cdot,a)\|_E \in C(V-E) \tag{46}$$

for any compact $E \subset V$.

3E. We retain the notation of 3A and let $S_a \subset V_a$, $S_b \subset V_b$ be hyperspheres concentric with V_a, V_b. Keeping S_a and S_b fixed we let the points a, b vary slightly about the centers. From Theorem 2D it follows that the continuity of $\|q_{V_a}(\cdot,a)\|_{S_a}$ in a and that of $\|q_{V_b}(\cdot,b)\|_{S_b}$ in b implies the uniform continuity of $\bar{h}_i$ and $\bar{k}_i$ in both a and b. The continuity of $\bar{k}_i(b)$ then implies the uniform continuity of $h_i = \bar{h}_i - \bar{k}_i(b)$ in V_a and consequently the continuity of the span $h_0(a) - h_1(a)$ in both a and b. In particular, we conclude that

$$\sup_{a \in E} S(\Omega, \{\beta_j\}_1^{j_\Omega}; a,b) < \infty \tag{47}$$

for any fixed compact $E \subset \Omega$.

Now (44) and (47) with $b = x_0$ imply the assertion of Theorem 3D, and the proof is complete.

3F. Noncompact regions. We next generalize (39) to arbitrary Riemannian spaces. Consider regular regions $\Omega \subset R$ containing $\{a,b\}$ and with a consistent system of partitions of the borders β_Ω. Let $\Omega \subset \Omega'$, and denote by $p_{\mu\lambda} = \mu p_0 + \lambda p_1$ and h_i quantities corresponding to Ω, by $p_{\mu\lambda}'$ and h_i' those corresponding to Ω'. Equation (39) gives for $p = p_0'|\bar{\Omega}$, $p_{\mu\lambda} = p_0$, $\beta = \beta_\Omega$, $\beta' = \beta_{\Omega'}$,

$$\int_\beta p_0' * dp_0' - h_0'(a) = -h_0(a) + D_\Omega(p_0' - p_0); \tag{48}$$

for $p = p_1'|\bar{\Omega}$, $p_{\mu\lambda} = p_1$,

$$\int_\beta p_1' * dp_1' + h_1'(a) = h_1(a) + D_\Omega(p_1' - p_1); \tag{49}$$

and for $p = p_1$, $p_{\mu\lambda} = p_0$,

$$\int_\beta p_1 * dp_1 - h_1(a) = -h_0(a) + D_\Omega(p_1 - p_0). \tag{50}$$

Since $\int_{\beta'-\beta} p_i' * dp_i' = D_{\Omega'-\Omega}(p_i) \geq 0$ and $\int_{\beta'} p_i' * dp_i' = 0$, we conclude that $\int_\beta p_i' * dp_i' \leq 0$.

$h_0(a)$ decreases, $h_1(a)$ increases with increasing Ω, and $h_1(a) \leq h_0(a)$ for every Ω.

Hence the directed limit $h_i(a) = \lim_{\Omega\to R} h_{i\Omega}(a)$ exists, as does $\lim_{\Omega\to R} D_\Omega(p_{i\Omega} - p_{i\Omega'}) = 0$. If we use this and the normalization $p_\Omega(b) - p_{\Omega'}(b) = 0$ Theorem 3D gives the harmonic directed limits $p_i = \lim_{\Omega\to R} p_{i\Omega}$ on $R-a-b$, the convergence being uniform on compacta. We then set

$$p_{\mu\lambda} = \mu p_0 + \lambda p_1$$

on $R-a-b$.

3G. The class $P_{\mu+\lambda}$ is defined for $R-a-b$ and for the given consistent system of partitions in obvious analogy with the case of Ω in 3A. To establish the extremal property of $p_{\mu\lambda}$ in $P_{\mu+\lambda}$ let $\Omega' \to R$ in (48) and (49). We obtain

$$\int_{\beta_\Omega} p_0 * dp_0 - h_0(a) = -h_{0\Omega}(a) + D_\Omega(p_0 - p_{0\Omega}),$$

$$\int_{\beta_\Omega} p_1 * dp_1 + h_1(a) = h_{1\Omega}(a) + D_\Omega(p_1 - p_{1\Omega}).$$

On letting $\Omega\to R$ we infer by $\int_{\beta_\Omega} p_i * dp_i \leq 0$ and the triangle inequality that

$$\lim_{\Omega\to R} D_\Omega(p_{\mu\lambda} - p_{\mu\lambda\Omega}) = 0.$$

From this and $D(p - p_{\mu\lambda}) = \lim_{\Omega\to R} D_\Omega(p - p_{\mu\lambda})$ one concludes, again by the triangle inequality, that

$$\lim_{\Omega\to R} D_\Omega(p - p_{\mu\lambda\Omega}) = D(p - p_{\mu\lambda}).$$

The deviation formula for Ω and $p \in P_{\mu+\lambda}$ on R reads

$$(51) \quad \int_{\beta_\Omega} p * dp + (\lambda - \mu) e(a) = \lambda^2 h_{1\Omega}(a) - \mu^2 h_{0\Omega}(a) + D_\Omega(p - p_{\mu\lambda\Omega}).$$

We introduce the symbolic expression

$$\int_{\beta} p * dp = \lim_{\Omega \to R} \int_{\beta_{\Omega}} p * dp.$$

On letting $\Omega \to R$ in (51) we obtain the generalization of (39) (Sario [28], Sario-Schiffer-Glasner [1]):

Theorem. *For every $p \in P_{\mu+\lambda}$ on a Riemannian space R*

$$\int_{\beta} p * dp + (\lambda - \mu) e(a) = \lambda^2 h_1(a) - \mu^2 h_0(a) + D(p - p_{\mu\lambda}). \tag{52}$$

In passing we observe that

$$\int_{\beta} p_0 * dp_0 = \int_{\beta} p_1 * dp_1 = 0. \tag{53}$$

This follows by choosing $p = p_{\mu\lambda} = p_i$, $i = 0,1$ in (52).

3H. The span for R. Again we have for $p_0 - p_1$ and the given system of partitions the minimum property

$$D(u) - 2u(a) = h_1(a) - h_0(a) + D(u - p_0 + p_1) \tag{54}$$

in the class P_0 of all harmonic functions u on R with $u(b) = 0$ and $\int_{\beta_j} * du = 0$, $j = 1, \cdots, j_{\Omega}$ for all $\Omega \subset R$. For the identity partition we are dealing with all harmonic functions, whereas for the canonical partition we are confronted with the analogue of real parts of analytic functions.

The *span* $S = S(R,P;a,b)$ of a Riemannian space R for a consistent partition P of the ideal boundary is defined as

$$S = h_0(a) - h_1(a). \tag{55}$$

The function $u \equiv 0$ in (54) gives the property

$$S = D(p_0 - p_1). \tag{56}$$

3I. In the class of univalent functions $p + ip^*$ with a singularity $1/(z - a)$ in a plane region R the integral $\int_{\beta} p\, dp^*$ gives the negative of the complementary area of the image of R, and hence is nonpositive (cf. II.2). The natural analogue for Riemannian spaces is the class $Q_1 \subset P_1$ of functions with $\int_\beta p * dp \leq 0$. Let h stand for e in the class P_1. Setting first $\mu = 1$, $\lambda = 0$, then $\mu = 0$, $\lambda = 1$ in (52), we conclude:

In the class Q_1 the quantity $h(a)$ is given its maximum $h_0(a)$ by p_0 and its minimum $h_1(a)$ by p_1. The span has the value $S = \max_{Q_1} h(a) - \min_{Q_1} h(a)$.

For $\mu = \lambda = \frac{1}{2}$ in (52) we have

$$\int_\beta p * dp = -\frac{S}{4} + D\left(p - \frac{p_0 + p_1}{2}\right). \tag{57}$$

If $S = 0$ then $p = (p_0 + p_1)/2 = p_0$. Conversely $p = p_0$ implies $e(a) = h_0(a)$ for all $p \in Q_1$, in particular for p_1, which gives $h_0(a) = h_1(a)$.

The functions p_0 and p_1 are identical if and only if $S = 0$. More generally, this condition is necessary and sufficient for every $p \in Q_1$ to coincide with $p_0 = p_1$.

Next denote by $A(p) = -\int_\beta p * dp$ the counterpart of the complementary area of the image under univalent functions. Define Q_2 by analogy with Q_1. Putting $\mu = \lambda = 1$ in (52) we obtain:

In $Q_2 \subset P_2$ the function $p_0 + p_1$ gives to $A(p)$ its maximum S. The condition $S = 0$ is necessary and sufficient for $A(p) = 0$ in all of Q_2.

Finally let $H(R;a,b) = \{u \in H(R) | u(a) = 1,\ u(b) = 0\}$. If $S \neq 0$ then for $\mu = 1/S$, $\lambda = -1/S$, the function $p_{\mu\lambda} = (p_0 - p_1)/S$ is in $H(R;a,b)$ and consequently (52) gives

$$D(u) = \frac{1}{S} + D\left(u - \frac{p_0 - p_1}{S}\right) \tag{58}$$

for $u \in H(R;a,b)$.

In $H(R;a,b)$ the function $(p_0 - p_1)/S$ gives to $D(u)$ its minimum $1/S$.

From the last two assertions we obtain the following interesting invariance relation:

$$\max_{Q_2} A(p) \cdot \min_{H(R;a,b)} D(u) = 1. \tag{59}$$

The above results were obtained in Sario-Schiffer-Glasner [1] (cf. also Sario [28]).

4. Classification of Riemannian spaces

4A. The class *HD*. As in the case of Riemann surfaces (cf. IV) we denote by O_{HD} the class of Riemannian spaces R for which the space HD reduces to a constant. In terms of the span, O_{HD} can be characterized as follows (Sario-Schiffer-Glasner [1], Sario [28]):

Theorem. *A Riemannian space R is not in O_{HD} if and only if $S \neq 0$ for some a, b and the identity partition.*

In fact, from (55) and (56) we conclude that $p_0 - p_1 \in HD$. Suppose there is a nonconstant $u \in HD$ on R. We may assume that $u(a) \neq 0$ and $u(b) = 0$ for some a,b. If $S = D(p_0 - p_1)$ were 0, then (54) would imply $u(a) = 0$. Therefore $S \neq 0$.

Conversely $S = D(p_0 - p_1) \neq 0$ entails the existence of a nonconstant HD-function $p_0 - p_1$ on R.

4B. The classes O_{HP}, O_{HB}, O_{HBD} are defined similarly. In addition to the property $p_0 - p_1 \in HD$, the functions $p_i = L_i p_i$ are bounded in a boundary neighborhood, and $p_0 - p_1 \in HBD$. Therefore

$$O_{HD} = O_{HBD} \tag{60}$$

(cf. Sario [28]), and we have the inclusion relations

$$O_{HP} \subset O_{HB} \subset O_{HD}. \tag{61}$$

4C. Green's function. Green's function $q_\Omega(\cdot, a)$ with pole at $a \in \Omega$ is defined by (D1) to (D4) on replacing V by Ω. By the maximum principle $q_\Omega \leq q_{\Omega'}$ for $\Omega \subset \Omega'$, and the directed limit $q = q_R = \lim_{\Omega \to R} q_\Omega$ either exists or is ∞ on R. In the former case it is called Green's function on R. A Riemannian space R is said to be *parabolic*, $R \in O_G$, if it has no Green's function; otherwise it is *hyperbolic*. The class O_G is contained in each of the above classes (see 4D below):

$$O_G \subset O_{HP} \subset O_{HB} \subset O_{HD}.$$

4D. Harmonic measures. Choose regular regions R_0, Ω with $\bar{R}_0 \subset \Omega \subset R$. The harmonic measure w_Ω of $\partial\Omega$ with respect to $\Omega - \bar{R}_0$ is, by definition, the harmonic function on $\Omega - \bar{R}_0$ with boundary values 0 on ∂R_0 and 1 on $\partial\Omega$. By the maximum principle $w_\Omega \geq w_{\Omega'}$ for $\Omega \subset \Omega'$, and the directed limit

$$w_R = \lim_{\Omega \to R} w_\Omega$$

exists. It is called the *harmonic measure* of the ideal boundary β of R with respect to $R - \bar{R}_0$. Using Theorem 2B one proves easily

$$w_R > 0 \quad \textit{in} \quad R - \bar{R}_0 \quad \textit{if and only if} \quad R \notin O_G.$$

We now return to the proof of $O_G \subset O_{HP}$. Let $u \in HP(R)$ be nonconstant and choose R_0 as above. Let $v = L_1 u - u$, where L_1 is the normal operator for $R - R_0$. Then $\int_{\partial R_0} {}^*dv = 0$, $v|\partial R_0 = 0$, and v is bounded from above. Furthermore, $M = \sup v > 0$. Indeed, if $v \leq 0$ on $R - R_0$ then the vanishing of the flux implies $\partial v/\partial n \equiv 0$ on ∂R_0. Thus v and its first order partial derivatives vanish on ∂R_0 and therefore (cf. 2C) $v \equiv 0$ by the uniqueness of the solution of the Cauchy problem for (21). This violates the assumption that u is nonconstant.

The harmonic measure w_Ω satisfies $M w_\Omega \geq v$ on $\Omega - R_0$, and a fortiori $w_R \not\equiv 0$. Hence $R \notin O_G$.

4E. Capacity functions. We shall introduce the capacity of the ideal boundary and of a boundary component of a Riemannian space R. Consider a regular region $\Omega \subset R$ with border $\beta = \gamma \cup \beta_1 \cup \cdots \cup \beta_{j_\Omega}$, where γ is a set of components of β and each β_j, $j = 1, \cdots, j_\Omega$, is a component of $\beta - \gamma$. Let V_a be a parametric ball centered at a with $\bar{V}_a \subset \Omega$. Denote by P the class of functions

$p \in H(\bar{\Omega} - a)$ such that

(62) $$p|\bar{V}_a = -q_{V_a}(\cdot, a) + h,$$

(63) $$\int_\gamma *dp = 1,$$

(64) $$\int_{\beta_j} *dp = 0, \qquad j = 1, \cdots, j_\Omega.$$

Here $h \in H(\bar{V}_a)$ and $h(a) = 0$. In P the capacity function p_γ of γ is singled out by the properties

(65) $$p_\gamma|\gamma = k_\gamma,$$

(66) $$p_\gamma|\beta_j = k_j,$$

k_γ, k_j being constants. The existence may be established using the Main Existence Theorem 2B as in 3A (cf. Ow [1]).

The capacity function has the following minimum property (Sario [28], Ow [1]):

(67) $$\min_P \int_\beta p * dp = k_\gamma + D_\Omega(p - p_\gamma).$$

In fact, on adding the quantity $\int_\beta p_\gamma * dp - \int_\beta p_\gamma * dp_\gamma = 0$ to the right side of $D_\Omega(p - p_\gamma) = \int_\beta p * d(p - p_\gamma)$ one obtains

$$D_\Omega(p - p_\gamma) = \int_\beta p * dp - \int_\beta p_\gamma * dp_\gamma + \int_\beta p_\gamma * dp - \int_\beta p * dp_\gamma.$$

By transferring the integral $\int_\beta p_\gamma * dp - p * dp_\gamma$ to an integral along ∂V_a one shows in the same manner as in 3B that its value is $h_\gamma(a) - h(a)$, hence 0.

4F. In passing we note that for $\gamma = \beta$, p_β also has the following extremal property (Sario [28], Ow [1]):

(68) $$\min_P \sup_\Omega p = \sup_\Omega p_\beta = k_\beta.$$

For the proof observe that

(69)
$$u(a) = \int_\beta u * dp_\beta$$

for any harmonic function u. In particular this is true for $u = p - p_\beta$. Since $\int_\beta p_\beta * dp_\beta = k_\beta$, it follows from $u(a) = 0$ that $\int_\beta p * dp_\beta = k_\beta$, and the possibility of $\sup_\Omega p < k_\beta$ is excluded.

4G. The capacity. Let $\{\Omega_m\}$ be a nested sequence of canonical regions of R with $\cup \Omega_m = R$. Consider a consistent system $\{\beta_{mj}\}$ of partitions of the $\{\partial\Omega_m\}$. A sequence $\{\gamma_m\} = \{\beta_{mj(m)}\}$ defines a subboundary γ of the ideal boundary β of R if $\beta_{m+1,j(m+1)}$ is in the component of $R - \Omega_m$ bordered by $\beta_{mj(m)}$. Equivalence in two exhaustions is defined in an obvious manner. For the identity partition γ is the ideal boundary β; for the canonical partition it is a boundary component.

For a canonical $\Omega \subset R$ let γ_Ω be the component of β_Ω that corresponds to a given γ. Let $p_{\Omega\gamma}$ be the capacity function of γ_Ω on Ω with $p_{\Omega\gamma}|\gamma_\Omega = k_{\Omega\gamma}$. Since

$$k_{\Omega\gamma} = \int_{\beta_\Omega} p_{\Omega\gamma} * dp_{\Omega\gamma} \leq \int_{\beta_\Omega} p_{\Omega'\gamma} * dp_{\Omega'\gamma} \leq \int_{\beta_{\Omega'}} p_{\Omega'\gamma} * dp_{\Omega'\gamma} = k_{\Omega'\gamma}$$

for $\bar{\Omega} \subset \Omega'$, we conclude that

(70)
$$k_{\Omega\gamma} \leq k_{\Omega'\gamma} \quad \text{for} \quad \bar{\Omega} \subset \Omega'.$$

Therefore the directed limit exists:

(71)
$$k_\gamma = \lim_{\Omega \to R} k_{\Omega\gamma}.$$

In the case $k_\gamma < \infty$ we could derive from this the uniform convergence of $p_{\Omega\gamma}$ to a unique limit p_γ on $R - a$, the capacity function for γ, in a manner similar to that in 3F. Here (67), (68) continue to hold in a class P defined in an obvious manner. If $k_\gamma = \infty$ limiting capacity functions still exist but uniqueness is lost. We shall, however, not use limiting functions in either case but introduce the following definitions (Sario [28], Ow [1]):

The capacity of the subboundary γ of a Riemannian space R is

$$c_\gamma = k_\gamma^{1/(2-n)} \qquad (n \geq 3), \tag{72}$$

$$c_\gamma = e^{-2\pi k_\gamma} \qquad (n = 2). \tag{72$'$}$$

A boundary component γ will be called weak if $c_\gamma = 0$.

We distinguish two classes of Riemannian spaces R:

$$C_\beta = \{R | c_\beta = 0\},$$

$$C_\gamma = \{R | \text{ each boundary component } \gamma \text{ is weak}\}.$$

Since $q_\Omega(\cdot, a) = k_{\beta_\Omega} - p_{\beta_\Omega}$ it is clear that

$$C_\beta = O_G. \tag{73}$$

4H. Completeness and degeneracy. A Riemannian space R is *complete* if the distance between any point in R and the ideal boundary of R is infinite. Therefore compact Riemannian spaces are complete. They also belong to all four of the following classes:

$$O_G \subset O_{HP} \subset O_{HB} \subset O_{HD}. \tag{74}$$

In this context the natural question arises: Does the completeness of a Riemannian space imply that it belongs to a null class? That the answer is in the negative in the 2-dimensional case is exemplified by the unit disk with the hyperbolic metric. The interest of the problem lies in the fact that the function-theoretic behavior of higher dimensional spaces is fundamentally different from that of Riemann surfaces. In particular, null classes of the latter are of course conformally invariant whereas those of the former are not (Nakai-Sario [6]). We shall show that, nevertheless, Riemannian spaces of any dimension share with Riemann surfaces the property that completeness does not imply degeneracy. Explicitly we have the following result (Nakai-Sario [6]):

Theorem. *For every dimension there exists a complete Riemannian space R which does not belong to any of the null classes in* (74).

For the proof take the noncompact space R obtained by puncturing a flat torus of dimension $n \geq 3$ at two points a_i, $i = 1, 2$. We may assume that the closed unit balls B_i about a_i and punctured at these points are disjoint. Take Euclidean coordinates $x_i = (x_i^1, \cdots, x_i^n)$ in B_i with $x_i = 0$ at a_i and set $|x_i| = (\sum_{j=1}^n (x_i^j)^2)^{1/2}$. We endow R with a C^∞ Riemannian metric g_{jk} such that $g_{jk}(x_i) = \lambda(x_i)\delta_{jk}$ in B_i. Then the Laplace-Beltrami operator takes the form

$$\Delta u = -\lambda^{-n/2} \sum_{j=1}^{n} \frac{\partial}{\partial x_i^j} \left(\lambda^{(n-2)/2} \frac{\partial u}{\partial x_i^j} \right) \tag{75}$$

in B_i. For the choice

$$\lambda(x_i) = |x_i|^{(2-2n)/(n-2)} \tag{76}$$

we can easily see by direct calculation that the function u on $B_1 \cup B_2$ given by

$$u|B_i = u_i = c_i(1 - |x_i|), \qquad i = 1, 2,$$

is harmonic in the interior of $B_1 \cup B_2$ and

$$D_{B_1 \cup B_2}(u) < \infty. \tag{77}$$

Clearly we can choose $c_i \neq 0$ so that $\int_{\partial B_1} * du_1 = -\int_{\partial B_2} * du_2$, or, what amounts to the same, $\int_{\partial A} * du = 0$ with $A = B_1 \cup B_2$. Let p be the principal function on R corresponding to L_1 for A and u. By (77) and the triangle inequality we conclude that $p \in HD$ on R. Since $L_1(u|\partial A) \equiv 0$, we have $u \not\equiv L_1(u|\partial A)$ and hence p is not constant. The fact that $R \notin O_{HD}$ implies with (74) that R does not belong to any of the null classes in (74).

On the other hand (76) shows that R is complete.

4I. Other null classes. We have considered six classes of Riemannian spaces R: O_G, O_{HP}, O_{HB}, O_{HD}, C_β, C_γ. We conclude this section by introducing other significant classes and by listing problems to which they lead.

In strict analogy with the concept of the real part of an analytic function in the 2-dimensional case we introduce for R the class F of harmonic functions u on R with vanishing flux $\int_{\gamma_{\Omega_j}} * du = 0$

across every component $\gamma_{\Omega j}$ of the boundary β_Ω of every regular region Ω of R. If the span S is defined for the canonical partition, the preceding reasoning for HD in 4B applies to FD and we obtain:

$$O_{FB} \subset O_{FD}. \tag{78}$$

A hypersurface α with the property that $R-\alpha$ consists of two components will be referred to as a dividing cycle. By definition, the class K for R is composed of harmonic functions on R with vanishing flux across every dividing cycle. In the same manner as above one obtains

$$O_{KB} \subset O_{KD}. \tag{79}$$

On a complex analytic manifold, $R^{2n} = \mathbf{C}^n$ for example, one can also consider the class A of complex analytic functions and the class of real parts of such functions.

Let p_Ω be the capacity function of $\partial\Omega = \beta_\Omega$ in Ω with singularity at a point $a \in \Omega$. The class HM_q in R consists of those $u \in H$ on R for which the mean $\int_{\beta_\Omega} |u|^q * dp_\Omega$ is bounded for all $\Omega \subset R$. More generally, for nonnegative Φ defined on $[0, \infty)$ the class $H\Phi$ in R is composed of those $u \in H$ on R for which $\Phi(|u|)$ admits a harmonic function on R as a majorant.

4J. In analogy with $\log|w|$ for a meromorphic function w on a plane region, we introduce the class L of harmonic functions on R with singularities $c_j q_{V_{x_j}}(\cdot, x_j)$ at isolated points x_j, $j = 1, 2, \cdots$, the coefficients being nonzero real numbers. Given $a \in R$ and $v \in L$ on R, take a regular region Ω containing a and decompose $v|\beta_\Omega$ into $v^+ = \max(v,0)$ and $v^- = \max(-v,0)$. Let x_Ω^+, x_Ω^- be the solutions on Ω of the Dirichlet problem with boundary values v^+, v^-, respectively.

Let $a_\mu (\mu = 1, \cdots, \mu_\Omega)$, $b_\nu (\nu = 1, \cdots, \nu_\Omega)$ be the positive and negative singularities of v in Ω. Set

$$y_\Omega^+ = \sum_{a_\mu \in \Omega} q_\Omega(\cdot, a_\mu), \quad y_\Omega^- = \sum_{b_\nu \in \Omega} q_\Omega(\cdot, b_\nu)$$

and $u_\Omega^+ = x_\Omega^+ + y_\Omega^+$, $u_\Omega^- = x_\Omega^- + y_\Omega^-$. The characteristic $C(\Omega)$

of $v \in L$ can be defined as

$$C(\Omega) = u_\Omega^+(a).$$

The class LC of functions of bounded characteristic on R consists of $v \in L$ with bounded $C(\Omega)$ for all $\Omega \subset R$.

4K. We have introduced classes IJ with $I = H, K, F, A, L$ and $J = P, B, D, M_q, \Phi, C$.

Given a Riemannian space R let $\bar{R}_1$ be the complement of a regular region with border α_1. With each nondegenerate class IJ we associate the class I_0J of functions $u \in IJ$ on $\bar{R}_1$ with $u|\alpha_1 = 0$. Such functions are useful in studying removability properties of the boundary.

4L. List of problems. A general classification theory can be developed for Riemannian spaces R. As special cases one can consider locally flat Riemannian spaces V, n-dimensional submanifolds V^n of a higher dimensional Euclidean space R^m, and regions $G \subset R^n$. Here we state 19 problems (Sario [28]), new and known, difficult and easy. For background and for partial solutions of some of the problems we refer to Ahlfors-Beurling [1], Sario [5], [12], [15], [28], Parreau [1], Tôki [1], [2], Nakai [1], [4], Nakai-Sario [7], Myers [1], and to the U.C.L.A. doctoral dissertations of Meehan [1], Larsen [1], Johnson [1], Smith [1], Glasner [1], Ow [1], and Breazeal [1].

(1) What are the inclusion relations between the various classes O_{IJ}, O_{I_0J}, C_β, C_γ? Do the classes O_{I_0J} for a fixed I coincide?

(2) Which inclusion relations are strict for G, which for V^n, which for V, and which for R? Do the classes O_{IJ} for a fixed I generally coincide in the first case? Is $O_{HM_q} = O_{HP}$ for $q = 1$, but $O_{HM_q} = O_{HB}$ for $q > 1$? What can be said about $O_{H\Phi}$? Can counterexamples be constructed by removing from the unit ball equidistant radial segments of "meridian" planes and by suitable identifying the "faces" of such segments?

(3) The modulus of a regular region Ω of R can be defined in analogy with that on Riemann surfaces. Can Ω be subdivided into two regular regions each with a modulus arbitrarily close to 1? Are there modular tests for a given R to belong to a given class?

(4) Can tests in terms of deep coverings or Riemannian metrics be formed?

(5) What metric properties do the boundaries of G, $V^n \in O_{IJ}$, C_β, C_γ possess?

(6) In what classes are the complements $R^n - C$ and $R^n - S$ of the n-dimensional analogues of Cantor sets C and Schottky sets S? What can be said about their complements with respect to a compact V (cf. Problem (8))?

(7) Is the complement of a generalized Cantor set in some class O_{IJ} if and only if the volume $\Pi_{i=1}^{\infty}(1-(1/p_i))^n$ vanishes?

(8) A compact V can be formed, for example, by identifying by pairs the opposite faces of an n-cube. Can unramified Abelian covering spaces of such spaces be formed and do they all belong to an O_{IJ}?

(9) Remove a disk D from R^n and take two copies V_1, V_2, of the remaining space. Identify the upper (lower) face of D in V_1 with the lower (upper) face of D in V_2 so as to form a locally flat covering space of R^n. More generally, construct covering spaces of the "cube" of (8) by removing several disks and using several duplicates of the remaining space, in the same manner as forming covering surfaces of R^2, with the branch points replaced by circles, the connecting line segments by encircled disks. Develop a classification of such covering spaces based on the ramification properties, in analogy with the classical type problem.

(10) If the potential p of a unit mass distribution $d\mu$ on a compact set E in R^n is defined as

$$p(z) = \int_E \frac{|z-\zeta|^{2-n}}{(n-2)\,\omega_n}\, d\mu(\zeta), \qquad \omega_n = 2(\sqrt{\pi})^n/\Gamma\left(\frac{n}{2}\right),$$

what is the relation between the equilibrium potential and our capacity function?

(11) Is the component γ of a compact set E in R^n a point if and only if $c_\gamma = 0$?

(12) Can an "equivalence" of R, V, V^n, G be defined in terms of isomorphisms of suitable function spaces, by quasiconformality, or by quasi-isometry?

(13) In the affirmative case, is a component γ of ∂V always a point or always a continuum or are there "unstable" components?

(14) Cover R^n with a set of cubes with side 1 and arrange the cubes in a sequence $\{Q_i\}$ such that the $R_j = \cup_1^j Q_i$, $j = 1, 2, \cdots$, form a nested sequence of regions exhausting R^n. For $\varepsilon > 0$ remove from Q_i a Cantor set C_i such that $Q_i - C_i$ has volume $2^{-i}\varepsilon$. Then the region $G = R^n - \cup_1^\infty C_i$ has an arbitrarily small volume ε, yet is dense in R^n. Does G have an equivalent G^* in R^n (at least if $n = 2m$) such that one boundary component of G^* is a continuum? Can G^* be a bounded region?

(15) Under what self-mappings of R^n is a class O_{IJ} preserved? In particular, what can be said about quasiconformally equivalent regions?

(16) Can the classification theory be extended to mappings of the complex space $\mathbf{C}^n$ into itself, with suitable modifications of properties P, B, D, M, C?

(17) To what extent can an analogue of the theory of meromorphic functions of bounded characteristic be developed for LC? In particular, can functions in LC be decomposed into extremal LP-functions? Do the Poisson-Stieltjes formula and the decompositions by Parreau [1] and Rao [1] generalize?

(18) Can a value distribution theory be developed for analytic functions suitably associated with harmonic functions in locally flat or nonflat spaces?

(19) Can the following interpolation problem be solved in terms of linear combinations of functions $p_0 - p_1$ with suitable singularities: given R, find a $u \in H(R)$ which takes prescribed values at a finite number of points and minimizes the Dirichlet integral (cf. 5 below).

5. Interpolation problem

5A. We shall consider the interpolation problem of constructing a harmonic function u on a Riemannian space R with given values r^i at given points $a^i \in R$, $i = 0, \cdots, m$, and with minimum $D_R(u)$.

Without loss of generality we may assume that $r^0 = 0$, for otherwise we consider functions $u - u(a^0)$ with values $r^i - r^0$ at the a^i. We set $a^0 = b$, and will understand that i ranges from 1 to m. We also put $a = \cup_i a^i$, $R' = R - a \cup b$.

Take parametric balls V^i, V^b about a^i, b with disjoint closures, and denote by q^i, q^b Green's functions $q_{V^i}(\cdot,a^i)$, $q_{V^b}(\cdot,b)$. Given real numbers $\alpha^1,\cdots,\alpha^m$, μ, λ, denote by $P^{\alpha}_{\mu+\lambda}$ the class of functions p with $p\in H(R')$,

$$p|\bar{V}^i = (\mu+\lambda)\alpha^i q^i+e^i,$$
(80)
$$p|\bar{V}^b = -(\mu+\lambda)\sum_1^m \alpha^i q^b+f$$

where $e^i\in H(\bar{V}^i)$, $f\in H(\bar{V}^b)$, $f(b) = 0$. As in 3 we can form functions $p_0, p_1\in P_1{}^{\alpha}$ defined as limits of functions $p_{0\Omega}$, $p_{1\Omega}$ on exhausting subregions Ω with (80) and with L_0 and L_1 behaviors at β_Ω. We set

(81)
$$p_{\mu\lambda} = \mu p_0+\lambda p_1.$$

For a fixed i take $\alpha^i = 1$, $\alpha^j = 0$ $(i\neq j)$, choose $\mu+\lambda = 1$, and let $p_0{}^i$, $p_1{}^i$ signify the corresponding functions p_0, p_1. Write $h_0{}^i$, $h_1{}^i$ for e^i of $p_0{}^i$, $p_1{}^i$. Clearly

(82)
$$p_0 = \sum_i \alpha^i p_0{}^i, \qquad p_1 = \sum_i \alpha^i p_1{}^i.$$

5B. Consider the system of m equations, $i = 1,\cdots,m$,

(83)
$$\sum_j \alpha^j(p_0{}^j(a^i) -p_1{}^j(a^i)) = r^i$$

where $p_0{}^i(a^i) -p_1{}^i(a^i)$ has the definite meaning $h_0{}^i(a^i) -h_1{}^i(a^i)$. For given r^i and a^i we can find a solution $\alpha^1,\cdots,\alpha^m$ of (83) if and only if

(84)
$$\operatorname{rank}\begin{pmatrix} p_{11} & \cdots & p_{1m} & r^1 \\ \cdot & & \cdot & \cdot \\ \cdot & & \cdot & \cdot \\ \cdot & & \cdot & \cdot \\ p_{m1} & \cdots & p_{mm} & r^m \end{pmatrix} = \operatorname{rank}\begin{pmatrix} p_{11} & \cdots & p_{1m} \\ \cdot & & \cdot \\ \cdot & & \cdot \\ \cdot & & \cdot \\ p_{m1} & \cdots & p_{mm} \end{pmatrix}$$

where $p_{ij} = p_0{}^j(a^i) -p_1{}^j(a^i)$.

The following solution of our problem is due to Sario-Fukuda [1]:

Theorem. *Condition* (84) *is necessary and sufficient for*

$$p_0 - p_1 = \sum_1^m \alpha^i (p_0{}^i - p_1{}^i) \tag{85}$$

to be the unique function minimizing $D_R(u)$ *among all harmonic functions* u *on* R *with* $u(a^i) = r^i$, $u(b) = 0$, *and* $D_R(u) < \infty$.

The value of the minimum is $\Sigma_i \alpha^i r^i$ *and the deviation for any admissible* u *from this minimum is* $D_R(u-(p_0-p_1))$:

$$D_R(u) = \sum_i \alpha^i r^i + D_R(u - p_0 + p_1). \tag{86}$$

It is clear that (84) is not always satisfied. For example, $R \in O_{HD}$ violates (84), as is seen from 4A. Therefore Problem (19) in 4L may be restated in the following form: characterize (84) in terms of a function-theoretic degeneracy of R.

For the proof it is sufficient to establish the deviation formula (86). First one derives it in the case where R is a regular region Ω and then passes to the limit as $\Omega \to R$. The proof is similar to that in 3, and we leave it to the reader.

6. Principal functions in physics

6A. Various natural phenomena are described by harmonic functions with suitable singularities and a specified boundary behavior, i.e. by principal functions in 2- and 3-dimensional spaces.

Examples which lead to previously introduced principal functions are:

(a) The velocity potential of a steady, irrotational ideal fluid flow in a region with an impenetrable boundary is the principal function p_0. If the streamlines are required to cross the boundary at right angles, the velocity potential is p_1.

(b) The electrostatic potential in a region bounded by an uncharged conductor influenced by poles outside the conductor is p_1.

(c) The steady-state temperature in a homogeneous body is the principal function p_0 when the body surface is insulated. It is p_1 when the surface is a perfect conductor.

(d) Suppose that there is a heat source at a, and a heat sink of equal strength at b. The heat energy exterior to small spheres about a and b is approximated by $\int_\beta p * dp$. Thus by II.2.2, the function $(p_0+p_1)/2$ describes the temperature distribution having minimum heat energy.

6B. Normal operators also have physical interpretations. The operator L_0 describes the temperature in an annulus or spherical shell in which the temperature is arbitrarily specified on a component of the boundary and has the rest of the boundary insulated. The operator L_0 minimizes the heat energy in the body. The operator L_1 provides us with the temperature when the remainder of the boundary is a perfect heat conductor held at a temperature which results in no net heat flow into the conductor.

6C. The main existence theorem for principal functions also applies to problems involving linear or areal distributions of sources and sinks with specified boundary conditions. Thus it yields the existence of single-layer potentials and also double-layer potentials.

§2. PRINCIPAL FORMS ON LOCALLY FLAT SPACES

The principal function problem may be restated as follows. Let A be the complement of a regular region of a Riemannian space R. Given a harmonic function (0-form) s on A, find a harmonic function p on R which "behaves like" s on A. The imitation of s by p has been given in terms of normal operators. We could have also described it by the boundedness of $|p-s|$ or by the finiteness of $D_A(p-s)$; in the case of a parabolic R, these amount to the same as $L(p-s) = p-s$, since there is only one normal operator $L = L_0 = L_1$ in this case.

We shall discuss the problem of constructing a harmonic p-form ρ on R such that, for a given harmonic p-form σ on A, $|\rho-\sigma|$ is bounded on A. In the present section we give a solution in the case of locally flat spaces. Our presentation also serves to indicate the nature of the solution one might expect on arbitrary Riemannian spaces.

In 1 we briefly explain fundamentals of p-forms. The construction of harmonic p-forms is given in 2. In 3 we introduce another approach, that of border reduction, which also seems to have some prospects for generalization.

The material in this section was given in Nakai-Sario [4], [10].

1. p-forms on regular regions

1A. Tangential and normal parts. Let R be a Riemannian n-space and G a regular subregion. We can find a parametric ball $U(x)$ for any $x \in \partial G$ such that the admissible coordinates $t = (t^1, \cdots, t^n)$ for $U(x)$ satisfy $t(x) = 0$ and

$$U(x) \cap G = \{t \mid |t| < 1, t^n > 0\},$$

the base $t^n = 0$ of the half-ball corresponding to the boundary of G. We suppose moreover that the t^n-curve is orthogonal to the t^i-curves, $i = 1, \cdots, n-1$. The coordinates t^i, $i = 1, \cdots, n$, are called *boundary coordinates.* Let

$$\varphi = \varphi_{\alpha_1 \cdots \alpha_p} dx^{\alpha_1} \wedge \cdots \wedge dx^{\alpha_p}$$

be a p-form (cf. 1.1B). Near the boundary of G we may express φ in boundary coordinates t^i as $\varphi = \varphi_{\alpha_1 \cdots \alpha_p} dt^{\alpha_1} \wedge \cdots \wedge dt^{\alpha_p}$, and we set

$$t\varphi = \varphi_{\alpha_1 \cdots \alpha_p} dt^{\alpha_1} \wedge \cdots \wedge dt^{\alpha_p}, \tag{87}$$

$$n\varphi = \varphi - t\varphi. \tag{88}$$

The differential p-forms $t\varphi$ and $n\varphi$ are called the *tangential part* and *the normal part* of φ respectively.

Since the t^n-curve is orthogonal to the boundary of G, we have

$$g_{in} = g^{in} = 0, \qquad i = 1, \cdots, n-1, \tag{89}$$

on the boundary of G. Here g_{ij} is the metric tensor corresponding to $t = (t^1,\cdots,t^n)$. It should be noted that $t\varphi$ and $n\varphi$ have an invariant meaning only on the boundary of G. For two p-forms φ and ψ we write $\varphi = \psi$ on ∂G if

$$t\varphi = t\psi, \qquad n\varphi = n\psi. \tag{90}$$

1B. The point norm. For two p-forms φ and ψ on R we define the *point inner product* $\langle\varphi,\psi\rangle$ and the *point norm* $|\varphi|$ by

$$\varphi \wedge *\psi = \langle\varphi,\psi\rangle dV \tag{91}$$

and

$$|\varphi|^2 = \langle\varphi,\varphi\rangle. \tag{92}$$

If we set

$$\varphi^{i_1\cdots i_p} = \sum_{l_1,\cdots,l_p} g^{i_1l_1}\cdots g^{i_pl_p}\varphi_{l_1\cdots l_p}, \tag{93}$$

where $\varphi_{l_1\cdots l_p}$ is extended to unordered indices by skew-symmetry, then by (6)

$$(*\varphi)_{\beta_1\cdots\beta_{n-p}} = \operatorname{sgn}\begin{pmatrix}1\cdots p & p+1\cdots n\\ i_1\cdots i_p & \beta_1\cdots\beta_{n-p}\end{pmatrix}\sqrt{g}\,\varphi^{i_1\cdots i_p}. \tag{94}$$

Therefore

$$\langle\varphi,\psi\rangle = \varphi_{i_1\cdots i_p}\psi^{i_1\cdots i_p}, \qquad |\varphi|^2 = \varphi_{i_1\cdots i_p}\varphi^{i_1\cdots i_p}, \tag{95}$$

where our summation convention of 1.1B applies. Each is a global function on R.

Observe from (90) that $\varphi = \psi$ on ∂G if and only if

$$|\varphi-\psi| = 0 \quad \text{on} \quad \partial G. \tag{96}$$

2. Bounded principal forms

2A. Locally flat spaces. Hereafter in this section the symbol V shall stand for a locally flat Riemannian n-space. Let U_a be a parametric ball about $a \in V$ with local coordinates $x_a = (x_a^1,\cdots,x_a^n)$ such that $x_a(a) = 0$ and

$$g_{ij}(x_a) \equiv \delta_{ij} \quad \text{on} \quad \bar{U}_a. \tag{97}$$

Such a U_a will be called a *distinguished coordinate neighborhood.* Since V is locally flat we can cover it by a union of such neighborhoods.

We call V *parallel* if we can find a covering $\{U_a|a \in V\}$ of distinguished coordinate neighborhoods U_a with local coordinates $x_a = (x_a^1, \cdots, x_a^n)$ such that in $U_a \cap U_b$

$$x_a^i = x_{ab}^i + c_{ab}^i \qquad (i = 1, \cdots, n) \tag{98}$$

where c_{ab}^i are constants. The covering $\{U_a|a \in V\}$ shall be referred to as the *parallel coordinate covering.* On such a space V we consider harmonic p-forms c with $|c| \equiv$ const.; they will be called *constant p-forms.*

2B. Problem. Let A be the complement of a regular subregion of V with border α. Suppose a harmonic p-form σ is given on A. If there exists a harmonic p-form ρ on V such that

$$|\rho - \sigma| \leq M < \infty \tag{99}$$

on A then $\rho = \rho[B, \sigma]$ will be called a *B-principal form* with singularity σ; here B refers to boundedness. The problem is to determine whether ρ exists.

The solution is given as follows (Nakai-Sario [4], [10]):

Theorem. *If* $V \notin O_G$ *then a B-principal form* $\rho = \rho[B, \sigma]$ *always exists. If* $V \in O_G$ *and if* V *is parallel then a necessary and sufficient condition for the existence of a B-principal form* ρ *is that*

$$\int_\alpha * d\langle \sigma, c \rangle = 0 \tag{100}$$

for every constant form c. In the latter case the B-principal form is unique up to an additive constant form.

Thus the problem is solved completely for $V \notin O_G$. However, for a $V \in O_G$ which is not parallel the problem is still open.

The proof for a parallel V will be given in 2C to 2G. For an arbitrary $V \notin O_G$ it will be postponed until 3.

2C. Hereafter in 2, *V is assumed to be parallel.* We fix a parallel coordinate covering $\{U_a|a \in V\}$ of V.

Take a p-form φ on V: $\varphi = {}_a\varphi_{i_1\cdots i_p} dx_a^{i_1} \wedge \cdots \wedge dx_a^{i_p}$. On $U_a \cap U_b$, $dx_a^i = dx_b^i$ and therefore

$$ {}_a\varphi_{i_1\cdots i_p} = {}_b\varphi_{i_1\cdots i_p}. $$

For this reason there exists a global function $\varphi_{i_1\cdots i_p}$ on V such that

$$ \varphi_{i_1\cdots i_p} \equiv {}_a\varphi_{i_1\cdots i_p} \tag{101} $$

on U_a. Conversely, given functions $\varphi_{i_1\cdots i_p}$ there exists a p-form $\varphi = {}_a\varphi_{i_1\cdots i_p} dx_a^{i_1} \wedge \cdots \wedge dx_a^{i_p}$ with $\varphi_{i_1\cdots i_p} \equiv {}_a\varphi_{i_1\cdots i_p}$ on each U_a.

Now take a harmonic form $\varphi = \varphi_{i_1\cdots i_p} dx^{i_1} \wedge \cdots \wedge dx^{i_p}$ with respect to a system of distinguished coordinates. It is easy to check that

$$ \Delta\varphi = (\Delta\varphi_{i_1\cdots i_p}) dx^{i_1} \wedge \cdots \wedge dx^{i_p}. \tag{102} $$

We conclude that $\varphi_{i_1\cdots i_p} \in H(V)$, and that this actually is the characterizing property of the harmonicity of φ.

We also note that

$$ |\varphi|^2 = \sum (\varphi_{i_1\cdots i_p})^2. \tag{103} $$

Lemma. *If φ is a harmonic p-form on V and $|\varphi|$ is constant on some open set $D \subset V$, then φ is a constant form and*

$$ \varphi = c_{i_1\cdots i_p} dx^{i_1} \wedge \cdots \wedge dx^{i_p} \tag{104} $$

with constants $c_{i_1\cdots i_p}$ independent of a distinguished coordinate neighborhood.

In fact, if $\varphi = \varphi_{i_1\cdots i_p} dx^{i_1} \wedge \cdots \wedge dx^{i_p}$, then $(\varphi_{i_1\cdots i_p})^2$ is subharmonic, and so is $|\varphi|^2$. However since $|\varphi|^2 = c$ (const.) on D we have there

$$ c - (\varphi_{i_1\cdots i_p})^2 = \sum_{j \neq i} (\varphi_{j_1\cdots j_p})^2. $$

The left-hand member is superharmonic and the right-hand member subharmonic on D. Therefore $(\varphi_{i_1\cdots i_p})^2$ is harmonic on D. But

$$ \Delta(\varphi_{i_1\cdots i_p})^2 = 2\,|\operatorname{grad} \varphi_{i_1\cdots i_p}|^2, $$

and for this reason $\varphi_{i_1\cdots i_p}$ is constant on D. Hence $\varphi_{i_1\cdots i_p}$ is constant on all of V.

2D. Normal operators. Let L^p be the linear transformation of the space of p-forms on α into the space of p-forms continuous on A and harmonic on its interior, which satisfies

(D1) $L^p\varphi|\alpha = \varphi$,

(D2) $|L^p\varphi| \leq \sup_\alpha |\varphi|$,

(D3) $\int_\alpha * d\langle L^p\varphi, c\rangle = 0$ *for every constant form* c.

We call L^p a *normal operator* for A.

For $\varphi = \varphi_{i_1\cdots i_p} dx^{i_1} \wedge \cdots \wedge dx^{i_p}$ we can write

$$L^p\varphi = ({}_{i_1\cdots i_p}L\varphi_{i_1\cdots i_p}) dx^{i_1} \wedge \cdots \wedge dx^{i_p}.$$

We express this by the notation

$$L^p = {}_{i_1\cdots i_p}L\, dx^{i_1} \wedge \cdots \wedge dx^{i_p}.$$

It is easy to see that L^p is a normal operator if and only if every ${}_{i_1\cdots i_p}L$ is normal for A in the sense of 1.2A. In particular, if ${}_{i_1\cdots i_p}L = L_0$ or L_1 for all $i = (i_1 \cdots i_p)$ we denote the corresponding L^p by $L_0{}^p$ or $L_1{}^p$.

2E. A q-lemma. Given a compact set E in V let F_E^p be the class of harmonic p-forms φ in V such that $\langle\varphi, c\rangle$ is not of constant sign in E except for being identically zero for every constant form c.

Lemma. *There exists a constant* q_E *such that* $0 < q_E < 1$ *and*

$$\max_E |\varphi| \leq q_E \sup_V |\varphi| \tag{105}$$

for all $\varphi \in F_E^p$.

For the proof we have only to consider forms φ with $\sup_V|\varphi| = 1$. Suppose there existed a sequence with $\max_E|\varphi_m| \to 1$ as $m \to \infty$. Since $\{\varphi | \varphi \in F_E^p, \sup_V|\varphi| = 1\}$ is a normal family, we would have $\varphi = \lim_m \varphi_m \in F_E^p$ with $\max_E|\varphi| = 1$. By the subharmonicity of $|\varphi|^2$, φ would be a constant form c on V. The contradiction $\langle\varphi, c\rangle = \langle c, c\rangle = 1$ completes the proof.

2F. The Main Existence Theorem. Let L^p be a normal operator for A, and σ a harmonic p-form on the interior of A, continuous on A. We have the following result (Nakai-Sario [10]):

Theorem. *In order that a harmonic p-form ρ on V exist with the property*

$$L^p(\rho-\sigma) = \rho-\sigma \tag{106}$$

it is necessary and sufficient that σ satisfy

$$\int_\alpha *d\langle\sigma,c\rangle = 0 \tag{107}$$

for all constant forms c. The solution ρ is unique up to an additive constant form.

The proof is analogous to that of Theorem 1.2B and we restrict ourselves to a brief outline.

Let $V_0 \subset V$ be a regular region with border α_0 such that $\overline{V-A} \subset V_0$. Denote by K the Dirichlet operator for V_0 (cf. 1.2C) and set

$$K^p\varphi = K\varphi_{i_1\cdots i_p}\, dx^{i_1}\wedge\cdots\wedge dx^{i_p}.$$

We have only to establish the convergence of $\varphi = \sum_0^\infty (L^pK^p)^m\sigma_0$ with $\sigma_0 = \sigma - L^p\sigma$.

Observe that condition (107) means that $\int_\gamma *d\langle\sigma,c\rangle = 0$ for every γ homologous to α. We conclude that

$$\int_\alpha \langle K^p(L^pK^p)^m\sigma_0,c\rangle *dh = 0,$$

with h the harmonic measure of α_0 in $\bar V_0 \cap A$. For this reason $K^p(L^pK^p)^m\sigma_0 \in F^p_\alpha(A)$, Lemma 2E applies in V_0, and we have the convergence.

2G. Proof of Theorem 2B for parallel V. Suppose $V \notin O_G$. Then σ need not satisfy (107). We set

$$\psi = \left(-\int_\alpha *d\sigma_{i_1\cdots i_p}\Big/\int_\alpha *d\omega\right)\omega\, dx^{i_1}\wedge\cdots\wedge dx^{i_p}$$

where $\sigma = \sigma_{i_1\cdots i_p} dx^{i_1} \wedge \cdots \wedge dx^{i_p}$ is the global expression in A and ω is the harmonic measure of the ideal boundary β of V with respect to A. Clearly $|\psi|$ is bounded in A. Consequently $\tilde{\sigma} = \sigma + \psi$ satisfies (107) and the solution ρ has the property

$$\rho - \sigma = L^p(\rho - \tilde{\sigma}) + \psi$$

on A. We infer that $|\rho - \sigma|$ is bounded.

The second part of Theorem 2B is nothing but Theorem 2F.

3. Border reduction

3A. Generalized Dirichlet operator. Given $f \in C(\alpha)$, we wish to construct a harmonic function u on the interior of A with boundary values f on α and "0 on the ideal boundary β of V." The function u is often referred to as the solution of the generalized Dirichlet problem for A (cf. Constantinescu-Cornea [1, p. 20]). Take an exhaustion $\{V_m\}_1^\infty$ of V and let $K_{0m} f$ be the harmonic function on the interior of $V_m \cap A$ with boundary values f on α and 0 on ∂V_m. Clearly

$$K_0 f = \lim_{m \to \infty} K_{0m} f$$

exists and is harmonic on the interior of A with boundary values f on α. The operator K_0 may again be called the *Dirichlet operator* for A, for if $\beta = \varnothing$ or $c_\beta = 0$ then K_0 is the genuine Dirichlet operator.

We next consider the linear operator $K_0{}^p$ from the space of continuous p-forms on α into the space of continuous p-forms on A which are harmonic on the interior of A with the following properties:

(A1) $K_0{}^p \varphi | \alpha = \varphi$,

(A2) $|K_0{}^p \varphi|^2 \leq \| \, |\varphi| \, \|_\alpha^2 u_\beta$.

Here u_β is the harmonic measure of α with respect to A. The operator $K_0{}^p$ will be again called the *Dirichlet operator* for A.

If V is parallel then $K_0{}^p$ is given simply by

$$K_0{}^p = \sum K_0 dx^{i_1} \wedge \cdots \wedge dx^{i_p}. \tag{108}$$

It is not normal in general. It is normal, however, if and only if $c_\beta = 0$, or what amounts to the same, $V \in O_G$.

For the general case we need the following rather deep result, for the proof of which we refer the reader to Duff-Spencer [2, p. 137]:

Let G be a regular region of a Riemannian space R. Suppose that $\Delta\varphi = 0$ and $\varphi = 0$ on the boundary imply $\varphi \equiv 0$. Then there exists a unique harmonic p-form φ on G with φ prescribed on the boundary.

For V, since $|\varphi|$ is subharmonic for φ with $\Delta\varphi = 0$, $\varphi = 0$ on the boundary implies $\varphi \equiv 0$. Therefore we can find a unique harmonic p-form $K_{0m}^p\varphi$ on the interior of $A \cap V_m$ with $K_{0m}^p\varphi|\alpha = \varphi$ and $K_{0m}^p\varphi|\partial V_m = 0$. It clearly satisfies (A2) for $A \cap V_m$. Therefore by the harmonicity of the coefficients of $K_{0m}^p\varphi$ and (103) for a general V, we can see that

$$K_0{}^p\varphi = \lim_{m\to\infty} K_{0m}^p\varphi$$

satisfying (A1) and (A2) exists on A.

Thus the existence of the Dirichlet operator $K_0{}^p$ for a general V is established.

3B. Suppose a continuous p-form σ is given on A which is harmonic on the interior of A. Let us try to find a harmonic p-form ρ on V such that

$$K_0{}^p(\rho - \sigma) = \rho - \sigma \tag{109}$$

on A.

3C. Border reduction theorem. Let A' be the complement of a regular region of V such that the interior of A' contains A. We set $\alpha' = \partial A'$. Typically we have in mind the case where $V - A'$ is a parametric ball. Suppose there exists a harmonic p-form σ on the interior of A which is continuous on A. Then we have the following (Nakai-Sario [4]):

Theorem. *There always exists a unique continuous p-form σ'*

on A which is harmonic on the interior of A' and satisfies

(110) $$\sigma'|\alpha' = 0,$$

(111) $$K_0^p(\sigma'-\sigma) = \sigma'-\sigma$$

on A, where K_0^p is the Dirichlet operator for A.

The meaning of the theorem is that in the principal function problem we may reduce α to a geometrically simpler configuration such as a sphere.

The unicity is simple. Indeed suppose there exist two required forms, σ' and $\bar{\sigma}'$. Since the function $|\bar{\sigma}'-\sigma'|^2$, subharmonic on the interior of A' and vanishing on α' also "vanishes" on the ideal boundary β of A', the maximum principle gives $|\bar{\sigma}'-\sigma'|^2 = 0$.

3D. For the existence proof take a regular region $V_0 \supset \overline{A'-A-\alpha'}$ of V contained in the interior of A' and set $\alpha_0 = \partial V_0 \cap A$. Let L^p be the Dirichlet operator for $V_0 \cup \alpha_0$, with α' considered as the ideal boundary of V_0. As in the proof of Theorem 2F we have only to establish the uniform convergence on α_0 of $\sigma' = \sum_0^\infty (K_0^p L^p)^m \sigma_0$ with $\sigma_0 = \sigma - K_0^p\sigma$.

To this end we first make the following observation. Let F_0 be the family of harmonic p-forms φ on V_0, continuous on $\bar{V}_0$ with $\varphi|\alpha' = 0$. Let ω be the harmonic measure of α_0 with respect to V_0 and set

$$q = \sqrt{\max_\alpha \omega}.$$

By the maximum principle $0<q<1$ and we have

(112) $$\max_\alpha |\varphi| \leq q \max_{\alpha_0} |\varphi|$$

for every $\varphi \in F_0$. In fact, $|\varphi|^2 \leq (\max_{\alpha_0}|\varphi|^2)\,\omega$ on V_0 implies (112).

Now set $M = \|\,|\sigma_0|\,\|_{\alpha_0}$. By (112) and the fact that the image of L^p is contained in F_0,

(113) $$\|\,|L^p(K_0^pL^p)^{m-1}\sigma_0|\,\|_\alpha \leq q\|\,|L^p(K_0^pL^p)^{m-1}\sigma_0|\,\|_{\alpha_0}$$
$$= q\|\,|(K_0^pL^p)^{m-1}\sigma_0|\,\|_{\alpha_0}.$$

By (A2) in 3A

(114) $$\|\,|(K_0^pL^p)^m\sigma_0|\,\|_{\alpha_0} \leq \|\,|L^p(K_0^pL^p)^{m-1}\sigma_0|\,\|_\alpha.$$

Hence by (113) and (114)

$$|| \, |(K_0^p L^p)^m \sigma_0| \, ||_{\alpha_0} \leq q || \, |(K_0^p L^p)^{m-1} \sigma_0| \, ||_{\alpha_0}$$

and a fortiori $|| \, |(K_0^p L^p)^m \sigma_0| \, ||_{\alpha_0} \leq q^n M$. This proves the uniform convergence of σ'.

3E. Solution to Problem 3B. If $V \in O_G$ and if V is parallel then K_0^p is the single normal operator for A, and Theorem 2F gives the solution to 3B, i.e. condition (107) is necessary and sufficient for the existence of a ρ satisfying (109).

In contrast, *if* $V \notin O_G$ *then* ρ *always exists.* We shall now prove this. By Theorem 3C we may suppose that A is connected and $\sigma|\alpha = 0$. This time let V_0 be a regular region of V with $V_0 \supset \overline{V - A}$ and set $\alpha_0 = \partial V_0$. We take the Dirichlet operator K^p for V_0. Again we have only to establish the uniform convergence of $\rho = \sum_0^\infty (K_0^p K^p)^m \sigma$ on α_0.

To this end consider the class F of continuous p-forms φ on A which satisfy $K_0^p \varphi = \varphi$. Take u_β for A. Since A *is connected* $V \notin O_G$ implies that $0 < u_\beta < 1$ on the interior of A. By (A2) we infer that $|\varphi|^2 = |K_0^p \varphi|^2 \leq || \, |\varphi| \, ||_\alpha^2 u_\beta$ on A. Hence on setting $q^2 = ||u_\beta||_{\alpha_0}$ we obtain $0 < q < 1$ and

$$|| \, |\varphi| \, ||_{\alpha_0} \leq q || \, |\varphi| \, ||_\alpha \tag{115}$$

for every $\varphi \in F$

Let $M = || \, |\sigma| \, ||_{\alpha_0}$. Since $|K^p \varphi|$ attains its maximum on α_0,

$$|| \, |K^p (K_0^p K^p)^{m-1} \sigma| \, ||_\alpha \leq || \, |K^p (K_0^p K^p)^{m-1} \sigma| \, ||_{\alpha_0}. \tag{116}$$

By (115) and (A1) in 3A

$$|| \, |(K_0^p K^p)^m \sigma| \, ||_{\alpha_0} \leq q || \, |K_0^p K^p (K_0^p K^p)^{m-1} \sigma| \, ||_\alpha$$

$$= q || \, |K^p (K_0^p K^p)^{m-1} \sigma| \, ||_\alpha.$$

This with (116) gives

$$|| \, |(K_0^p K^p)^m \sigma| \, ||_{\alpha_0} \leq q || \, |(K_0^p K^p)^{m-1} \sigma| \, ||_{\alpha_0},$$

and we conclude that $|| \, |(K_0^p K^p)^m \sigma| \, ||_{\alpha_0} \leq q^m M$.

3F. The solution to Problem 3B is automatically the one to Problem 2B. Thus the proof of the first part of Theorem 2B for a general V is also herewith complete.

§3. PRINCIPAL FORMS ON RIEMANNIAN SPACES

We turn to harmonic p-forms on arbitrary Riemannian spaces. After a brief preliminary on the classification of p-form, we discuss principal harmonic fields in 2 and then principal harmonic forms in 3. Some in-between classes will be considered in 4. A generalization which contains the classes in 2–4 as special cases will be given in 5.

The results in this section were obtained in Nakai-Sario [3], [5], [9], [12], and Larsen-Nakai-Sario [1].

We turn to harmonic p-forms on arbitrary Riemannian spaces.

1. Classes of p-forms

1A. Weak derivatives. We denote the space of all p-forms on a Riemannian space R with locally Lebesgue square integrable coefficients by E^p_{loc}. For $\varphi, \psi \in E^p_{\text{loc}}$ we set

$$(\varphi,\psi) = \int_R \varphi \wedge *\psi, \qquad ||\varphi||^2 = (\varphi,\varphi) \tag{117}$$

whenever the integrals are meaningful. The subspace of all p-forms with C^∞ coefficients and compact supports is denoted by $D_\infty{}^p$.

Besides the exterior derivative d, coderivative δ, and the Laplace-Beltrami operator $\Delta = d\delta + \delta d$, we shall also consider the operators

$$\Delta' = \delta d, \qquad \Delta'' = d\delta. \tag{118}$$

Let T be one of these operators and T^* its adjoint, i.e. $d^* = \delta$, $\delta^* = d$, $\Delta^* = \Delta$, $(\Delta')^* = \Delta'$, and $(\Delta'')^* = \Delta''$. For a given $\alpha \in E^p_{\text{loc}}$ we set $T\alpha = \beta$ if there exists a $\beta \in E_{\text{loc}}$ such that

$$(\beta,\varphi) = (\alpha,T^*\varphi) \tag{119}$$

for every $\varphi \in D_\infty^p$ when $T = \Delta, \Delta', \Delta''$; for every $\varphi \in D_\infty^{p+1}$ when $T = d$; and for every $\varphi \in D_\infty^{p-1}$ when $T = \delta$. The generalized operation $T\alpha$ for $\alpha \in E_{\text{loc}}^p$ is called the weak derivative. If α is of class C^∞ then $T\alpha$ has its conventional meaning.

The following Weyl-Kodaira-de Rham theorem is fundamental in the sequel. For the proof we refer the reader to the monograph of de Rham [2].

If $\Delta\alpha = 0$ for $\alpha \in E_{\text{loc}}^p$ then α is C^∞, i.e., there exists a C^∞ harmonic form φ with $\|\alpha - \varphi\| = 0$.

1B. Subclasses of harmonic forms. A p-form φ satisfying

$$d\varphi = 0, \qquad \delta\varphi = 0 \tag{120}$$

is called a *harmonic p-field.* Clearly harmonic fields are harmonic forms. A harmonic 0-field is a constant and a harmonic n-field is a constant multiple of dV.

A closed harmonic p-form φ is characterized by

$$\Delta''\varphi = 0, \qquad d\varphi = 0 \tag{121}$$

and a coclosed harmonic p-form by

$$\Delta'\varphi = 0, \qquad \delta\varphi = 0. \tag{122}$$

The forms which satisfy (121) or (122) are called *p-tensor potentials.*

We shall also consider p-forms φ with

$$\Delta'\varphi = \Delta''\varphi = 0, \tag{123}$$

which will be referred to as *harmonic semifields.*

The following inclusion relations are obviously valid:

$$\{\text{harmonic forms}\} \supset \{\text{harmonic semifields}\} \supset \{p\text{-tensor potentials}\} \supset \{\text{harmonic fields}\}. \tag{124}$$

1C. Green's formulas. Let G be a regular region of R. For a

p-form φ and a $(p+1)$-form ψ we have by (12)

$$\int_G d(\varphi \wedge *\psi) = \int_G (d\varphi \wedge *\psi + (-1)^p \varphi \wedge d*\psi)$$

$$= \int_G (d\varphi \wedge *\psi - \varphi \wedge *\delta\psi) = \int_{\partial G} \varphi \wedge *\psi.$$

Therefore we can write

$$(d\varphi,\psi)_G - (\varphi,\delta\psi)_G = \int_{\partial G} \varphi \wedge *\psi. \tag{125}$$

From this we immediately have Green's formulas:

$$(d\varphi,d\psi)_G - (\varphi,\Delta'\psi)_G = \int_{\partial G} \varphi \wedge *d\psi, \tag{126}$$

$$(\Delta''\varphi,\psi)_G - (\delta\varphi,\delta\psi)_G = \int_{\partial G} \delta\varphi \wedge *\psi, \tag{127}$$

$$(\Delta'\varphi,\psi)_G - (\varphi,\Delta'\psi)_G = \int_{\partial G} (\varphi \wedge *d\psi - \psi \wedge *d\varphi), \tag{128}$$

$$(\Delta''\varphi,\psi)_G - (\varphi,\Delta''\psi)_G = \int_{\partial G} (\delta\varphi \wedge *\psi - \delta\psi \wedge *\varphi), \tag{129}$$

$$(\Delta\varphi,\psi)_G - (\varphi,\Delta\psi)_G = \int_{\partial G} (\varphi \wedge *d\psi - \psi \wedge *d\varphi) + (\delta\varphi \wedge *\psi - \delta\psi \wedge *\varphi). \tag{130}$$

2. Principal harmonic fields

2A. Problem. Let A be the complement of a regular region of a Riemannian space R. Set $\alpha = \partial A$. Let σ be a harmonic p-field in the interior of A and C^∞ on R. We are to find a harmonic p-field ρ on R which "imitates" σ in the neighborhood of the ideal boundary β of R. It is in the nature of the problem that only the case $0<p<n$ offers any interest.

We start by describing our choice of the imitation. We denote by $E_d{}^p$ and $E_\delta{}^p$ the spaces

$$E_d{}^p = \overline{dD_\infty^{p-1}}, \qquad E_\delta{}^p = \overline{\delta D_\infty^{p+1}} \tag{131}$$

where the closure is taken with respect to the norm $||\cdot|| = (\cdot,\cdot)^{1/2}$. Clearly $E_d{}^p$ and $E_\delta{}^p$ are mutually orthogonal closed subspaces of the Hilbert space $E^p = \overline{D_\infty{}^p}$. We also consider the direct sum

$$E_{d\delta}^p = E_d^p \oplus E_\delta^p = \overline{dD_\infty{}^{p-1} + \delta D_\infty{}^{p+1}}. \tag{132}$$

By the Weyl-Kodaira-de Rham theorem the orthogonal complement of $E_{d\delta}^p$ in E^p is the space F^p of harmonic p-fields in E^p

$$E^p = E_{d\delta}^d \oplus F^p. \tag{133}$$

Denote by $L_{d\delta}^p$ the projection operator from E^p onto $E_{d\delta}^p$. For the required imitation of σ by ρ we choose

$$L_{d\delta}^p(\rho - \sigma) = \rho - \sigma. \tag{134}$$

This means two things. First, $||\rho - \sigma|| < \infty$. Second, in view of the elements of $E_{d\delta}^p$ being limits of forms in $dD_\infty{}^{p-1} + \delta D_\infty{}^{p+1}$, $\rho - \sigma$ has in some sense zero data on the ideal boundary β of R. Thus the imitation is close indeed.

The solution ρ will be called an *$L_{d\delta}^p$-principal field* for σ.

2B. Main theorem. A complete solution of our problem is as follows (Nakai-Sario [5]):

Theorem. *A necessary and sufficient condition for the existence of an $L_{d\delta}^p$-principal field is that*

$$\int_\alpha \sigma \wedge *\varphi = O(||\delta\varphi||) \tag{135}$$

for all $\varphi \in D_\infty^{p+1}$ and

$$\int_\alpha *\sigma \wedge \psi = O(||d\psi||) \tag{136}$$

for all $\psi \in D_\infty^{p-1}$.

It is important to observe that the conditions are given in terms of σ on α only.

For the proof note that $(\sigma,\delta\varphi) = (d\sigma,\varphi) = (d\sigma,\varphi)_{R-A}$. By (125) we have

$$|(\sigma,\delta\varphi)| = \left|O(||\delta\varphi||) + \int_\alpha \sigma \wedge *\varphi\right|. \tag{137}$$

Similarly we obtain

$$|(\sigma,d\psi)| = \left|O(||d\psi||) + \int_\alpha *\sigma \wedge \psi\right|. \tag{138}$$

Therefore (135) and (136) are equivalent to

$$(\sigma,\delta\varphi) = O(||\delta\varphi||) \tag{135$'$}$$

and

$$(\sigma,d\psi) = O(||d\psi||). \tag{136$'$}$$

Since $\delta\varphi$ and $d\psi$ are orthogonal, (135) and (136) together are equivalent to

$$(\sigma,\eta) = O(||\eta||) \tag{139}$$

for every $\eta \in dD_\infty^{p-1} + \delta D_\infty^{p+1}$.

Suppose the principal field ρ exists. Then by setting $\rho - \sigma = \tau \in E_{d\delta}^p$ we see that $(\sigma,\eta) = (\rho,\eta) - (\tau,\eta) = O(||\eta||)$, since ρ is orthogonal to $dD_\infty^{p-1} + \delta D_\infty^{p+1}$.

Conversely if (139) is valid then $(\sigma,\cdot)$ can be continuously extended to $E_{d\delta}^p$, and there exists a form $-\tau \in E_{d\delta}^p$ such that $(\sigma,\eta) = (-\tau,\eta)$ for every $\eta \in dD_\infty^{p-1} + \delta D_\infty^{p+1}$. If we set $\rho = \sigma + \tau$ then ρ is orthogonal to $dD_\infty^{p-1} + \delta D_\infty^{p+1}$ and hence is a harmonic field on R.

This completes the proof.

2C. Specialization. Since the conditions (135) and (136) are in terms of $\sigma|\alpha$ alone, they can be satisfied by forms with strikingly simple properties. For instance (Nakai-Sario [5]):

Theorem. *An $L^p_{d\delta}$-principal field ρ exists if*

$$\sigma|\alpha = d\theta, \qquad *\sigma|\alpha = d\tau \tag{140}$$

for some forms θ, τ defined in a neighborhood U of α.

For the proof take neighborhoods U_0, U_1 of α such that $\bar{U}_1$ is compact and $\bar{U}_0 \subset U_1 \subset \bar{U}_1 \subset U$. Let $f \in C^\infty(R)$ be 1 on $\bar{U}_0$ and 0 on $R - U_1$. Then $\theta_0 = f\theta$ and $\tau_0 = f\tau$ are C^∞ forms on R.

We conclude that

$$\left|\int_\alpha \sigma \wedge *\varphi\right| = \left|\int_\alpha d\theta_0 \wedge *\varphi\right| = \left|\int_{R-A} d\theta_0 \wedge *\delta\varphi\right| = O(\|\delta\varphi\|),$$

i.e. condition (135) is met. Similarly (136) is satisfied.

2D. Sphere-like components. Perhaps somewhat surprisingly, no conditions whatever are needed on σ if the components α_i of α are *sphere-like.* By this we mean that they have disjoint neighborhoods, each homeomorphic to a spherical shell $S_i = S^{n-1} \times [-1,1]$ such that α_i goes to $S^{n-1} \times 0$, where S^{n-1} is a sphere.

By the Poincaré - de Rham theorem (see e.g. Whitney [1, Ch. IV], de Rham [2, Ch. IV]), a closed form γ on $S^{n-1} \times [-1,1]$ is an exact form, i.e. $\gamma = d\eta$, if the degree of γ is different from $n-1$. If it is $n-1$ then $\int_{S^{n-1}\times 0} \gamma = 0$ assures the same conclusion.

We thus have the following result (Nakai-Sario [5]):

Theorem. *Let the α_i be sphere-like. Then an $L^p_{d\delta}$-principal field ρ exists for every σ if $1 < p < n-1$; for every σ with $\int_{\alpha_i} *\sigma = 0$ if $p = 1$; for every σ with $\int_{\alpha_i} \sigma = 0$ if $p = n-1$.*

2E. It is clear that the α_i are not always sphere-like, as is exemplified by the solid torus. But for $n = 2$ this condition is automatically fulfilled, and we have (Nakai-Sario [5]):

Theorem. *On Riemannian 2-spaces (hence on Riemann surfaces) an $L^1_{d\delta}$-principal field ρ exists if*

$$\int_{\alpha_i} \sigma = \int_{\alpha_i} *\sigma = 0. \tag{141}$$

Note that a harmonic 1-field is a harmonic differential of order 1. For exact differentials, i.e. for differentials of functions, the first part of (141) is satisfied by definition. In the second part we recognize the flux condition.

It is known that in the 2-dimensional case, however, no condition on σ is needed for the existence of ρ if $R \notin O_G$. On the other hand, the criterion $\int_\alpha \sigma = \int_\alpha * \sigma = 0$ is necessary and sufficient if $R \in O_G$ (Nakai [6]).

2F. Point singularities. On a Riemannian space R_0 let K be a compact discrete set contained in a finite set B of disjoint balls B_i. Let σ be a harmonic p-field on $\bar{B} - K$ with singularities at K. The problem is to find a harmonic field ρ on $R_0 - K$ with behavior σ on $B - K$. We also require that $\|\rho\|_{R_0 - B} < \infty$.

Let B' be a set of slightly larger disjoint balls $B_i' \supset B_i$. For the space R of our main theorem we take $R_0 - K$ and set $A = (\bar{B} - K) \cup (R_0 - B')$. Then the α_i are sphere-like. For σ we take the given σ in $\bar{B} - K$ and set $\sigma = 0$ in $R_0 - B'$. Theorem 2D applies and we conclude (Nakai-Sario [5]):

Theorem. *An $L^p_{d\delta}$-principal field ρ with given singularities σ at $K \subset R$ exists for every σ if $1 < p < n-1$; for every σ with $\int_{\alpha_i} * \sigma = 0$ if $p = 1$; for every σ with $\int_{\alpha_i} \sigma = 0$ if $p = n-1$.*

2G. Ahlfors' method. Ahlfors' method (Ahlfors-Sario [1, p. 300]) may also be applied for the solution of Problem 2F. It actually gives a sharper result, verbally communicated to the authors by Royden:

Theorem. *There exists a harmonic p-field ρ such that $\rho - \sigma$ is harmonic in B and $\|\rho\|_{R_0 - B} < \infty$ for every σ if $1 < p < n-1$. If $p = 1$ or $n-1$ then $\int_{\alpha_i} * \sigma = 0$ or $\int_{\alpha_i} \sigma = 0$ guarantees the same conclusion.*

Take B' of 2F so small that σ is still defined on $\overline{B'}$. The restrictions on σ imply the existence of a $C^\infty(p-1)$-form θ and a C^∞ $(n-p-1)$-form τ on $B' - K$ such that

$$\sigma = d\theta, \qquad * \sigma = d\tau. \tag{142}$$

Let $f \in C^\infty(R_0)$ such that $f \equiv 1$ on $\bar{B}$ and the support of f is contained in B'. Then

$$\theta_0 = f\theta, \qquad \tau_0 = f\tau$$

are C^∞-forms on R_0. Clearly

$$d\theta_0 - (-1)^{np+p} * d\tau_0 = d\theta_0 - (-1)^{np+n}\delta * \tau_0$$

vanishes on $\bar{B} \cup (R_0 - B')$ and therefore belongs to $D_\infty{}^p$ on R_0.

From (132) and (133) it follows that

$$d\theta_0 - \varepsilon\delta * \tau_0 = \xi + \eta + \omega \tag{143}$$

with $\xi \in E_d{}^p$, $\eta \in E_\delta{}^p$, $\omega \in F^p$. Here $\varepsilon = (-1)^{np+n}$. Let

$$\rho \equiv d\theta_0 - \xi = \varepsilon\delta * \tau_0 + \eta + \omega. \tag{144}$$

If we apply d, δ in the weak sense to ρ then

$$d\rho = d(d\theta_0 - \xi) = 0, \qquad \delta\rho = \delta(\varepsilon\delta * \tau_0 + \eta + \omega) = 0.$$

Therefore ρ is a harmonic field on $R = R_0 - K$ and, since $d\theta_0 | B = \sigma | B$, $(\sigma - \rho) | B = \xi | B$. In (143), ξ and η are C^∞ along with $d\theta_0 - \varepsilon\delta * \tau_0$ (see de Rham [2]). Thus the conditions $d\xi = 0$, and $\delta\xi = 0$ which held on $B - K$ remain valid on all of B, i.e. $\xi = \sigma - \rho$ is harmonic on B. The finiteness of $\|\rho\|_{R_0 - B}$ is obvious. Actually $\rho \in E_d{}^p \subset E_{d\delta}^p$ on R_0 if we suitably modify ρ about K.

3. Principal harmonic forms

3A. Problem. A discussion analogous to that in 2 can be given for harmonic forms. In fact, let σ be a harmonic p-form in the interior of A and C^∞ on R. We are interested in finding a harmonic p-form ρ on R which imitates the behavior of σ in A. This time we include in our consideration every p with $0 \leq p \leq n$.

In order to specify the imitation, set

$$E_\Delta{}^p = \overline{\Delta D_\infty{}^p}, \tag{145}$$

and denote by H^p the space of harmonic p-forms in $E^p = \overline{D_\infty{}^p}$. It follows easily, by the Weyl-Kodaira-de Rham theorem, that

$$E^p = E_\Delta{}^p \oplus H^p. \tag{146}$$

In the same manner as in 2A we let $L_\Delta{}^p$ signify the projection of E^p onto $E_\Delta{}^p$. The imitation is then in the sense

$$L_\Delta{}^p(\rho-\sigma) = \rho-\sigma. \tag{147}$$

The solution ρ will be called the *$L_\Delta{}^p$-principal form* for σ.

3B. Main theorem. A complete solution can again be given in terms of σ on α alone (Larsen-Nakai-Sario [1]):

Theorem. *A necessary and sufficient condition for the existence of an $L_\Delta{}^p$-principal form ρ is that*

$$\int_\alpha \sigma\wedge *d\varphi-\delta\varphi\wedge *\sigma+\delta\sigma\wedge *\varphi-\varphi\wedge *d\sigma = O(\|\Delta\varphi\|) \tag{148}$$

for every $\varphi\in D_\infty{}^p$.

In fact, $(\sigma,\Delta\varphi) = (\Delta\sigma,\varphi) = (\Delta\sigma,\varphi)_{R-A}$. By (130) the left-hand member of (148) is equal to $(\sigma,\Delta\varphi)-(\sigma,\Delta\varphi)_{R-A}$. Thus (148) is equivalent to

$$(\sigma,\Delta\varphi) = O(\|\Delta\varphi\|) \tag{148}$$

for every $\varphi\in D_\infty{}^p$.

If a principal form ρ with $\rho-\sigma = \tau\in E_\Delta$ exists then, since $(\rho,\Delta\varphi) = (\Delta\rho,\varphi) = 0$, we have $(\sigma,\Delta\varphi) = (-\tau,\Delta\varphi) = O(\|\Delta\varphi\|)$.

Conversely, if (148)′ is valid then $(-\sigma,\cdot)$ is bounded on $\Delta D_\infty{}^p$ and there exists a $\tau\in E_\Delta{}^p$ with $(-\sigma,\cdot) = (\tau,\cdot)$ on $\Delta D_\infty{}^p$. Set $\rho = \sigma+\tau$. Since $(\rho,\Delta\varphi) = 0$ for all $\varphi\in D_\infty{}^p$, ρ is harmonic on R and $\rho-\sigma = \tau\in E_\Delta^p$.

3C. Specialization. As a counterpart of Theorem 2C we state (Larsen-Nakai-Sario [1]):

Theorem. *If there exists an exact p-form θ and an exact $(n-p)$-form τ defined in a neighborhood of α such that*

$$\sigma|\alpha = \theta, \qquad *\sigma|\alpha = \tau \tag{149}$$

then an $L_\Delta{}^p$-principal form ρ exists.

There exist forms $\theta_0 \in D_\infty^{p-1}$ and $\tau_0 \in D_\infty^{n-p-1}$ such that $\theta = d\theta_0$ and $\tau = d\tau_0$ in a neighborhood of α. Thus the left-hand member of (148) is equal to

$$\int_\alpha d\theta_0 \wedge *d\varphi - \delta\varphi \wedge d\tau_0 = (-1)^p \int_{R-A} d\theta_0 \wedge d*d\varphi - \int_{R-A} d\delta\varphi \wedge d\tau_0$$

which in turn equals $O(\|\delta d\varphi\|) + O(\|d\delta\varphi\|) = O(\|\Delta\varphi\|)$, since $(d\delta\varphi, \delta d\varphi) = 0$.

3D. Counterparts of 2D to 2G are not valid for harmonic forms unless we add the additional assumption that σ and $*\sigma$ are closed. From the heuristic discussion in §2 it is plausible to reason that a condition for complete solvability might be hyperbolicity or a condition similar to (100). To obtain the analogue of (100) for arbitrary Riemannian spaces seems to be the most important problem in this direction.

For harmonic fields we have reached a rather satisfactory result in the sense that at least in the case of point singularities the problem is solved. It is indicative of the present early state of the theory that even this concrete case is not settled for harmonic forms. The problem and its ramifications seem to offer considerable interest.

4. Principal semifields

4A. Semifields. We next consider the case in which σ is a semifield on the interior of A and C^∞ on R. We set

$$E^p_{\Delta'} = \overline{\Delta' D_\infty{}^p}, \qquad E^p_{\Delta''} = \overline{\Delta'' D_\infty{}^p}, \qquad E^p_{\Delta'\Delta''} = E^p_{\Delta'} \oplus E^p_{\Delta''} \tag{150}$$

and denote by SF^p the class of semifields in E^p. It is easy to see that

$$E^p = E^p_{\Delta'\Delta''} \oplus SF^p. \tag{151}$$

We are to find an $L^p_{\Delta'\Delta''}$-*principal semifield* ρ where $L^p_{\Delta'\Delta''}$ is the projection of E^p onto $E^p_{\Delta'\Delta''}$: $L^p_{\Delta'\Delta''}(\rho-\sigma) = \rho-\sigma$.

In analogy with Theorem 2B we obtain (Nakai-Sario [9]):

Theorem. *An* $L^p_{\Delta'\Delta''}$-*principal semifield* ρ *exists if and only if*

$$\int_\alpha \sigma\wedge *d\varphi-\varphi\wedge *d\sigma = O(||\Delta'\varphi||) \tag{152}$$

and

$$\int_\alpha \delta\sigma\wedge *\varphi-\delta\varphi\wedge *\sigma = O(||\Delta''\varphi||) \tag{153}$$

for every $\varphi\in D_\infty{}^p$.

From (128) and (129) it follows that (152) and (153) are equivalent to

$$(\sigma,\Delta'\varphi) = O(||\Delta'\varphi||) \tag{152$'$}$$

and

$$(\sigma,\Delta''\varphi) = O(||\Delta''\varphi||) \tag{153$'$}$$

respectively. These two together are equivalent to

$$(\sigma,\eta) = O(||\eta||) \tag{154}$$

for every $\eta\in\Delta' D_\infty{}^p+\Delta'' D_\infty{}^p$. The same argument as in 2B provides us with a proof of our theorem.

4B. Tensor potentials. Finally we consider the case in which σ is a closed harmonic p-form (resp. coclosed harmonic p-form) in A and C^∞ on R. Let

$$E^p_{\delta\Delta''} = E_\delta{}^p\oplus E^p_{\Delta''}, \qquad E^p_{d\Delta'} = E_d{}^p\oplus E^p_{\Delta'}. \tag{155}$$

If $P_d{}^p$ (resp. $P_\delta{}^p$) denotes the class of all closed (resp. coclosed) harmonic p-forms in E^p then we can easily conclude that

$$E^p = E^p_{\delta\Delta''}\oplus P_d{}^p = E^p_{d\Delta'}\oplus P_\delta{}^p. \tag{156}$$

Let $L^p_{\delta\Delta''}$ (resp. $L^p_{d\Delta'}$) be the projection operator from E^p onto $E^p_{\delta\Delta''}$ (resp. $E^p_{d\Delta'}$). We are to find an $L^p_{\delta\Delta''}$-(resp. $L^p_{d\Delta'}$-) *principal tensor potential* ρ which is a closed (resp. coclosed) harmonic p-form ρ on R with $L^p_{\delta\Delta''}(\rho-\sigma) = \rho-\sigma$ (resp. $L^p_{d\Delta'}(\rho-\sigma) = \rho-\sigma$). We claim (Larsen-Nakai-Sario [1]):

Theorem. *An* $L^p_{\delta\Delta''}$-*(resp.* $L^p_{d\Delta'}$-*) principal tensor potential exists if and only if both* (135) *and* (152) *(resp.* (136) *and* (153)*) are valid.*

We shall give the proof for closed harmonic forms only. The case of coclosed harmonic forms is similar.

Assume that (135) and (152) are valid. As before $(-\sigma,\cdot)$ is bounded on δD^{p+1}_∞ and $\Delta'' D_\infty{}^p$, and consequently on $\delta D^{p+1}_\infty \oplus \Delta'' D_\infty{}^p$. Therefore there exists a $\tau \in E^p_{\delta\Delta''}$ such that $(\tau,\cdot) = (-\sigma,\cdot)$. Setting $\rho = \tau+\sigma$ we see that $d\rho = 0$ and $\Delta''\rho = 0$ in the weak sense. However, $\Delta\rho = \delta(d\rho)+\Delta''\rho = 0$ in the weak sense, and a fortiori ρ is C^∞. We conclude that $d\rho = 0$ and $\Delta''\rho = 0$ in the strict sense, and that $\rho-\sigma = \tau \in E_{\delta\Delta''}$.

Conversely assume the existence of such a ρ. Then on setting $\rho-\sigma = \tau \in E_{\delta\Delta''}$, we have (135) as in 2B. Formula (152) is obtained similarly.

5. Generalization

5A. L_{T^*}-principal forms. We close by considering the more general situation where no assumptions are made on σ other than that it belong to the space E_{loc} of locally square integrable differential p-forms. Here and in the sequel we drop the upper index p from the notation, as the degree will always be clear from the context.

We also unify the treatment by dealing with a more general differential operator T, e.g. a polynomial of operators d and δ, acting on D_∞. It is extended to those elements α of E_{loc} for which there exists a $\beta \in E_{\text{loc}}$ with

$$(\beta,\varphi) = (\alpha,T^*\varphi)$$

for all $\varphi \in D_\infty$; then $\beta = T\alpha$.

Let $T^*D_\infty = \{T^*\varphi | \varphi \in D_\infty\}$ and $E_{T^*} = \overline{T^*D_\infty}$ where the closure is taken with respect to the norm $||\cdot||$. The L_{T^*}-principal form problem consists in finding, for a given $\sigma \in E_{loc}$, a ρ with $T\rho = 0$ such that

$$L_{T^*}(\rho - \sigma) = \rho - \sigma. \tag{157}$$

Here L_{T^*} is the projection operator from E_{loc} to E_{T^*}, defined by $(\alpha,\varphi) = (L_{T^*}\alpha,\varphi)$ for all $\varphi \in T^*D_\infty$. The form ρ, if it exists, is called the *L_{T^*}-principal form.*

5B. Existence. We maintain (Nakai–Sario [12]):

Theorem. *Given an arbitrary locally square integrable differential form σ on a noncompact Riemannian space and a differential operator T, there exists an L_{T^*}-principal form ρ if and only if*

$$(\sigma,T^*\varphi) = O(||T^*\varphi||) \tag{158}$$

for all $\varphi \in D_\infty$.

For the necessity, we shall prove the slightly stronger result that (158) already follows from the existence of a ρ' with properties $T\rho' = 0$ and $||\rho' - \sigma|| < \infty$. In fact, then

$$(\sigma,T^*\varphi) = (T\sigma,\varphi) = (T(\sigma - \rho'),\varphi) = ((\sigma - \rho'),T^*\varphi)$$

and therefore

$$|(\sigma,T^*\varphi)| \leq ||\sigma - \rho'|| \cdot ||T^*\varphi|| = O(||T^*\varphi||).$$

To prove the sufficiency of (158), we observe that $(\sigma,T^*\varphi)$ has a continuous extension to E_{T^*}. For this reason there exist a $\tau \in E_{T^*}$ such that

$$(-\sigma,T^*\varphi) = (\tau,T^*\varphi).$$

The form $\rho = \tau + \sigma$ then has the required property $(\rho,T^*\varphi) = 0$, that is, $T\rho = 0$.

5C. System of operators. From the above theorem we can now easily derive the following more general result Nakai–Sario [12]:

Theorem. *Let $T_1,\cdots,T_m$ be differential operators such that $T_1{}^*D_\infty,\cdots,T_m{}^*D_\infty$ are orthogonal by pairs.*

Given $\sigma\in E_{\text{loc}}(R)$, the conditions

$$(\sigma,T_i{}^*\varphi) = O(\|T_i{}^*\varphi\|), \tag{159}$$

$i = 1,\cdots,m$, for all $\varphi\in D_\infty$ are necessary and sufficient for the existence of an $L_{T_1{}^*\cdots T_m{}^*}$-principal form ρ characterized by $T_i\rho = 0$, $i = 1,\cdots,m$, and

$$\sigma-\rho\in E_{T_1{}^*}\oplus\cdots\oplus E_{T_m{}^*}. \tag{160}$$

The sufficiency is contained in Theorem 5B. For the necessity, let $\tau_i\in E_{T_i{}^*}$ be the form with

$$(-\sigma,T_i{}^*\varphi) = (\tau_i,T_i{}^*\varphi) \tag{161}$$

for all $\varphi\in D_\infty$; its existence is also given by Theorem 5B. By summing the (161) for $i = 1,\cdots,m$ we obtain for $\varphi\in D_\infty$

$$(-\sigma, \sum_1^m T_i{}^*\varphi) = (\sum_1^m \tau_i, \sum_1^m T_i{}^*\varphi),$$

because $(\tau_i,T_j{}^*\varphi) = 0$ for $i\neq j$. We conclude that $\rho = \sigma+\sum_1^m\tau_i$ has the required properties.

5D. Special cases. We now impose upon $\sigma\in E_{\text{loc}}$ the restriction that $T_i\sigma = 0$, $i = 1,\cdots,m$, in the complement A of a regular region with boundary $\alpha = -\partial A$. Since the support of $\varphi\in D_\infty$ is compact, Green's formulas give, for commonly used operators T, the existence on α of a form $\gamma(\sigma,\varphi)$ such that

$$\int_\alpha \gamma(\sigma,\varphi) = \int_A \sigma\wedge *T^*\varphi-T\sigma\wedge *\varphi. \tag{162}$$

Some expressions for γ are listed in the following table with the operator T indicated on the left:

$$\begin{array}{ll} d: & \sigma \wedge *\varphi \\ \delta: & -\varphi \wedge *\sigma \\ \Delta': & \sigma \wedge *d\varphi - \varphi \wedge *d\sigma \\ \Delta'': & -\delta\varphi \wedge *\sigma + \delta\sigma \wedge *\varphi \\ \Delta: & \sigma \wedge *d\varphi + \delta\sigma \wedge *\varphi - \varphi \wedge *d\sigma - \delta\varphi \wedge *\sigma. \end{array}$$

Theorem. *Let* $\sigma \in E_{\text{loc}}$ *and* $T_1, \cdots, T_m$ *be as in Theorem* 5C. *Suppose* $T_i\sigma|A = 0$, $i = 1, \cdots, m$. *Then there exists an* $L_{T_1{}^* \cdots T_m{}^*}$-*principal form if and only if*

$$\int_\alpha \gamma(\sigma, \varphi) = O(\|T_i{}^*\varphi\|) \tag{163}$$

for $i = 1, \cdots, m$ *and all* $\varphi \in D_\infty$.

Most of the preceding statements on principal forms appear as special cases of this theorem.

CHAPTER VII
PRINCIPAL FUNCTIONS ON HARMONIC SPACES

In this chapter we shall discuss the problem of finding on a given harmonic space a harmonic function which imitates the behavior of a given harmonic function on a neighborhood of the ideal boundary of the harmonic space.

Harmonic spaces are the most general spaces on which harmonicity can be considered. Consequently the problem we are going to discuss presents the most general form of the principal function problem.

Fundamentals of harmonic spaces are compiled in §1. The complete solution of the problem, when the imitation is understood to be bounded deviation, is given in §2. In the final §3 the generalized main existence theorem is established which includes the main existence theorem of Chapter I as a special case.

From a methodological viewpoint the main features of the present chapter are: (1) the use of the q-lemma is avoided and instead the complete continuity of the normal operator is exploited, (2) the discussion of the convergence of the Neumann series is replaced by the Riesz-Schauder theory for the abstract Fredholm equation of completely continuous operators. These replacements are inevitable since no differentiable structure or the like exists on a harmonic space. Yet one can even define the flux and the method seems to supply a new weapon to attack the principal form problem.

§1. HARMONIC SPACES

A harmonic space is a certain topological space on which "harmonic functions" are distinguished. Riemann surfaces and Riemannian spaces are harmonic spaces with their usual harmonic functions. Any differentiable manifold with an elliptic linear differential operator L of order 2 becomes a harmonic space when one takes the solutions of $Lu = 0$ as the harmonic functions (see 3B below). The concept of harmonic space was introduced by Brelot [2]. In this section we shall briefly describe fundamentals of harmonic spaces.

1. Harmonic structures

1A. Regularity of open sets. Let R be a locally compact Hausdorff space which is connected and locally connected. Suppose that to every open set Ω there corresponds a linear space $H(\Omega)$ of finitely continuous real-valued functions defined on Ω; they give rise to a family $H = \{H(\Omega)\}_\Omega$ of functions with domains in R.

A relatively compact open set Ω is called *regular* for H if for each bounded continuous function f on $\partial\Omega$ there exists a unique continuous function h_f on $\bar{\Omega}$ such that $h_f|\Omega \in H(\Omega)$, $h_f|\partial\Omega = f$, and $h_f \geq 0$ for $f \geq 0$.

If Ω is regular and $x \in \Omega$ then there exists a positive Radon measure $\mu(\cdot,x,\Omega)$ on $\partial\Omega$ such that $h_f(x) = \int f(y)d\mu(y,x,\Omega)$. If D is a component of Ω then D is also regular and $\mu(\cdot,x,\Omega)|\partial D = \mu(\cdot,x,D)$ for $x \in D$.

Consider the class $\bar{H}(\Omega)$ of functions s with the following properties:

(α) s is lower semicontinuous on Ω,

(β) $s > -\infty$ and $s \not\equiv \infty$ on any component of Ω,

(γ) for any regular region D such that $\bar{D} \subset \Omega$ and for any $x \in D$, $s(x) \geq \int s(y)d\mu(y,x,D)$.

1B. Definition of harmonic space. A pair (R,H) is called a *harmonic space* if the following axioms are satisfied:

Axiom I. *H forms a presheaf of* $\mathbf{R}$*-modulus on R, i.e. $\Omega_1 \subset \Omega_2$ and $h \in H(\Omega_2)$ imply that $h \mid \Omega_1 \in H(\Omega_1)$, and if h is a function on $\Omega = \bigcup \Omega_i$ with $h|\Omega_i \in H(\Omega_i)$ for every i then $h \in H(\Omega)$.*

Axiom II. *The regular regions in R for H form a basis of open sets.*

Axiom III. *The upper envelope of an upper directed family of functions in $H(D)$ is either ∞ or in $H(D)$ for any region D in R.*

Axiom IV. $1 \in \bar{H}(R)$.

R and H are referred to as the base space and the harmonic structure of the harmonic space (R,H). We shall sometimes loosely call R itself the harmonic space if H is well understood.

Functions in H are called *harmonic*; those in $\bar{H} = \{\bar{H}(\Omega)\}_\Omega$, *superharmonic*. If $-s$ is superharmonic then s itself is said to be *subharmonic*; the family of subharmonic functions is denoted by $\underline{H}$.

Remark. Axioms I–III were introduced by Brelot [2], Axiom IV by Loeb [1].

1C. Basic properties. We list some fundamental properties of harmonic and superharmonic functions. Except for the property singled out by an asterisk they can be easily deduced from the definition (cf. e.g. Brelot [2], [3]):

(a) If $h \in H(D)$ with a region D, $h \geq 0$ on D, and $h(x) = 0$ for some $x \in D$ then $h \equiv 0$.

(b) If $f \geq 0$ is lower semicontinuous on ∂D and if D is regular then the function $h(x) = \int f(y)\,d\mu(y,x,D)$ is either ∞ or in $H(D)$.

(c) For any $x \in D$ and any compact set $K \subset D$ there exists a constant M such that $h(y) \leq Mh(x)$ for all $y \in K$ and $h \in H(D)$ with $h \geq 0$ (Constantinescu–Cornea [2]).

(d)* For any $x \in D$ the family $\{h \in H(D) | h \geq 0,\ h(x) = 1\}$ is equicontinuous at x (Loeb-Walsh [1]).

(e) $\{h \in H(D) | h \geq 0,\ h(x) = 1\}$ is compact with respect to the compact convergence topology.

(f) If $s \in \bar{H}(\Omega)$ and $\alpha > 0$ then $\alpha s \in \bar{H}(\Omega)$.

(g) If s_1, $s_2 \in \bar{H}(\Omega)$ then $s_1 + s_2$, $\min(s_1,s_2) \in \bar{H}(\Omega)$.

(h) If $\{s_i\}$ is an upper directed family of superharmonic functions on a region D then the upper envelope of $\{s_i\}$ is either ∞ or in $\bar{H}(D)$.

(i) No superharmonic function assumes ∞ over a nonempty open set.

(j) If D is a regular region in an open set Ω and $s \in \bar{H}(\Omega)$ then $h(x) = \int s(y)\,d\mu(y,x,D) \in H(D)$; if $s' = s$ in $\Omega - D$ and $s' = h$ in D then $s' \in \bar{H}(\Omega)$.

(k) Let a and b be constants with $a \leq 0 \leq b$. If $s \in \bar{H}(D)$ and $s \geq a$ then either $s = a$ or $s > a$; if $s \in \underline{H}(D)$ and $s \leq b$ then either $s = b$ or $s < b$; if $h \in H(D)$ is not constant then h takes neither a nonnegative maximum nor a nonpositive minimum in D.

1D. Perron family. Let Ω be an open set and $\mathfrak{F}$ a family of functions in $\bar{H}(\Omega)$. $\mathfrak{F}$ is called a Perron family if (1) $\mathfrak{F} \neq \emptyset$ and $\mathfrak{F}$ is lower directed, (2) for any regular region D and $s \in \mathfrak{F}$, the superharmonic function s' defined by $s'|\Omega - D = s|\Omega - D$ and $s'|D = \int s(y) d\mu(y, \cdot, D)$ also belongs to $\mathfrak{F}$, and (3) $\mathfrak{F}$ is locally uniformly bounded from below.

Perron's theorem (see e.g. Brelot [2]) states:

If $\mathfrak{F}$ is a Perron family on Ω, then the lower envelope of $\mathfrak{F}$ belongs to $H(\Omega)$.

2. Dirichlet's problem

2A. Regular points. Let Ω be an open set in R. Denote by $\delta\Omega$ the boundary of Ω relative to the one point compactification of R. If Ω is relatively compact then $\delta\Omega = \partial\Omega$, and if Ω is not relatively compact, then $\delta\Omega$ consists of points in $\partial\Omega$ and the Alexandroff ideal boundary point.

Given a bounded continuous function f on $\delta\Omega$ consider the family

$$\mathfrak{U}(f) = \{s \in \bar{H}(\Omega) \mid \liminf_{x \in \Omega, x \to x_0} s(x) \geq f(x_0) \quad \text{for all} \quad x_0 \in \delta\Omega\}.$$

Clearly it is a Perron family and therefore

$$h_f(x) = \inf_{s \in \mathfrak{U}(f)} s(x)$$

is harmonic on Ω.

A point $x_0 \in \partial\Omega$ is called *regular* if

$$\lim_{x \in \Omega, x \to x_0} h_f(x) = f(x_0)$$

for every bounded continuous function f on $\delta\Omega$. In particular every boundary point of a regular open set is regular.

A point x_0 on the relative boundary of an open set Ω is regular if there exists a strictly positive superharmonic function s defined in the intersection of Ω and an open neighborhood of x_0 such that $\lim_{x\to x_0} s(x) = 0$. Such a function is called a barrier for Ω at x_0.

For a detailed account of these facts we refer the reader to Loeb [1].

2B. Outer-regular sets. A compact set $K \subset R$ is called *outer-regular* if there is a barrier for $R-K$ at each point of ∂K. The following result of R.-M. Hervé and Loeb [1] will be fundamental in the sequel:

Theorem. *For any compact set K and any region D with $K \subset D$ there always exist an outer-regular compact set F and a regular region G such that $K \subset F - \partial F \subset F \subset G \subset \bar{G} \subset D$.*

As a corollary of this the existence of an *exhaustion* $\{\Omega\}$ of R follows: there exists an upper directed family $\{\Omega\}$ of regular regions in R such that $R = \cup \Omega$.

3. Classification

3A. The operator B. Let A be the closure of the complement of an outer-regular compact set F with respect to R and let α be the relative boundary ∂A of A. Denote by $C(\alpha)$ the space of bounded continuous functions on α. For $f \in C(\alpha)$ define $f^* \in C(\delta A)$ by $f^*|\alpha = f$ and $f^*|\delta A - \alpha = 0$. Here $\delta A - \alpha$ may be empty, i.e. R may be compact. Since every point in α is regular with respect to $A-\alpha$, a function Bf on A can be defined by

$$Bf|\alpha = f, \qquad Bf|A-\alpha = h_{f^*}$$

where h_{f^*} is considered with respect to $(A-\alpha, f^*)$ (cf. 2A).

Bf can also be defined as follows. Let $\{\Omega\}$ be an exhaustion of R with $\partial\Omega \subset A-\alpha$. Then $\Omega \cap (A-\alpha)$ is a regular open set and hence we can find a unique continuous function $B_\Omega f$ on $\overline{\Omega \cap (A-\alpha)}$ such that $B_\Omega f|\Omega\cap(A-\alpha) \in H(\Omega\cap(A-\alpha))$, $B_\Omega f|\alpha = f$, and $B_\Omega f|\partial\Omega = 0$. In view of 1C we have the existence of $\lim_{\Omega\to R} B_\Omega f$, and we infer that

$$Bf = \lim_{\Omega\to R} B_\Omega f. \tag{1}$$

The correspondence $f \to Bf$ is an operator from $C(\alpha)$ into $H(A) = H(A-\alpha) \cap C(A)$ with the following properties:

(B1) $B(f+g) = Bf+Bg,\ B(cf) = cB(f)$,
(B2) $Bf|\alpha = f$,
(B3) $\min(0, \min_\alpha f) \leq Bf \leq \max(0, \max_\alpha f)$.

Intuitively speaking Bf is the harmonic function on $A-\alpha$ with boundary values f at α and "0" at the ideal boundary of R.

3B. Parabolicity. A harmonic space is called *parabolic* if $B1$ is the constant 1 on A for every choice of $A \subset R$. Otherwise it is called *hyperbolic*. If there is an open set in R on which the constant function 1 is not harmonic then R is hyperbolic.

Example 1. Let R be a Riemann surface or a Riemannian space and H the presheaf of (ordinary) harmonic functions on open sets of R. Then (R,H) is a harmonic space and R is parabolic if and only if $R \in O_G$ (cf. Chapters IV and VI).

Example 2. Let R be a C^2-manifold with an invariantly defined elliptic differential operator

$$L = \sum a_{ij}(x) \frac{\partial^2}{\partial x_i \partial x_j} + \sum b_i(x) \frac{\partial}{\partial x_i} + c(x), \qquad c(x) \leq 0,$$

where the coefficients are Lipschitz continuous and $\sum a_{ij}(x)\xi_i\xi_j$ is a positive definite quadratic form. Let

$$H = \{u | Lu = 0 \quad \text{on open sets}\}.$$

Then (R,H) is a harmonic space (cf. e.g. Duff [5]). It is hyperbolic if $c(x) \not\equiv 0$.

Example 3. Let R be the real line and let H consist of linear functions. Then (R,H) is a harmonic space which is parabolic.

§2. HARMONIC FUNCTIONS WITH GENERAL SINGULARITIES

We shall study the problem of finding a harmonic function on the entire space which behaves like a given harmonic function in a neighborhood of the ideal boundary. The problem is always solvable if the space is hyperbolic. If the space is parabolic then for the solvability it is necessary and sufficient that the given function has vanishing "flux" across the ideal boundary. The result obtained in this section is due to Nakai [11].

1. Problem and its reduction

1A. Problem. Let N and N' be open sets in R with compact complements. They can be considered as neighborhoods of the ideal boundary of R. Suppose that $\bar{N}' \subset N$ and a function $\sigma \in H(N)$ is given. We shall consider the following problem:

Find a $\rho \in H(R)$ *such that*

$$\sup_{x \in N'} |\rho(x) - \sigma(x)| < \infty. \tag{2}$$

It is in the nature of the problem that the domain N of σ may be replaced by any smaller open set. For this reason we take a set A which is the closure of the complement of an outer-regular compact F. We denote by α the relative boundary ∂A of A. By Theorem 1.2B we may assume that $A \subset N$. To solve our original problem it is thus necessary and sufficient to solve the following

Problem. *Given a* $\sigma \in H_1(A) = H(A - \alpha) \cap C(A)$, *find a* $\rho \in H(R)$ *with*

$$\sup_{x \in A} |\rho(x) - \sigma(x)| < \infty. \tag{3}$$

1B. Reformulation. Let B be the operator introduced in 1.3A for the set A. We observe that Problem 1A is equivalent to:

Problem. *Given a* $\sigma \in H_1(A) = H(A - \alpha) \cap C(A)$, *find a* $\rho \in H(R)$ *with*

$$B(\rho - \sigma) = \rho - \sigma. \tag{4}$$

Strictly speaking (4) should be written as $B(\rho|\alpha - \sigma|\alpha) = \rho|A - \sigma$. The function σ will be referred to as a *singularity* at the ideal boundary, and the function ρ with (4), if it exists, will be called a *principal function* with respect to (σ, B), or simply a (σ, B)-principal function.

Clearly a (σ,B)-principal function ρ is a solution of Problem 1A. Conversely suppose there exists a solution ρ of Problem 1A. Let $\{\Omega\}$ be an exhaustion of R with $\partial\Omega \subset A$ and set $u = \rho - \sigma$. Then u is a bounded function in $H_1(A)$, and we put $c = \sup_A |u|$. Using the notation in 1.2A we consider functions v_Ω and w_Ω defined by

$$v_\Omega = h_{u|\partial\Omega}, \qquad w_\Omega = h_c$$

for the regular region Ω. By 1.1C.(k), $\{w_\Omega\}$ is monotone decreasing and thus $w = \lim_\Omega w_\Omega$ exists and is in $H(R)$; moreover, $0 < v_\Omega + w_\Omega < 2c$. By 1.1C.(e), we may assume, by choosing a suitable subexhaustion, that $\{v_\Omega + w_\Omega\}_\Omega$ is convergent. In summary, we have seen that $v = \lim_\Omega v_\Omega$ exists and belongs to $H(R)$.

From the construction we have $u - v_\Omega = 0$ on $\partial\Omega$ and $u - v_\Omega \in H(\Omega \cap (A - \alpha))$. Hence in terms of B_Ω considered in 1.3A we obtain

$$B_\Omega(u - v_\Omega) = u - v_\Omega. \tag{5}$$

Observe that

$$|B_\Omega v_\Omega - Bv| \leq |B_\Omega(v_\Omega - v)| + |B_\Omega v - Bv|$$

$$\leq \sup_\alpha |v_\Omega - v| + |B_\Omega v - Bv|.$$

This implies that $\lim_\Omega B_\Omega v_\Omega = Bv$. On letting $\Omega \to R$ in (5) we now conclude that $B(u - v) = u - v$. Set $\bar{\rho} = \rho - v \in H(R)$. Then we see that

$$B(\bar{\rho} - \sigma) = \bar{\rho} - \sigma,$$

i.e. (4) is solved by $\bar{\rho}$.

Therefore we only need discuss problem (4).

1C. Reduction. Let D be a regular region with $D \supset \overline{R - A}$. We denote by δ the relative boundary of D; $\delta = \partial D$. For each $f \in C(\delta)$, let Kf be the function on $\bar{D}$ such that $(Kf)|\delta = f$ and $Kf \in H_1(D) = H(D) \cap C(\bar{D})$. Then $f \to Kf$ is an operator from $C(\delta)$ into $H_1(\bar{D})$ with the following properties:

(K1) $K(f+g) = Kf + Kg, \qquad K(cf) = cK(f)$,
(K2) $Kf|\delta = f$,
(K3) $\min(0, \min_\delta f) \leq Kf \leq \max(0, \max_\delta f)$.

Define a third operator $T = T_B$, from $C(\delta)$ into itself, by

$$Tf = B((Kf)|\alpha)|\delta. \tag{6}$$

It satisfies

(T1) $T(f+g) = Tf+Tg, \qquad T(cf) = cTf,$
(T2) $\min(0,\min_\delta f) \leq Tf \leq \max(0,\max_\delta f)$.

Problem 1B is reduced to the following:

Problem. *Given a* $\sigma \in H_1(A)$, *set* $\sigma_0 = (\sigma - B\sigma)|\delta \in C(\delta)$. *Find a* $\varphi \in C(\delta)$ *with*

$$(I-T)\varphi = \sigma_0 \tag{7}$$

where I is the identity operator on $C(\delta)$.

If (4) has a solution ρ then $\varphi = \rho|\delta$ is a solution of (7). Conversely suppose that $\varphi \in C(\delta)$ is a solution of (7) and set

$$\rho = \begin{cases} K\varphi & \text{on } \bar{D}, \\ B((K\varphi)|\alpha) + \sigma - B\sigma & \text{on } \bar{A}. \end{cases} \tag{8}$$

Since $K\varphi = B((K\varphi)|\alpha) + \sigma - B\sigma$ on $\alpha \cup \delta$ and the same is true on $\bar{D} \cap \bar{A}$, ρ is well defined on R and $\rho \in H(R)$. Moreover

$$\begin{aligned} B(\rho - \sigma) &= B[(B((K\varphi)|\alpha) + \sigma - B\sigma) - \sigma] \\ &= B((K\varphi)|\alpha) - B\sigma = \rho - \sigma, \end{aligned}$$

i.e. ρ is a solution of (4).

1D. The space $C(\delta)$ forms a Banach space with the norm $\|\varphi\| = \sup_\delta|\varphi|$. Thus (T2) with 1.1C.(k) implies that

(T2)₁ $\|T\varphi\| \leq \|\varphi\|$,

(T2)₂ $T\varphi \geq 0$ for $\varphi \geq 0$ and $T\varphi > 0$ for $\varphi \geq 0$ with $\varphi \not\equiv 0$.

Moreover we have:

Lemma. *T is completely continuous, i.e.* $\{T\varphi \mid \|\varphi\| \leq 1\}$ *is relatively compact in* $C(\delta)$.

Since $\{T\varphi|\ ||\varphi||\leq 1,\varphi\geq 0\} = \frac{1}{2}[\{T\varphi|\ ||\varphi||\leq 1\}+T1]$, we have only to show that $\{T\varphi|\ ||\varphi||\leq 1,\varphi\geq 0\}$ is relatively compact in $C(\delta)$. By 1.1C.(e), $\{K\varphi|\ ||\varphi||\leq 1,\varphi\geq 0\}$ is relatively compact in $C(D)$ with respect to the compact convergence topology, and therefore $\{K\varphi|\alpha|\ ||\varphi||\leq 1,\varphi\geq 0\}$ is relatively compact in $C(\alpha)$. By the fact that $f\to Bf|\delta$ is continuous from $C(\alpha)$ into $C(\delta)$, we conclude that $B\{K\varphi|\alpha|\ ||\varphi||\leq 1,\varphi\geq 0\}|\delta = \{T\varphi|\ ||\varphi||\leq 1,\varphi\geq 0\}$ is relatively compact in $C(\delta)$.

1E. In summary, we are given a completely continuous positive operator T on the Banach space $C(\delta)$ and a $\sigma_0\in C(\delta)$. The problem is to solve the abstract Fredholm integral equation $(I-T)\varphi=\delta_0$, $\varphi\in C(\delta)$. To this end we shall apply the Riesz-Schauder theory.

2. Riesz-Schauder theory

2A. Dual operator T^*. Let $C(\delta)^*$ be the dual space of $C(\delta)$, i.e. the space of continuous linear functionals on $C(\delta)$. Then $C(\delta)^*$ is nothing but the linear space of all signed regular Borel measures μ on δ with the norm $||\mu||$ = the total variation of μ (cf. e.g. Yosida [1, p. 119]). The dual operator T^* of T is the bounded linear operator from $C(\delta)^*$ into itself given by

$$\langle T\varphi,\mu\rangle = \langle\varphi,T^*\mu\rangle \tag{9}$$

for every $\varphi\in C(\delta)$, $\mu\in C(\delta)^*$. Here $\langle\psi,\nu\rangle = \int\psi d\nu$.

It is easy to see that T^* is again completely continuous along with T (see e.g. Yosida [1, p. 282]).

2B. Eigenvalues. A number λ is an *eigenvalue* of T if $(\lambda I-T)\varphi = 0$ for some nonzero $\varphi\in C(\delta)$. The dimension of the linear space $\{\varphi|(\lambda I-T)\varphi = 0\}$ is called the *multiplicity* of the eigenvalue λ and each $\varphi\in\{\varphi|(\lambda I-T)\varphi = 0\}$ is an *eigenvector* belonging to λ. These notions are similarly defined for T^*.

We shall make essential use of the following:

Riesz-Schauder Theory. (a) *If $\lambda\neq 0$ is not an eigenvalue of T, then the operator $\lambda I-T$ has the continuous inverse $(\lambda I-T)^{-1}$.*

(b) *$\lambda\neq 0$ is an eigenvalue of T if and only if it is an eigenvalue of T^*, and the multiplicities of λ are the same for T and T^*.*

(c) *The equation $(\lambda I-T)\varphi = \psi$ admits a solution $\varphi\in C(\delta)$ if and only if $\psi\in C(\delta)$ satisfies $\langle\psi,\mu\rangle = 0$ for every $\mu\in C(\delta)^*$ with $(\lambda I^*-T^*)\mu = 0$ where I^* is the identity operator on $C(\delta)^*$.*

Actually the above results constitute only that part of the Riesz-Schauder theory which is needed in the sequel. Moreover, the Riesz-Schauder theory is concerned with general completely continuous operators on arbitrary Banach spaces. For the proofs we refer the reader to Yosida [1, pp. 282–285].

2C. The eigenvalue 1. In addition to the above facts of a general nature we shall need the following specific observation:

Lemma. *If $T1 \neq 1$, then 1 is not an eigenvalue of T. If $T1 = 1$, then 1 is an eigenvalue of T with multiplicity 1.*

Suppose $T\varphi = \varphi$ for $\varphi \in C(\delta)$. As in 1C we see that the function ρ given by

$$\rho = \begin{cases} K\varphi & \text{on } \bar{D}, \\ B((K\varphi)|\alpha) & \text{on } \bar{A} \end{cases}$$

is well defined on R and belongs to $H(R)$. Since $T\varphi = \varphi$, $\rho|\delta = \varphi$. Note that

$$\min(0, \min_{\delta} \varphi) \le \rho|\bar{D} \le \max(0, \max_{\delta} \varphi),$$

$$\min(0, \min_{\alpha} K\varphi) \le \rho|\bar{A} \le \max(0, \max_{\alpha} K\varphi)$$

and consequently

$$\min(0, \min_{\delta} \varphi) \le \rho \le \max(0, \max_{\delta} \varphi)$$

on R. This with $\rho|\delta = \varphi$ implies that $\rho \in H(R)$ takes either a positive maximum or a negative minimum unless $\rho \equiv 0$. By 1.1C. (k), ρ must be a constant on R.

If $T\varphi = \varphi$ for T with $T1 \neq 1$, then since φ is constant, $\varphi = 0$ on R. This means that 1 is not an eigenvalue of T.

Next suppose that $T1 = 1$. Then $1 \in \{\varphi | (1 \cdot I - T)\varphi = 0\}$ and thus 1 is an eigenvalue of T. Moreover $\{\varphi | (1 \cdot I - T)\varphi = 0\}$ consists of only constants and its dimension is 1, i.e. the multiplicity of 1 is 1.

2D. Invariant measure. Suppose $T1 = 1$. A signed measure ν on δ such that

$$(10) \qquad \langle \varphi, \nu \rangle = \langle T\varphi, \nu \rangle \quad \text{for every} \quad \varphi \in C(\delta), \qquad \nu(\delta) = 1$$

will be called an *invariant measure* associated with T. Clearly (10) is equivalent to

$$(11) \qquad T^*\nu = \nu, \qquad \nu(\delta) = 1.$$

We shall show:

Lemma. *Let* $T1 = 1$. *There exists a unique invariant measure* $\nu = \nu_T = \nu_{T_B}$ *associated with* T. *It is a positive regular Borel measure whose support coincides with* δ.

By the Riesz-Schauder theory, 1 is also an eigenvalue of T^* with multiplicity 1 (see 2B, 2C). Therefore $\{\mu | \mu \in C(\delta)^*, T^*\mu = \mu\}$ is a vector space of dimension 1. Let $\nu \in \{\mu | \mu \in C(\delta)^*, T^*\mu = \mu\}$ with $\nu(\delta) = 1$. Then $\{\mu | \mu \in C(\delta)^*, T^*\mu = \mu\} = \{c\nu | c \text{ constants}\}$.

The proof of the existence and the unicity of the invariant measure is herewith complete. (For another proof, see Walsh [1].)

2E. Next we prove that $\nu = \nu_T$ is a positive measure with $S_\nu = \delta$ where S_ν is the support of ν.

To this end let $\nu = \nu_1 - \nu_2$ be the Hahn decomposition of ν, i.e. ν_1 and ν_2 are positive measures on δ such that there is a decomposition of δ: $\delta = \delta_1 \cup \delta_2$, $\delta_1 \cap \delta_2 = \varnothing$ into Borel sets with the property $\nu_1(\delta_2) = \nu_2(\delta_1) = 0$. Since $\nu(\delta) = \nu_1(\delta) - \nu_2(\delta) = 1$, we have $\nu_1(\delta) = \nu_1(\delta_1) > 0$.

First we show that $S_{\nu_1} = \delta$. Note that $|\nu| = \nu_1 + \nu_2$ is a positive regular measure on δ. Suppose that the open set $\delta - S_{\nu_1}$ in δ were not empty. Choose closed subsets $\theta_n \subset \delta - S_{\nu_1}$ with $|\nu|(\delta - S_{\nu_1} - \theta_n) < 1/n$ and functions $\varphi_n \in C(\delta)$ such that $0 \leq \varphi_n \leq 1$, $\varphi_n | \theta_n = 1$, $\varphi_n | S_{\nu_1} = 0$, and $\varphi_1 \leq \cdots \leq \varphi_n \leq \cdots$.

By the complete continuity of T, a subsequence of $\{T\varphi_n\}$, say $\{T\varphi_{n'}\}$, is uniformly convergent on δ. Thus $\psi = \lim_{n'} T\varphi_{n'} \in C(\delta)$. Moreover by $(T2)_2$, $\psi \geq T\varphi_{n'} > 0$ and $\{T\varphi_{n'}\}$ is monotone increasing. Therefore

$$(12) \qquad \langle \psi, \nu_1 \rangle > 0.$$

By the construction of $\varphi_{n'}$ it is easily seen that

$$(13) \qquad \lim_{n'\to\infty} \langle \varphi_{n'}, \nu \rangle = -\nu_2(\delta_2).$$

Since $1 \geq \psi \geq T\varphi_{n'}$, we also conclude that

$$(14) \qquad \lim_{n'\to\infty} \langle T\varphi_{n'}, \nu \rangle = \langle \psi, \nu_1 \rangle - \langle \psi, \nu_2 \rangle \geq \langle \psi, \nu_1 \rangle - \nu_2(\delta_2).$$

Observe that $\langle \varphi_{n'}, \nu \rangle = \langle T\varphi_{n'}, \nu \rangle$. On letting $n' \to \infty$ we infer from (13) and (14) that

$$-\nu_2(\delta_2) \geq \langle \psi, \nu_1 \rangle - \nu_2(\delta_2).$$

This contradicts (12) and we have $S_{\nu 1} = \delta$.

2F. We shall complete the proof by showing that $\nu_2(\delta_2) = 0$. Contrary to the assertion, suppose that $\nu_2(\delta_2) > 0$. Take open subsets ϵ_n of δ and closed subset θ_n of δ with $\epsilon_n \supset \delta_2 \supset \theta_n$ and $|\nu|(\epsilon_n - \theta_n) < 1/n$ $(n = 1,2,\cdots)$. Choose also functions $\varphi_n \in C(\delta)$ such that $0 \leq \varphi_n \leq 1$, $\varphi_n|\theta_n = 1$, and $\varphi_n|\delta - \epsilon_n = 0$. As in 2E, the complete continuity of T implies the existence of a subsequence of $\{T\varphi_n\}$, say $\{T\varphi_{n'}\}$, which is uniformly convergent on δ. Clearly $\psi = \lim_{n'} T\varphi_{n'} \geq 0$ on δ and belongs to $C(\delta)$.

In the same manner as in 2E

$$(15) \qquad \lim_{n'} \langle \varphi_{n'}, \nu \rangle = -\nu_2(\delta_2)$$

and also in view of $0 \leq \psi \leq 1$

$$(16) \qquad \lim_{n'} \langle T\varphi_{n'}, \nu \rangle = \langle \psi, \nu_1 \rangle - \langle \psi, \nu_2 \rangle \geq \langle \psi, \nu_1 \rangle - \nu_2(\delta_2).$$

We let $n' \to \infty$ in $(\varphi_{n'}, \nu) = (T\varphi_{n'}, \nu)$ and conclude that

$$(17) \qquad -\nu_2(\delta_2) = \langle \psi, \nu_1 \rangle - \langle \psi, \nu_2 \rangle \geq \langle \psi, \nu_1 \rangle - \nu_2(\delta_2).$$

This shows that $\langle \psi, \nu_1 \rangle \leq 0$. On the other hand, $\langle \psi, \nu_1 \rangle \geq 0$ since $\psi \geq 0$. Therefore $\langle \psi, \nu_1 \rangle = 0$. This with $S_{\nu_1} = \delta$ implies that $\psi \equiv 0$ on δ. Again by (17) we deduce $\nu_2(\delta_2) = 0$, a contradiction.

The proof of Lemma 2D is herewith complete.

3. Solution of Problem 1C

3A. Result. We are now in a position to prove (Nakai [11]):

Theorem. *Let* $\sigma \in H_1(A)$ *and* $\sigma_0 = (\sigma - B\sigma)|\delta$. *If* $T1 \neq 1$ *then there is a unique* $\varphi \in C(\delta)$ *such that* $(I-T)\varphi = \sigma_0$. *If* $T1 = 1$ *then a necessary and sufficient condition for the existence of a* $\varphi \in C(\delta)$ *with* $(I-T)\varphi = \sigma_0$ *is that* $\langle \sigma_0, \nu_T \rangle = 0$. *In this case* φ *is unique up to an additive constant.*

3B. First we shall prove the existence. If $T1 \neq 1$, then by Lemma 2C, 1 is not an eigenvalue of T. By the Riesz-Schauder theory $(I-T)^{-1}$ exists and $\varphi = (I-T)^{-1}\sigma_0 \in C(\delta)$ is a desired solution.

Next suppose that $T1 = 1$. Then by the fact that

$$\{\mu \in C(\delta)^* | T^*\mu = \mu\} = \{c\nu_T\},$$

$\langle \sigma_0, \nu_T \rangle = 0$ is equivalent to

$$\langle \sigma_0, \mu \rangle = 0 \quad \text{for all} \quad \mu \in \{\mu \in C(\delta)^* | T^*\mu = \mu\}. \tag{18}$$

Again by the Riesz-Schauder theory the solvability of $(I-T)\varphi = \sigma_0$ is equivalent to (18).

3C. Suppose that $(I-T)\varphi_j = \sigma_0$ for $\varphi_j \in C(\delta)$ $(j = 1,2)$. Set $\psi = \varphi_1 - \varphi_2$. Then $T\psi = \psi$. By the proof of Lemma 2C, ψ is a constant which, in particular, is zero if $T1 \neq 1$. Thus the unicity of the solution is proved.

4. Solution of Problem 1B

4A. Although Problems 1A, 1B, and 1C are completely settled by Theorem 3A, the formulation is somewhat implicit, and more direct statements for the solutions of Problems 1A and 1B are preferable. To derive them we first observe:

Lemma. $T1 = 1$ *(resp.* $T1 \neq 1$*) if and only if* R *is parabolic (hyperbolic).*

Suppose first that R is parabolic. Then $1 \in H(R)$, because otherwise $B1 \neq 1$ for some choice of A. Therefore $K1 = 1$ and $B1 = 1$ implies that $T1 = 1$.

Conversely if $T1 = 1$, then $K1 = 1$ and $B1 = 1$. This in particular implies that $1 \in H(R)$. In this case $B1 = 1$ does not depend on the choice of A, i.e. R is parabolic.

4B. Flux. Suppose that R is parabolic. We define the *flux* of $\sigma \in H_1(A)$ across the ideal boundary of R by

$$\text{flux}(\sigma) = \int_\delta (\sigma - B\sigma) d\nu_{T_B}. \tag{19}$$

One can easily prove by using Theorem 3A that $\text{flux}(\sigma) = 0$ does not depend on the choice of A and D and also that $\text{flux}(\sigma) = 0$ if σ is bounded. Details of the proof will be left to the reader.

A justification of our terminology is exemplified by the following concrete model:

Example. *Let R be a parabolic Riemann surface or Riemannian space and let α and δ be smooth. Take $w \in H(A \cap D - \alpha) \cap C(A \cap \bar{D})$ with $w|\delta = 1$ and $w|\alpha = 0$. Then*

$$d\nu_{T_B} = \frac{*dw}{\int_\delta *dw}, \tag{20}$$

$$\text{flux}(\sigma) = \frac{\int_\delta *d\sigma}{\int_\delta *dw}. \tag{21}$$

In fact, for any real analytic φ in $C(\delta)$

$$\langle T\varphi, *dw \rangle = \int_\delta BK\varphi *dw = \int_\alpha BK\varphi *dw = \int_\alpha K\varphi *dw,$$

$$\langle \varphi, *dw \rangle = \int_\delta \varphi *dw = \int_\delta K\varphi *dw = \int_\alpha K\varphi *dw.$$

Hence $\langle T\varphi, *dw/\int_\delta *dw \rangle = \langle \varphi, *dw/\int_\delta *dw \rangle$ for every real analytic $\varphi \in C(\delta)$ and as is seen on passing to the limit, for every $\varphi \in C(\delta)$. Since the total mass of $*dw/\int_\delta *dw$ is one, we obtain (20).

The proof of (21) is immediate:

$$\text{flux}(\sigma) = \int_\delta (\sigma - B\sigma) *dw \Big/ \int_\delta *dw$$

$$= \int_\delta *d(\sigma - B\sigma) \Big/ \int_\delta *dw = \int_\delta *d\sigma \Big/ \int_\delta *dw.$$

4C. Result. We have reached the complete solution of Problem 1B (Nakai [11]):

Theorem. *Let* $\sigma \in H_1(A)$. *If* R *is hyperbolic then there always exists a unique* (σ,B)*-principal function. If* R *is parabolic then for the existence of a* (σ,B)*-principal function it is necessary and sufficient that* flux$(\sigma) = 0$. *In this case the* (σ,B)*-principal function is unique up to an additive constant.*

This result is merely a restatement of Theorem 3A.

4D. Solution of the original problem. Again by reformulating Theorem 4C we obtain the following complete solution of Problem 1A (Nakai [11]):

Theorem. *Let* σ *be a harmonic function in a neighborhood of the ideal boundary of a harmonic space* R. *If* R *is hyperbolic then there always exists a harmonic function* ρ *on* R *such that* $\rho - \sigma$ *is bounded. If* R *is parabolic then such a* ρ *exists if and only if* flux$(\sigma) = 0$.

§3. GENERAL PRINCIPAL FUNCTION PROBLEM

Instead of the operator B we shall consider more general operators L and study the existence problem for principal functions with respect to L.

1. Principal functions

1A. Quasinormal operators. Let $A \subset R$ and α be as in 2.1A. Consider operators L from $C(\alpha)$ into $H_1(A) = H(A - \alpha) \cap C(A)$ satisfying the following three conditions:

(L1) $L(f+g) = Lf + Lg, \qquad L(cf) = cLf,$
(L2) $Lf|\alpha = f,$
(L3) $\min(0, \min_\alpha f) \leq Lf \leq \max(0, \max_\alpha f).$

Such an L will be called a *quasinormal operator*. Normal operators in Chapters I and VI are examples of quasinormal operators. The operator B in §2 is also a typical example.

Given a singularity $\sigma \in H_1(A)$ we wish to find a $\rho \in H(R)$ with the property

$$L(\rho-\sigma) = \rho-\sigma. \tag{22}$$

Such a ρ, if it exists, will be called a *principal function* with respect to (σ,L) or simply a (σ,L)-principal function.

1B. Associated operator. Let $D \supset \overline{R-A}$, $\delta = \partial D$, and K be as in 2.1C. In analogy with (6) we consider the operator T_L defined by

$$T_L f = L((Kf)|\alpha)|\delta. \tag{23}$$

It is an operator from $C(\delta)$ into itself. As in 2.1D, T_L is completely continuous. We say that L is of *the first* (resp. *second*) *kind* if $T_L 1 = 1$ (resp. $T_L 1 \neq 1$). Normal operators in Chapters I and VI are all of the first kind. Quasinormal operators on parabolic harmonic spaces are all of the first kind. The operator B on a hyperbolic harmonic space is of the second kind.

1C. *L*-flux. Let L be a quasinormal operator of the first kind. As in 2.2D one can prove the existence of a positive regular Borel measure ν_{T_L} on δ such that

$$T_L^* \nu_{T_L} = \nu_{T_L}, \qquad \nu_{T_L}(\delta) = 1, \qquad S_{\nu_{T_L}} = \delta. \tag{24}$$

It is again called the *invariant measure* associated with T_L.

For $\sigma \in H_1(A)$ we then define the *L-flux* of σ across the ideal boundary of R by

$$L\text{-flux}(\sigma) = \int_\delta (\sigma - L\sigma)\, d\nu_{T_L}. \tag{25}$$

If R is a Riemann surface or a Riemannian space and L is a normal operator then as in (20) and (21) we see that

$$dv_{T_L} = \frac{*dw}{\int_\delta *dw}, \tag{26}$$

$$L\text{-flux}(\sigma) = \frac{\int_\delta *d\sigma}{\int_\delta *dw}. \tag{27}$$

2. Generalized main existence theorem

2A. Result. A complete solution of the problem in 1A is given by the following (Nakai [11])

Generalized Main Existence Theorem. *Let $\sigma \in H_1(A)$ and let L be a quasinormal operator. If L is of the second kind, then there exists a unique (σ,L)-principal function. If L is of the first kind, then for the existence of a (σ,L)-principal function it is necessary and sufficient that L-flux$(\sigma) = 0$. In this case the (σ,L)-principal function is unique up to an additive constant.*

In view of (27) this is a generalization of the Main Existence Theorem of Chapter I.

On replacing B by L the entire argument in 2.1C–4C can be applied verbatim to prove the above theorem; the details are left to the reader.

APPENDIX

SARIO POTENTIALS ON RIEMANN SURFACES

by

MITSURU NAKAI

Three kernels are considered in classical potential theory: the hyperbolic kernel $\log[(1-\bar{a}\zeta)/(\zeta-a)]$ for the disk, the parabolic kernel $\log(1/|\zeta-a|)$ for the plane, and the elliptic kernel $\log(1/[\zeta,a])$ for the sphere. The first one can be extended to Green's kernel on hyperbolic Riemann surfaces; the corresponding potential theory has been given a comprehensive discussion in the monograph of Constantinescu-Cornea [1]. The second kernel generalizes to the Evans kernel on parabolic Riemann surfaces (Nakai [9]); the corresponding potential theory may be developed in the same fashion as that of logarithmic potentials presented in Tsuji's monograph [5].

These two kernels, however, suffer from the restrictions imposed upon their domain surfaces. In contrast, the third kernel can be extended to the Sario kernel, i.e. the proximity function in the terminology of Chapter V, on an *arbitrary* Riemann surface (Sario [21], [25], [26]). Thus the corresponding theory of the Sario potentials has the advantage of full generality.

In this appendix we shall systematically develop the theory of Sario potentials from the viewpoint of potential-theoretic principles. We shall establish the continuity principle, unicity principle, Frostman's maximum principle, capacitary principle (the fundamental theorem of potential theory), and energy principle (Nakai [7], [8], [10]). Their applications e.g. to value distribution theory (cf. Nakai [7], [8]) are not included.

§1. CONTINUITY PRINCIPLE

We shall study Sario potentials $s_\mu(\zeta) = \int s(\zeta,a)d\mu(a)$ with respect to the Sario kernel $s(\zeta,a)$. In this section we shall establish the continuity principle for these potentials (Nakai [7]). This will follow from the joint continuity of $s(\zeta,a)$. The unicity principle will also be established.

1. Joint continuity of $s(\zeta,a)$

1A. Definition of $s(\zeta,a)$. We shall first review the definition of $s(\zeta,a)$. On an arbitrary Riemann surface S take arbitrary but then fixed points ζ_j $(j = 0,1)$ and disjoint parametric disks D_j with centers ζ_j $(j = 0,1)$. Let $t_0(\zeta) = t(\zeta,\zeta_0,\zeta_1)$ be a harmonic function on $S - \{\zeta_0,\zeta_1\}$ such that $t_0(\zeta) + 2\log|\zeta - \zeta_0|$ and $t_0(\zeta) - 2\log|\zeta - \zeta_1|$ are harmonic in D_0 and D_1 respectively. Moreover we require that $t_0 = (I)L_1t_0$ in a neighborhood of the ideal boundary of S, with $(I)L_1$ the normal operator of Chapter I. Such a function t_0 is unique up to an additive constant, and we assume that $t_0(\zeta) + 2\log|\zeta - \zeta_0| \to 0$ as $\zeta \to \zeta_0$ in D_0. The functions $s_0(\zeta) = \log(1 + e^{t_0(\zeta)})$ and $s_0(\zeta) + 2\log|\zeta - \zeta_0|$ are finitely continuous on $S - \{\zeta_0\}$ and D_0 respectively.

For an arbitrary point a in $S - \{\zeta_0\}$ we construct the function $t(\zeta,a) = t(\zeta,a,\zeta_0)$ in the same manner as $t(\zeta,\zeta_0,\zeta_1)$ but this time we choose the normalization $t(\zeta,a) - 2\log|\zeta - \zeta_0| \to s_0(a)$ as $\zeta \to \zeta_0$ in D_0.

Let $s_1(\zeta,a) = s_0(\zeta) + t(\zeta,a)$. Then the functions $s_1(\zeta,a)$ and $s_1(\zeta,a) + 2\log|\zeta - a|$ are finitely continuous on $S - \{a\}$ and a parametric disk about a respectively. We also put $s_1(\zeta,\zeta_0) = s_0(\zeta)$, i.e. $t(\zeta,\zeta_0) = 0$.

The function $s_1(\zeta,a)$ thus defined on $S \times S$ is essentially the same as the proximity function constructed in V.1. By Theorem V.1.1F it is therefore bounded from below. We finally define $s(\zeta,a) = s_1(\zeta,a) + c$ where the constant c is so chosen that

$$s(\zeta,a) > 0 \tag{1}$$

for all $(\zeta,a) \in S \times S$. By Theorem V.1.1D

$$s(\zeta,a) = s(a,\zeta) \tag{2}$$

for all (ζ,a) in $S \times S$. This function s was introduced and properties

(1) and (2) were established by Sario [21], [25], [26]. From a potential-theoretic viewpoint, we shall refer to it as the *Sario kernel* on S. It is a positive symmetric kernel.

1B. Continuity outside the diagonal set. For efficient development of a potential theory for the Sario kernel a task of compelling importance is to prove the continuity of s on $S \times S$. As the first step we shall establish the finite continuity of s on $S \times S$ outside the diagonal set, i.e. we shall show that $s(\zeta,a)$ is finitely continuous at $(\zeta',a') \in S \times S$ with $\zeta' \neq a'$. In view of (2) we may assume that $\zeta' \neq \zeta_0$. Take parametric disks U and V about ζ' and a' respectively, such that $\bar{U} \cap \bar{V} = \emptyset$ and $\zeta_0 \notin \bar{U}$. Since $(\zeta,a) \to s_0(\zeta)$ is finitely continuous on $U \times V$ and $s(\zeta,a) = s_0(\zeta) + t(\zeta,a) + c$, we have only to show that $(\zeta,a) \to t(\zeta,a)$ is finitely continuous on $U \times V$.

We note that $t(\zeta,a)$ has the following properties:

(α) $\zeta \to t(\zeta,a)$ is harmonic on U for fixed $a \in V$,
(β) $a \to t(\zeta,a)$ is finitely continuous on V for fixed $\zeta \in U$,
(γ) $(\zeta,a) \to t(\zeta,a)$ is bounded from below on $U \times V$.

Property (α) is a direct consequence of the definition of $t(\zeta,a)$, and (β) follows from (2) on observing that $t(\zeta,a) = s(a,\zeta) - c - s_0(\zeta)$. By virtue of this equality and (1) we deduce (γ).

From (α), (β), (γ), and the Harnack inequality it follows that $(\zeta,a) \to t(\zeta,a)$ is finitely continuous on $U \times V$.

1C. Decomposition of $s(\zeta,a)$. Let Ω be a regular region with $a \in \Omega$, and $g_\Omega(\zeta,a)$ be Green's kernel on Ω. It is easily seen that g_Ω is continuous on $\Omega \times \Omega$.

Consider the 2-form $\lambda^2(\zeta)dS_\zeta$ on $S - \{\zeta_0,\zeta_1\}$ defined by

$$\lambda^2(\zeta) = \Delta_\zeta s_0(\zeta) = \frac{e^{t_0}|\operatorname{grad} t_0(\zeta)|^2}{(1+e^{t_0(\zeta)})^2}$$

with $\lambda(\zeta) \geq 0$ and the local Euclidean area element dS_ζ on S. It is readily verified that $\lambda^2(\zeta)dS)_\zeta$ can be continued to a nonnegative finitely continuous 2-form on S and that

$$\Delta_\zeta s(\zeta,a) = \lambda^2(\zeta) \tag{3}$$

on $S - \{a,\zeta_0,\zeta_1\}$.

We introduce the functions on Ω and $\Omega\times\Omega$

$$G_\Omega(\zeta) = \frac{1}{2\pi}\int_\Omega \lambda^2(b)g_\Omega(b,\zeta)dS_b,$$

$$H_\Omega(\zeta,a) = \frac{1}{2\pi}\int_{\partial\Omega} v_\Omega(b,a) * d_b g_\Omega(b,\zeta)$$

where

$$v_\Omega(\zeta,a) = s(\zeta,a) - 2g_\Omega(\zeta,a).$$

We shall prove the continuity of G_Ω and H_Ω. We start with G_Ω. Let $\zeta' \in \Omega$ and let U be a disk with center ζ' and radius 1 such that $\bar{U}\subset\Omega$. Denote by U_r the disk $|b-\zeta'|<r$ in U with $0<r<1$, and by g Green's kernel on U. Then

$$g(b,\zeta) = \log\left|\frac{1-\bar{\zeta}b}{b-\zeta}\right| \le \log\frac{1}{|b-\zeta|} + \log 2.$$

Since $g_\Omega(b,\zeta)-g(b,\zeta)>0$ is finitely continuous on $U\times U$, its supremum M for $(b,\zeta)\in U_{1/2}\times U_{1/2}$ is finite. For $0<\varepsilon<\frac{1}{6}$ and $\zeta\in U_\varepsilon$, $g_\Omega(b,\zeta)\le -\log|b-\zeta|+M'$ in $|b-\zeta|<2\varepsilon$ with $M' = M+\log 2$.

On setting $m = \sup\{\lambda^2(b)\,|\,b\in U\}<\infty$ we obtain

$$(4)\quad \int_{|b-\zeta|<2\epsilon} \lambda^2(b)g_\Omega(b,\zeta)dS_b \le m\int_0^{2\pi}\int_0^{2\epsilon}\left(\log\frac{1}{r}+M'\right) r\,dr\,d\theta = O(\epsilon).$$

For any $\zeta''\in U_\epsilon$ we have

$$|G_\Omega(\zeta')-G_\Omega(\zeta'')| \le \int_{\Omega-U_\epsilon} \lambda^2(b)\,|g_\Omega(b,\zeta')-g_\Omega(b,\zeta'')|\,dS_b + \sum_{\zeta=\zeta',\zeta''}\int_{|b-\zeta|<2\epsilon}\lambda^2(b)g_\Omega(b,\zeta)dS_b.$$

In view of (4) the second term of the right-hand side is $O(\varepsilon)$, and since $g_\Omega(b,\zeta')\to g_\Omega(b,\zeta'')$ uniformly on $\Omega-U_\varepsilon$ as $\zeta''\to\zeta'$, we see that

$$\limsup_{\zeta''\to\zeta'} |G_\Omega(\zeta')-G_\Omega(\zeta'')|<O(\varepsilon).$$

Thus

$$\lim_{\zeta''\to\zeta'} G_\Omega(\zeta'') = G_\Omega(\zeta').$$

We turn to the continuity of H_Ω. By 1B, $(b,a)\to v_\Omega(b,a) = s(b,a)-2g_\Omega(b,a) = s(b,a)$ is finitely continuous on $\partial\Omega\times\Omega$ and the same is true of the coefficients of $*d_b g_\Omega(b,\zeta)$ as functions of the pair (b,ζ) of local parameters b and ζ on $\partial\Omega$ and Ω. Therefore we can easily conclude that $H_\Omega(\zeta,a)$ is finitely continuous on $\Omega\times\Omega$.

1D. Next we prove that

$$v_\Omega(\zeta,a) = -G_\Omega(\zeta)-H_\Omega(\zeta,a) \tag{5}$$

for all $(\zeta,a)\in\Omega\times\Omega$ and consequently v_Ω is finitely continuous on $\Omega\times\Omega$.

For a fixed $a\in\Omega$, the function $b\to v_\Omega(b,a)$ is bounded and continuous on $\bar\Omega$, of class C^∞ on $\bar\Omega-\{a,\zeta_0,\zeta_1\}$, and $\Delta_b v_\Omega(b,a) = \lambda^2(b)$. Assume that a is different from ζ_0 and ζ_1. Remove from Ω disjoint small closed disks with centers ζ_0 and ζ_1 and radii $1/n$ such that the resulting region Ω_n contains a. Let g_n be Green's kernel on Ω_n and let U_ε be a disk about $\zeta\in\Omega_n$ of radius ε such that $\bar U_\varepsilon\subset\Omega_n$.

By Green's formula

$$\int_{\partial\Omega_n-\partial U_\varepsilon} v_\Omega(b,a)*d_b g_n(b,\zeta)-\int_{\partial\Omega_n-\partial U_\varepsilon} g_n(b,\zeta)*d_b v_\Omega(b,a)$$

$$= \int_{\Omega_n-U_\varepsilon} (v_\Omega(b,a)\,\Delta_b g_n(b,\zeta)-g_n(b,\zeta)\,\Delta_b v_\Omega(b,a))\,dS_b.$$

On letting $\varepsilon\to0$ we obtain

$$2\pi v_\Omega(\zeta,a)+\int_{\partial\Omega} v_\Omega(b,a)*d_b g_n(b,\zeta)+\int_{\beta_n} v_\Omega(b,a)*d_b g_n(b,\zeta) \tag{6}$$

$$= -\int_\Omega g_n(b,\zeta)\lambda^2(b)\,dS_b$$

where $\beta_n = \partial\Omega_n - \partial\Omega$. We set $g_n(b,\zeta) = 0$ on $\Omega - \Omega_n$ and note that $g_\Omega(b,\zeta) - g_n(b,\zeta) \geq 0$ converges to 0 uniformly on $\bar{\Omega}$ with respect to b as $n \to \infty$. Therefore the coefficients of $*d_b g_n$ converge uniformly to those of $*d_n g_\Omega$ locally, and

$$\int_{\beta_n} *d_b g_n(b,\zeta) = 2\pi - \int_{\partial\Omega} *d_b g_n(b,\zeta)$$

$$= \int_{\partial\Omega} *d_b g_\Omega(b,\zeta) - *d_b g_n(b,\zeta)$$

converges to zero as $n \to \infty$. Since $*dg_n > 0$ on β_n and $|v_\Omega(b,a)| \leq K < \infty$ on Ω for a fixed a, we have

$$\left| \int_{\beta_n} v_\Omega(b,a) *d_b g_n(b,\zeta) \right| \leq K \int_{\beta_n} *d_b g_n(b,\zeta),$$

and the right-hand side tends to 0 as $n \to \infty$. Thus on letting $n \to \infty$ in (6) we obtain (5) provided ζ and a are different from ζ_0 and ζ_1. By the separate continuity of v_Ω, H_Ω, and G_Ω on Ω, we deduce the validity of (5) for every ζ and a.

1E. We have arrived at the following conclusion (Nakai [7]):

Theorem. *The Sario kernel $s(\zeta,a)$ is continuous on $S \times S$ and finitely continuous on $S \times S$ outside the diagonal set. Moreover, for every regular region Ω of S the decomposition*

$$s(\zeta,a) = 2g_\Omega(\zeta,a) + v_\Omega(\zeta,a) \tag{7}$$

is valid, where g_Ω is the Green's kernel on Ω and v_Ω is a finitely continuous function on $\Omega \times \Omega$.

2. Sario potentials

2A. Potential-theoretic principles. Let μ be a nonnegative regular Borel measure with compact support S_μ in S. Unless specified otherwise we consider only such measures μ. The Sario potential s_μ of the measure μ is defined by

$$s_\mu(\zeta) = \int s(\zeta,a)\,d\mu(a).$$

By (1) it is nonnegative, and positive unless $\mu = 0$. As a consequence of Theorem 1E, it is lower semicontinuous on S and finitely continuous on $S - S_\mu$. By virtue of (3), s_μ is subharmonic on $S - S_\mu$.

The object of potential theory is to find properties independent of μ of the family $\{s_\mu\}_\mu$. Such properties are customarily called principles of potential theory.

2B. Local maximum principle. Since s_μ is subharmonic in $S - S_\mu$, its magnitude is determined by its behavior at the ideal boundary of S and at S_μ. Regarding the latter we shall show:

Theorem (local maximum principle). *Let F be a closed subset of S containing S_μ. Then for any $\zeta' \in F$*

$$\limsup_{\zeta \in S-F, \zeta \to \zeta'} s_\mu(\zeta) \le \limsup_{\zeta \in F, \zeta \to \zeta'} s_\mu(\zeta). \tag{8}$$

For the proof take a parametric disk D about ζ', let $\mu' = \mu|D$, i.e. $\mu'(\cdot) = \mu(\cdot \cap D)$, and set $\mu'' = \mu - \mu'$. Then $s_\mu = s_{\mu'} + s_{\mu''}$ and $s_{\mu''}$ is continuous on D. Thus if we can prove (8) for $s_{\mu'}$ it will follow for s_μ. Therefore we may suppose that $F \subset D$. Let $u(\zeta,a) = s(\zeta,a) - 2\log(1/|\zeta - a|)$ on $D \times D$. By Theoerm 1E, $u(\zeta,a)$ is finitely continuous on $D \times D$ and a fortiori $\zeta \to \int u(\zeta,a) d\mu(a)$ is finitely continuous on D. Hence the proof of (8) is reduced to that of

$$\limsup_{\zeta \in D-F, \zeta \to \zeta'} \int \log \frac{1}{|\zeta - a|}\, d\mu(a) \le \limsup_{\zeta \in F, \zeta \to \zeta'} \int \log \frac{1}{|\zeta - a|}\, d\mu(a),$$

which is elementary (see e.g. Tsuji [5, p. 53]).

2C. Continuity principle. As a consequence of the local maximum principle we are now able to state:

Theorem (continuity principle). *If $s_\mu | S_\mu$ is continuous (resp. finitely continuous) at $\zeta \in S_\mu$ on S_μ, then the same is true for s_μ at ζ on S.*

In fact, by the lower semicontinuity of s_μ on S and (8)

$$s_\mu(\zeta) \le \liminf_{\zeta' \in S, \zeta' \to \zeta} s_\mu(\zeta') \le \limsup_{\zeta' \in S, \zeta' \to \zeta} s_\mu(\zeta') \le \limsup_{\zeta' \in S_\mu, \zeta' \to \zeta} s_\mu(\zeta').$$

But the last term is $s_\mu(\zeta)$ since $s_\mu|S_\mu$ is continuous, and the assertion follows.

3. Unicity principle

3A. Uniqueness. The potentials s_μ determine a linear operator $\mu \to s_\mu$ from the measure space into a function space. We now see that this operator is injective (Nakai [10]):

Theorem (unicity principle). $s_\mu = s_\nu$ *implies that* $\mu = \nu$.

We shall actually prove more: if $s_\mu = s_\nu + u$, with u a harmonic function on S, then $\mu = \nu$.

3B. Let $C_0^2(S)$ be the space of twice continuously differentiable functions with compact supports in S. If C_ϵ is the clockwise oriented boundary of a disk D_ϵ about $b \in S$ with radius $\epsilon > 0$, then Green's formula yields

$$(9)\quad \int_{S-D_\epsilon} (f(a)\,\Delta_a s(a,b) - s(a,b)\,\Delta_a f(a))\,dS_a$$

$$= \int_{C_\epsilon} (f(a) * d_a s(a,b) - s(a,b) * d_a f(a))$$

for every $f \in C_0^2(S)$. In view of (3), we obtain on letting $\epsilon \to 0$ in

$$(10)\qquad f(b) = \frac{1}{4\pi}\int_S f(a)\lambda^2(a)dS_a - \frac{1}{4\pi}\int_S s(a,b)\Delta_a f(a)dS_a.$$

Let σ be the signed measure $\mu - \nu$. Then for every $f \in C_0^\infty(S)$

$$(11)\quad \int f(b)d\sigma(b) = \frac{\sigma(S)}{4\pi}\int f(a)\lambda^2(a)dS_a - \frac{1}{4\pi}\int s_\sigma(a)\Delta_a f(a)dS_a,$$

as is seen from (10) on integrating both sides with respect to σ. Since $s_\sigma \equiv u$,

$$(12)\qquad \Delta_a s_\sigma(a) = 0,$$

and $\int s_\sigma(a)\Delta_a f(a)dS_a - \int f(a)\Delta_a s_\sigma(a)dS_a = 0$, (11) implies that

$$\int f(b)(d\sigma(b) - \frac{\sigma(S)}{4\pi}\lambda^2(b)dS_b) = 0$$

for every $f \in C_0^\infty(S)$. From this we conclude that

$$d\sigma(b) = \frac{\sigma(S)}{4\pi}\lambda^2(b)dS_b.$$

If $\sigma(S)$ were not zero, then σ would be of constant sign and therefore $s_\sigma \neq 0$, i.e. $s_\mu \neq s_\nu$, a contradiction. Thus $\sigma(S) = 0$ and consequently $\sigma = 0$, i.e. $\mu = \nu$.

3C. *Remark.* In Theorem 3A the assumption $s_\mu = s_\nu$ on S can be weakened as follows: $s_\mu = s_\nu$ on S except for a set on which there is no unit measure τ with $\int s(a,b)d\tau(a)d\tau(b) < \infty$ (see 2.2A below).

In fact, by (7) we see that

$$s_\xi(a) = \lim_{\varepsilon \to 0} \frac{1}{2\pi} \int s_\xi(a + \varepsilon e^{i\theta})d\theta$$

for $\xi = \mu, \nu$. From this it follows that $s_\mu = s_\nu$ on all of S without exception.

§2. MAXIMUM PRINCIPLE

Frostman's maximum principle will be established for Sario potentials (Nakai [8]). As a consequence the fundamental theorem of potential theory will be proved. We shall also deduce the validity of the energy principle.

1. Frostman's maximum principle

1A. Global maximum. By the local maximum principle the behavior of s_μ near S_μ is regulated by that of $s_\mu|S_\mu$. Moreover the construction suggests that $\max_{S-S_\mu} s_\mu$ is ruled by the behavior of s_μ at the relative boundary of $S - S_\mu$. Therefore it is plausible to state (Nakai [8]):

Theorem (Frostman's maximum principle). *$s_\mu \mid S_\mu \leq M$ implies $s_\mu \leq M$ on S.*

1B. Let K be a compact subset in S. We then have

$$M(K) = \sup_{a \in K} \limsup_{\zeta \to \beta} s(\zeta, a) < \infty \tag{13}$$

where β stands for the ideal boundary of S. If $\beta = \emptyset$ we understand that $M(K) = 0$. For the proof of (13) take a regular region S_0 containing $K \cup \{\zeta_0\}$. By 1.1B, $(\zeta,a) \to t(\zeta,a)$ is finitely continuous on $\partial S_0 \times K$ and hencc it takes a finite maximum M_0; $t(\zeta,a) \leq M_0$ on ∂S_0 for any fixed $a \in K$. Since $L_1 t = t$ on $S - S_0$, we conclude that $t(\zeta,a) \leq M_0$ for $(\zeta,a) \in (S - S_0) \times K$. Clearly $s_0(\zeta)$ is bounded on $S - S_0$, and therefore (13) follows from the definition of $s(\zeta,a)$.

Set $B(M,\mu) = \max\{M, \mu(S_\mu) M(S_\mu)\}$. By the local maximum principle and the maximum principle for subharmonic functions we obtain the following partial result:

(14) $s_\mu | S_\mu \leq M$ *implies that* $s_\mu \leq B(M,\mu)$ *on* S.

1C. If S is closed then $B(M,\mu) = M$ and Theorem 1A follows. If S is open but parabolic then, since s_μ is bounded by $B(M,\mu)$ and subharmonic on $S - S_\mu$,

$$\sup_{S - S_\mu} s_\mu = \sup_{\zeta' \in \partial S_\mu} \limsup_{\zeta \to \zeta'} s_\mu | S - S_\mu .$$

By the local maximum principle the right-hand side is dominated by

$$\sup_{\zeta' \in S_\mu} \limsup_{\zeta \in S_\mu, \zeta \to \zeta'} s_\mu | S_\mu \leq M.$$

Thus Theorem 1A again follows. Therefore the proof will be complete if we give it for a hyperbolic S.

1D. Hereafter in 1 we always suppose that S is hyperbolic, i.e. Green's kernel $g(\zeta,a)$ exists on S. In this case we conclude by the unicity of t_0 on S that

(15) $$t_0(\zeta) = 2g(\zeta,\zeta_0) - 2g(\zeta,\zeta_1) + k$$

where k is a suitable constant. Therefore

(16) $$s_0(\zeta) = \log(e^{-2g(\zeta,\zeta_0)} + e^{-2g(\zeta,\zeta_1)+k}) + 2g(\zeta,\zeta_0).$$

Similarly we obtain

(17) $$t(\zeta,a) = 2g(\zeta,a) - 2g(\zeta,\zeta_0) + s_0(a) - k.$$

We set

$$(18) \qquad u(\zeta) = \log(e^{-2g(\zeta,\zeta_0)} + e^{-2g(\zeta,\zeta_1)+k}) - \log(1+e^k).$$

Then Sario's kernel is expressed in terms of Green's kernel as follows:

$$(19) \qquad s(\zeta,a) = 2g(\zeta,a) + u(\zeta) + u(a) + m$$

with a suitable constant m.

1E. We are to prove that $s_\mu \leq M$ on S under the assumption $s_\mu | S_\mu \leq M$. For this purpose we may assume without loss of generality that $\mu(S) = 1$. By (19) what we have to show is that the validity of

$$(20) \qquad 2g_\mu(\zeta) + u(\zeta) \leq M'$$

on S_μ implies that on all of S, where $g_\mu(\zeta) = \int g(\zeta,a)d\mu(a)$ and

$$M' = M - m - \int u(a)d\mu(a).$$

We now assume that (20) holds on S_μ, and first prove that $M' \geq 0$.

To this end take the unit measure ν on S such that $S_\nu \subset S_\mu$ and

$$\int g(\zeta,a)d\nu(\zeta)d\nu(a) = V_g(S_\mu) = \inf_\theta \int g(\zeta,a)d\theta(\zeta)d\theta(a)$$

where the θ's are unit measures with $S_\theta \subset S_\mu$. Such a measure ν always exists and has the properties $g_\nu \leq V_g(S_\mu)$ on S and $g_\nu = V_g(S_\mu)$ on S_μ except for a set which carries no measure $\theta \neq 0$ with $\int g d\theta d\theta < \infty$. For this well-known fact we refer the reader to e.g. Constantinescu-Cornea [1, p. 48] (see also the proof of Theorem 2B below).

Since $\int g d\mu d\mu < \infty$ in view of (20), the μ-measure of the subset of S_μ on which $g_\nu \neq V_g(S_\mu)$ is zero, and hence $\int g_\nu d\mu = \int V_g(S_\mu) d\mu = V_g(S_\mu)$. Fubini's theorem now yields $\int g_\mu d\nu =$

$V_g(S_\mu)$. On integrating both sides of (20) with respect to ν we obtain

$$2V_g(S_\mu) + \int u(\zeta)d\nu(\zeta) \le M'. \tag{21}$$

By (18) we see that

$$\int u(\zeta)d\nu(\zeta) = -2g_\nu(\zeta_0) + \int \varphi(\zeta)d\nu(\zeta)$$

with

$$\varphi(\zeta) = \psi(2g(\zeta,\zeta_0) - 2g(\zeta,\zeta_1)),$$

$$\psi(\xi) = \log(1+e^{\xi+k}) - \log(1+e^k).$$

Since $\psi(\xi)$ is a convex function, Jensen's inequality $\psi(\int \xi d\nu(\zeta)) \le \int \psi(\xi)d\mu(\zeta)$ is valid for any ν-integrable function $\xi = \xi(\zeta)$ and in particular for $\xi(\zeta) = 2g(\zeta,\zeta_0) - 2g(\zeta,\zeta_1)$. Hence

$$\log \frac{1+e^{2g_\nu(\zeta_0)-2g_\nu(\zeta_1)+k}}{1+e^k} \le \int \varphi(\zeta)d\nu(\zeta)$$

and therefore

$$\log \frac{e^{-2g_\nu(\zeta_0)} + e^{-2g_\nu(\zeta_1)+k}}{1+e^k} \le \int u(\zeta)d\nu(\zeta).$$

Since $g_\nu(\zeta_0)$ and $g_\nu(\zeta_1)$ are dominated by $V_g(S_\mu)$, we now conclude that

$$\int u(\zeta)d\nu(\zeta) \ge -2V_g(S_\mu).$$

This with (21) implies $M' \ge 0$, as desired.

1F. Consider the family $\mathfrak{F}$ of sequences $\{\zeta_n\} \subset S$ converging to the ideal boundary β of S, and the family $\mathfrak{F}^+ \subset \mathfrak{F}$ consisting of $\{\zeta_n\}$ with $\liminf_n g(\zeta_n,a) > 0$ for some and hence for all $a \in S$. There exists a positive superharmonic function v on S such that $\lim_n v(\zeta_n) = \infty$ for $\{\zeta_n\} \in \mathfrak{F}^+$ (see e.g. Constantinescu-Cornea

[1, p. 27]). Observe that

$$w_m(\zeta) = M' - (g_\mu(\zeta) + u(\zeta)) + \frac{v(\zeta)}{m}$$

is a superharmonic function on $S - S_\mu$ for any $m = 1, 2, \cdots$. By the local maximum principle and (20),

$$\liminf_{\zeta \in S - S_\mu, \zeta \to \zeta'} w_m(\zeta) \geq 0 \tag{22}$$

for every $\zeta' \in \partial S_\mu$. We also have

$$\liminf_n w_m(\zeta_n) \geq 0 \tag{23}$$

for every $\{\zeta_n\} \in \mathfrak{F}$. In fact, if $\liminf_n w_m(\zeta_n) < 0$ for some $\{\zeta_n\} \in \mathfrak{F}$ then since g_μ and u are bounded near β, $\{\zeta_n\}$ cannot be in $\mathfrak{F}^+$. Therefore we can find a subsequence $\{\zeta_n'\}$ of $\{\zeta_n\}$ such that $\lim_n w_m(\zeta_n') < 0$ and $\lim_n g(\zeta_n', a) = 0$ for every $a \in S$. Clearly this implies that

$$\lim_n g_\mu(\zeta_n') = \lim_n u(\zeta_n') = 0.$$

Thus we would obtain

$$0 > \lim_n w_m(\zeta_n') \geq \lim_n (M' - (g_\mu(\zeta_n') + u(\zeta_n')) = M',$$

in violation of $M' \geq 0$.

By (22), (23), and the minimum principle for superharmonic functions we see that $w_m(\zeta) \geq 0$ on $S - S_\mu$. On letting $m \to \infty$ we obtain (20) on $S - S_\mu$ and hence on S.

The proof of Theorem 1A is herewith complete.

2. Fundamental theorem

2A. Capacity. We define a set function $V(K) = V_s(K)$ first for compact sets $K \subset S$ by

$$V(K) = \inf_\mu \int s(\zeta, a) d\mu(\zeta) d\mu(a)$$

where μ runs over all unit measures with $S_\mu \subset K$. For a general set $X \subset S$ we write

$$V(X) = \sup_K V(K)$$

where K runs over all compacta $K \subset X$.

The quantity

$$c(X) = c_s(X) = \frac{1}{V(K)}$$

will be referred to as the (inner) *Sario capacity* of X. From the definition it is easy to verify that for Borel sets X, $c(X) = 0$ is characterized by

$$(24) \quad \mu(X) = 0 \text{ for every } \mu \text{ with } \int s(\zeta,a)d\mu(\zeta)d\mu(a) < \infty.$$

Using this we can prove:

Theorem. *A set X is of Sario capacity zero if and only if X is locally of logarithmic capacity zero.*

For the proof we may suppose that X is compact. Take a parametric disk D. By virtue of (7), $\mu(D \cap X) = 0$ for every μ in D with $\int_D \log(1/|\zeta - a|)d\mu(\zeta)d\mu(a) < \infty$ provided (24) is valid. This means that $X \cap D$ has logarithmic capacity zero.

2B. Capacitary measure. Let K be a compact set with $c(K) > 0$. Since s is jointly continuous, by the selection theorem for a sequence of measures we can find a unit measure μ with $S_\mu \subset K$ such that $\int s d\mu d\mu = V(K)$. Such a measure μ is called the *capacitary measure* for K. We shall see in 3B that μ is unique. For this measure we prove the following *capacitary principle*:

Theorem (fundamental theorem of potential theory). *Let K be a compact subset of S with positive Sario capacity, and μ its capacitary measure. Then $s_\mu \le V(K)$ on S, and $s_\mu = V(K)$ on K except for an F_σ-set of Sario capacity zero.*

2C. As the first step we shall prove that $s_\mu \geq V(K)$ on K except for an F_σ-set of Sario capacity zero. Let A and A_n be the subsets of K on which $s_\mu < V(K)$ and $s_\mu \leq V(K) - 1/n$ ($n = 1,2,\cdots$) respectively. Then A_n is compact and

$$A_1 \subset \cdots \subset A_n \subset \cdots, \qquad A = \cup A_n.$$

This indicates that A is an F_σ-set. We shall show that $c(A) = 0$.

Suppose that this were not the case. Then there would exist an n with $c(A_n) > 0$. This means that for a suitable $\varepsilon > 0$ there exists a compact subset $K_1 \subset K$ with

$$s_\mu | K_1 < V(K) - 2\varepsilon, \qquad c(K_1) > 0. \tag{25}$$

Note that $\int s_\mu s_\mu d\mu = V(K)$ implies the existence of a point $\zeta_0 \in S_\mu$ with $s_\mu(\zeta_0) > V(K) - \varepsilon$. By (25), $\zeta_0 \notin K_1$. Therefore we can choose an open disk U about ζ_0 with $\bar{U} \cap K_1 = \emptyset$ and

$$s_\mu | U > V(K) - \varepsilon. \tag{26}$$

Moreover since $\zeta_0 \in S_\mu$

$$\mu(U) > 0. \tag{27}$$

By virtue of $c(K_1) > 0$ there exists a measure ν with $S_\nu \subset K_1$ such that

$$\nu(K_1) = \mu(U), \qquad \int s d\nu d\nu < \infty. \tag{28}$$

Using this ν we construct a new signed measure ν_1 by

$$\nu_1 | K_1 = \nu | K_1, \qquad \nu_1 | U = -\mu | U, \qquad \nu_1 | S - K_1 \cup U = 0. \tag{29}$$

Clearly $\mu_t = \mu + t\nu_1$ is a unit measure for every $t \in (0,1)$ with $S_{\mu_t} \subset K$. Therefore

$$\int s d\mu_t d\mu_t \geq \int s d\mu d\mu = V(K). \tag{30}$$

On the other hand a simple calculation shows that

$$\int s d\mu_t d\mu_t - \int s d\mu d\mu = 2t\left(\int_{K_1} s_\mu d\nu_1 + \int_U s_\mu d\nu_1\right) + t^2 \int s d\nu_1 d\nu_1$$

$$< 2t(\mu(U)(V-2\epsilon) - \mu(U)(V-\epsilon)) + t^2 \int s d\nu_1 d\nu_1$$

$$= -t(2\mu(U)\epsilon - t\int s d\nu_1 d\nu_1).$$

The last member can be made negative by taking t sufficiently small. This contradicts (30) and we have $c(A) = 0$.

2D. As the second step we shall show that $s_\mu | S_\mu \leq V(K)$. Contrary to the assertion assume that $s_\mu(\zeta_1) > V(K)$ for a $\zeta_1 \in S_\mu$. Take an open disk U_1 about ζ_1 such that

$$s_\mu | U_1 > V(K) + \varepsilon, \qquad \varepsilon > 0.$$

Note that $\mu(U_1) > 0$. By 2C we see that

$$V(K) = \int_{U_1} s_\mu d\mu + \int_{S-U_1} s_\mu d\mu$$

$$> (V(K)+\epsilon)\mu(U_1) + V(K)\mu(S-U_1)$$

$$= V(K) + \epsilon\mu(U_1) > V(K),$$

a contradiction.

By the maximum principle $s_\mu | S_\mu \leq V(K)$ implies that $s_\mu \leq V(K)$ on all of S. This with 2C proves Theorem 2B.

2E. Subadditivity. As an application of the fundamental theorem we prove the subadditivity of the Sario capacity:

Theorem. *Let X_n $(n = 1,2,\cdots)$ be sets in S and $X = \cup_1^\infty X_n$. Then*

$$c(X) \leq \sum_1^\infty c(X_n). \tag{31}$$

We may assume that $V(X) < \infty$ and X and X_n are compact.

Let μ and μ_n be capacitary measures for X and X_n respectively. Then

$$V(X) = \int_X s(\zeta,a)d\mu(a) \geq \int_{X_n} s(\zeta,a)d\mu(a) \tag{32}$$

for $\zeta \in X_n \subset X$ except for a set of Sario capacity zero. Since $\int s_{\mu_n}(a)d\mu_n(a) = V(X_n)$, integration of both sides of (32) with respect to μ_n and Fubini's theorem give

$$V(X) \geq \int_{X_n} s_{\mu_n}(a)d\mu(a) = V(X_n)\mu(X_n),$$

i.e.

$$c(X)\mu(X_n) \leq c(X_n).$$

Therefore

$$c(X) = c(X) \sum_1^\infty \mu(X_n) \leq \sum_1^\infty c(X_n).$$

3. Energy principle

3A. Ninomiya's theorem. Let Ω be a locally compact Hausdorff space and $k(x,y)$ a continuous positive function on $\Omega \times \Omega$ with $k(x,x) = \infty$ and $k(x,y) = k(y,x)$. One can consider potentials $k_\mu(x) = \int k(x,y)d\mu(y)$. Ninomiya [1, Lemma 6] proved that if k_μ satisfies Frostman's maximum principle and the unicity principle in the form of Remark 1.3C then

$$\int k(x,y)d\sigma(x)d\sigma(y) > 0$$

for any nonzero signed measure σ.

Since the Sario potentials enjoy both Frostman's maximum principle and unicity principle, Ninomiya's theorem can be applied to obtain:

Theorem (energy principle). *For any measures μ and ν with $\sigma = \mu - \nu \neq 0$,*

$$\int s(\zeta,a)d\sigma(\zeta)d\sigma(a) > 0. \tag{33}$$

For signed measures $\sigma = \mu - \nu$ set

$$(\sigma_1, \sigma_2) = \int s\,d\sigma_1 d\sigma_2, \qquad \|\sigma\| = \sqrt{(\sigma,\sigma)}.$$

$\|\sigma\|^2$ is referred to as the *energy* of σ. Let $\mathcal{E} = \{\sigma \mid \|\sigma\| < \infty\}$. The energy principle assures that $\mathcal{E}$ is a pre-Hilbert space with (σ_1, σ_2) inner product and $\|\sigma\|$ norm. Thus we have the Schwarz inequality $|(\sigma_1,\sigma_2)| \leq \|\sigma_1\|\,\|\sigma_2\|$ and the triangle inequality $\|\sigma_1+\sigma_2\| \leq \|\sigma_1\| + \|\sigma_2\|$.

3B. Unicity of capacitary measure. We are now in a position to prove the unicity of the capacitary measure as anticipated in 2B.

Let μ_1 and μ_2 be capacitary measures for a compact set K with positive Sario capacity so that $\|\mu_1\|^2 = \|\mu_2\|^2 = V(K)$. By Theorem 2B

$$s_{\mu_i}(\zeta) = V(K)$$

on K except for a set of Sario capacity zero, and hence

$$(\mu_i, \mu_j) = \int s_{\mu_i}(\zeta)\,d\mu_j(\zeta) = V(K).$$

Observe that

$$\begin{aligned}\|\mu_1 - \mu_2\|^2 &= \|\mu_1\|^2 - 2(\mu_1,\mu_2) + \|\mu_2\|^2 \\ &= V(K) - 2V(K) + V(K) = 0.\end{aligned}$$

The energy principle yields $\mu_1 = \mu_2$.

BIBLIOGRAPHY

ACCOLA, R.

[1] *The bilinear relation on open Riemann surfaces.* Trans. Amer. Math. Soc. 96 (1960), 143–161.

[2] *Differentials and extremal length on Riemann surfaces.* Proc. Nat. Acad. Sci. U.S.A. 46 (1960), 540–543.

[3] *On semi-parabolic Riemann surfaces.* Trans. Amer. Math. Soc. 108 (1963), 437–448.

[4] *On a class of Riemann surfaces.* Proc. Amer. Math. Soc. 15 (1964), 607–611.

[5] *Some classical theorems on open Riemann surfaces.* Bull. Amer. Math. Soc. 73 (1967), 13–26.

AHLFORS, L.

[1] *Bounded analytic functions.* Duke Math. J. 14 (1947), 1–11.

[2] *Open Riemann surfaces and extremal problems on compact subregions.* Comment. Math. Helv. 24 (1950), 100–134.

[3] *Remarks on the classification of open Riemann surfaces.* Ann. Acad. Sci. Fenn. Ser. AI No. 87 (1951), 8 pp.

[4] *Remarks on Riemann surfaces.* Lectures on functions of a complex variable, Univ. of Mich. Press, Ann Arbor, Mich., 1955. 45–48.

[5] *Abel's theorem for open Riemann surfaces.* Seminars on analytic functions II. Institute for Advanced Study, Princeton, N.J., 1958. 7–19.

[6] *The method of orthogonal decomposition for differentials on open Riemann surfaces.* Ann. Acad. Sci. Fenn. Ser. AI No. 249/7 (1958), 15 pp.

[7] *Complex analysis.* Second edition. McGraw-Hill Book Company, New York, 1966. 317 pp.

AHLFORS, L.; BEURLING, A.

[1] *Conformal invariants and function-theoretic null-sets.* Acta Math. 83 (1950), 101–129.

AHLFORS, L.; SARIO, L.

[1] *Riemann surfaces.* Princeton Univ. Press, Princeton, N.J., 1960. 382 pp.

AKAZA, T.

[1] *On the weakness of some boundary component.* Nagoya Math. J. 17 (1960), 219–223.

AKAZA, T.; OIKAWA, K.

[1] *Examples of weak boundary components.* Nagoya Math. J. 18 (1961), 165–170.

ALLENDOERFER, C.; EELLS, J.

[1] *On the cohomology of smooth manifolds.* Comment. Math. Helv. 32 (1958), 165–179.

BADER, R.; SÖRENSEN, W.

[1] *Formes harmoniques sur une surface de Riemann.* Comment. Math. Helv. 34 (1960), 140–174.

BEHNKE, H.; STEIN, K.

[1] *Entwicklung analytischer Funktionen auf Riemannschen Flächen.* Math. Ann. 120 (1949), 430–461.

BERGMAN, S.

[1] *The kernel functions and conformal mapping.* Math. Surveys 5, Amer. Math. Soc., New York, 1950. 161 pp.

BESICOVITCH, A.

[1] *On two problems of Loewner.* J. London Math. Soc. 27 (1952), 141–144.

BEURLING, A.

[1] Cf. AHLFORS, L.; BEURLING, A. [1].

BIDAL, P.; DE RHAM, G.

[1] *Les formes différentielles harmoniques.* Comment. Math. Helv. 19 (1946), 1–49.

BIEBERBACH, L.

[1] *Über einen Riemannschen Satz aus der Lehre von der konformen Abbiddung.* Sitzungsber. Preuss. Akad. Wiss. Berlin 24 (1925), 6–9.

BLATTER, C.

[1] *Une inéqualité de géométrie différentielle.* C. R. Acad. Sci. Paris 250 (1960), 1167.

[2] *Über Extremallängen auf geschlossenen Flächen.* Comment. Math. Helv. 35 (1961), 153–168.

[3] *Zur Riemannschen Geometrie im Grossen auf dem Möbiusband.* Compositio Math. 15 (1961), 88–107.

BOURBAKI, N.

[1] *Éléments de mathématique, Livre VI, Intégration. Chapitres 1–4.* Second edition, Hermann, Paris, 1965. 283 pp.

BREAZEAL, N.

[1] *The class O_{AD} of Riemannian 2-spaces.* Doctoral dissertation, Univ. of Calif., Los Angeles, Calif., 1966. 43 pp.

BRELOT, M.

[1] *Éléments de la théorie classique du potentiel.* 3e édition. Centre de Documentation Universitaire, Paris, 1965. 209 pp.

[2] *Lectures on potential theory.* Tata Inst. Fund. Res., Bombay, 1960. 153 pp.

[3] *Axiomatique des fonctions harmoniques.* Sem. Math. Sup., Univ. de Montréal, 1965. 140 pp.

BRELOT, M.; CHOQUET, G.

[1] *Espaces et lignes de Green.* Ann. Inst. Fourier 3 (1951), 199–263.

BROWDER, F.

[1] *Principal functions for elliptic systems of differential equations.* Bull. Amer. Math. Soc. 71 (1965), 342–344.

BRUCKNER, J.

[1] *Triangulations of bounded distortion in the classification theory of Riemann surfaces.* Doctoral dissertation, Univ. of Calif., Los Angeles, Calif., 1960. 64 pp.

CARTAN, E.

[1] *Les systèmes différentielles extérieurs et leurs applications géométriques.* Hermann, Paris, 1945. 214 pp.

[2] *Leçons sur la géométrie des espaces de Riemann.* Deuxième edition, Gauthier-Villars, Paris, 1951. 378 pp.

CHEVALLEY, C.

[1] *Introduction to the theory of algebraic functions of one variable.* Math. Surveys 6, Amer. Math. Soc., New York, 1951, 188 pp.

CHOQUET, G.

[1] *Le théorème de representation intégrale dans les ensembles convexes compacts.* Ann. Inst. Fourier 10 (1960), 333–344.

[2] Cf. BRELOT, M.; CHOQUET, G. [1].

CONNER, P.

[1] *The Green's and Neumann's problems for differential forms on Riemannian manifolds.* Proc. Nat. Acad. Sci. U.S.A. 40 (1954), 1151–1155.

[2] *The Neumann's problem for differential forms on Riemannian manifolds* Mem. Amer. Math. Soc. No. 20, 1956. 56 pp.

CONSTANTINESCU, C.; CORNEA, A.

[1] *Ideale Ränder Riemannscher Flächen.* Springer-Verlag, Berlin-Göttingen-Heidelberg, 1963. 244 pp.

[2] *Compactifications of harmonic spaces.* Nagoya Math. J. 25 (1965), 1–57.

CORNEA, A.

[1] Cf. CONSTANTINESCU, C.; CORNEA, A. [1].

COURANT, R.; HILBERT, D.

[1] *Methods of mathematical physics.* Vol. II. Interscience Publishers, New York, 1962. 830 pp.

DUFF, G.

[1] *Differential forms in manifolds with boundary.* Ann. of Math. (2) 56 (1952), 115–127.

[2] *A tensor equation of elliptic type.* Canad. J. Math. 5 (1953), 524–535.

[3] *On the potential theory of coclosed harmonic forms.* Ibid. 7 (1955), 126–137.

[4] *On the Neumann and dual-adjoint problems of generalized potential theory.* Trans. Roy. Soc. Canada 50 (1956), 23–31.

[5] *Partial differential equations.* Univ. of Toronto Press, Toronto, 1956. 248 pp.

[6] *Hyperbolic mixed problems for harmonic tensors.* Canad. J. Math. 9 (1957), 161–179.

DUFF, G.; SPENCER, D.

[1] *Harmonic tensors on manifolds with boundary.* Proc. Nat. Acad. Sci. U.S.A. 37 (1951), 614–619.

[2] *Harmonic tensors on Riemannian manifolds with boundary.* Ann. of Math. (2) 56 (1952), 128–156.

EELLS, J.

[1] Cf. ALLENDOERFER, C.; EELLS, J.

EMIG, R.

[1] *Meromorphic functions and the capacity function on abstract Riemann surfaces.* Doctoral dissertation, Univ. of Calif., Los Angeles, Calif., 1962. 81 pp.

EVANS, G.

[1] *Potentials and positively infinite singularities of harmonic functions.* Monatsh. Math. Phys. 43 (1936), 419–424.

FELLER, W.

[1] *Über die Lösungen der linearen partiellen Differentialgleichungen zweiter Ordung vom elliptischen Typus.* Math. Ann. 102 (1930), 633–649.

FRIEDRICHS, K.

[1] *Differential forms on Riemannian manifolds.* Comm. Pure Appl. Math. 8 (1955), 551–590.

FUGLEDE, B.

[1] *Extremal length and functional completion.* Acta Math. 98 (1957), 171–219.

FUKUDA, N.
[1] Cf. SARIO, L.; FUKUDA, N. [1].

FULLER, D.
[1] *Mappings into abstract Riemann surfaces.* Doctoral dissertation, Univ. of Calif., Los Angeles, Calif., 1963. 74 pp.

GAFFNEY, M.
[1] *The harmonic operator for exterior differential forms.* Proc. Nat. Acad. Sci. U.S.A. 37 (1951), 48–50.
[2] *A special Stokes's theorem for complete Riemannian manifolds.* Ann. of Math. (2) 60 (1954), 140–145.
[3] *The heat equation method of Milgram and Rosenbloom for open Riemannian manifolds.* Ibid. (2) 60 (1954), 458–466.
[4] *Hilbert space methods in the theory of harmonic integrals.* Trans. Amer. Math. Soc. 78 (1955), 426–444.

GILLIS, P.
[1] *Sur des formes différentielles et la formule de Stokes.* Acad. Roy. Belgique. Cl. Sci. Mém. Coll. (2)20 (1943), 95 pp.

GIRAUD, G.
[1] *Problèmes mixtes et problèmes sur des variétés closes, relativement aux équations linéares du type elliptique.* Polskie Towarzstow Matematyczue 12 (1933), 35–54.

GLASNER, M.
[1] *Harmonic functions with prescribed boundary behavior in Riemannian spaces.* Doctoral dissertation, Univ. of Calif., Los Angeles, Calif., 1966. 57 pp.
[2] Cf. SARIO, L.; SCHIFFER, M.; GLASNER, M. [1].

GOLDSTEIN, M.
[1] *L- and K-kernels on an arbitrary Riemann surface.* Doctoral dissertation, Univ. of Calif., Los Angeles, Calif., 1963. 71 pp.
[2] *Normal operators* (to appear).

GOWRISANKARAN, K.
[1] *Extreme harmonic functions and boundary value problems.* Ann. Inst. Fourier 13 (1963), 307–356.

GRÖTZSCH, H.
[1] *Das Kreisbogenschlitztheorem der konformen Abbildung schlichter Bereiche.* Ber. Verh. Sächs. Akad. Wiss. Leipzig 83 (1931), 238–253.

GRUNSKY, H.
[1] *Neue Abschätzungen zur konformen Abbildung ein- und mehrfach zusammenhängender Bereiche.* Schriften Sem. Univ. Berlin 1 (1932), 95–140.

HARMON, S.
[1] *Regular covering surfaces of Riemann surfaces.* Doctoral dissertation, Univ. of Calif., Los Angeles, Calif., 1958. 74 pp.
[2] *Regular covering surfaces of Riemann surfaces.* Pacific J. Math. 10 (1960), 1263–1289.

HEINS, M.
[1] *On the continuation of a Riemann surface.* Ann. of Math. (2) 43 (1942), 280–297.
[2] *Riemann surfaces of infinite genus.* Ibid. (2) 55 (1952), 296–317.
[3] *A problem concerning the continuation of Riemann surfaces.* Contributions to the theory of Riemann surfaces, Princeton Univ. Press, Princeton, N.J., 1953. 55–62.

HEINS, M.
[4] *Studies in the conformal mapping of Riemann surfaces I.* Proc. Nat. Acad. Sci. U.S.A. 39 (1953), 322–324.
[5] *Studies in the conformal mapping of Riemann surfaces II.* Ibid. 40 (1954), 302–305.
[6] *On the Lindelöf principle.* Ann. of Math. (2) 61 (1955), 440–473.
[7] *Lindelöfian maps.* Ibid. (2) 62 (1955), 418–446.
[8] *Functions of bounded characteristic and Lindelöfian maps.* Internat. Congr. Math. Edinburgh, 1958. 376–388.
[9] *On the boundary behavior of a conformal map of the open unit disk into a Riemann surface.* J. Math. Mech. 9 (1960), 573–581.

HERSCH, J.
[1] *Longueurs extrémales et théorie des fonctions.* Comment. Math. Helv. 29 (1955), 301–337.

HILBERT, D.
[1] Cf. COURANT, R.; HILBERT, D. [1].

HILLE, E.
[1] *Analytic function theory.* Ginn and Co., New York. Vol. I, 1959, 308 pp. Vol. II, 1962, 496 pp.

HODGE, W.
[1] *A Dirichlet problem for harmonic functionals, with applications to analytic varieties.* Proc. London Math. Soc. 36 (1934), 257–303.
[2] *Harmonic functionals in a Riemannian space.* Ibid. 38 (1935), 72–95.
[3] *The existence theorem for harmonic integrals.* Ibid. 41 (1936), 483–496.
[4] *The theory and applications of harmonic integrals.* Second edition, Cambridge Univ. Press, Cambridge, 1952. 282 pp.

HÖRMANDER, L.
[1] *Linear partial differential operators.* Springer-Verlag, Berlin-Göttingen-Heidelberg, 1963. 287 pp.

JENKINS, J.
[1] *Univalent functions and conformal mapping.* Springer-Verlag, Berlin-Göttingen-Heidelberg, 1958. 169 pp.
[2] *On some span theorems.* Illinois J. Math. 7 (1963), 104–117.

JOHNSON, W.
[1] *Harmonic functions of bounded characteristic in locally Euclidean spaces.* Doctoral dissertation, Univ. of Calif., Los Angeles, Calif., 1964. 52 pp.

JURCHESCU, M.
[1] *Modulus of a boundary component.* Pacific J. Math. 8 (1958), 791–804.
[2] *Bordered Riemann surfaces.* Math. Ann. 143 (1961), 264–292.

KELLEY, J.
[1] *General topology.* D. Van Nostrand Company, Inc., Princeton, N.J., 1955. 298 pp.

KLOTZ, T.; SARIO, L.
[1] *Gaussian mapping of arbitrary minimal surfaces.* J. Analyse Math. 17 (1966), 209–217.

KOBORI, A.; SAINOUCHI, Y.
[1] *On the Riemann's relation on open Riemann surfaces.* J. Math. Kyoto Univ. 2 (1962), 11–23.

KODAIRA, K.

[1] *Über die Harmonischen Tensorfelder in Riemannschen Mannigfaltigkeiten I, II, III.* Proc. Imp. Acad. Tokyo 20 (1944), 186–198, 257–261, 353–358.

[2] *Harmonic fields in Riemannian manifolds (generalized potential theory).* Ann. of Math. 50 (1949), 587–665.

[3] Cf. DE RHAM, G.; KODAIRA, K. [1].

KORN, A.

[1] *Zwei Anwendungen der Methode der sukzessiven Anwendungen.* Schwarz Festschrift, 1914. 215–229.

KURAMOCHI, Z.

[1] *Mass distributions on the ideal boundaries of abstract Riemann surfaces I.* Osaka Math. J. 8 (1956), 119–137.

KUSUNOKI, Y.

[1] *On Riemann's period relations on open Riemann surfaces.* Mem. Coll. Sci. Univ. Kyoto Ser. A. Math. 30 (1956), 1–22.

[2] *Notes on meromorphic covariants.* Ibid. 30 (1957), 243–249.

[3] *Contributions to Riemann-Roch's theorem.* Ibid. 31 (1958), 161–180.

LARSEN, K.

[1] *Extremal harmonic functions in locally Euclidean n-spaces.* Doctoral dissertation, Univ. of Calif., Los Angeles, Calif., 1964. 49 pp.

LARSEN, K.; NAKAI, M., SARIO, L.

[1] *Principal harmonic forms on Riemannian spaces.* Bull. Sci. Math. (to appear).

LEHTO, O.

[1] *On the existence of analytic functions with a finite Dirichlet integral.* Ann. Acad. Sci. Fenn. Ser. AI No. 67 (1949), 7 pp.

LEVI, E.

[1] *I problemi dei valori al contorno per le equazioni totalmente ellitique all derivate parziali.* Memorie di Matematica e di Fisica della Societa italianà delle Scienze, 16 (1909). 112 pp.

LEVI-CIVITA, T.

[1] *The absolute differential calculus.* Blackie and Son, Ltd., London, 1961. 452 pp.

LICHNEROWICZ, A.

[1] *Eléments de calcul tensoriel.* Armand Colin, Paris, 1950. 216 pp.

[2] *Courbure, nombres de Betti, et espaces symétriques.* Internat. Congr. Math. Cambridge, 1950. 216–223.

LICHTENSTEIN, L.

[1] *Zur Theorie der konformen Abbildung nichtanalytischer, singularitätenfreier Flächenstücke auf ebene Gebiete.* Bull. Internat. Acad. Sci. Gracovie, Cl. Sci. Math. Nat. Ser. A. (1916), 192–217.

LOEB, P.

[1] *An axiomatic treatment of pairs of elliptic differential equations.* Ann. Inst. Fourier 16 (1966), 167–208.

LOEB, P.; WALSH, B.

[1] *The equivalence of Harnack's principle and Harnack's inequality in the axiomatic system of Brelot.* Ann. Inst. Fourier 15 (1965), 597–600.

Lokki, O.

[1] *Beiträge zur Theorie der analytischen und harmonischen Funktionen mit endlichem Dirichletintegral.* Ann. Acad. Sci. Fenn. Ser. AI No. 92 (1951), 11 pp.

Maeda, F.-Y.

[1] *On spectral representations of generalized spectral operators.* J. Sci. Hiroshima Univ. Ser. A-I Math. 27 (1963), 137–149.

[2] *Axiomatic treatment of full-superharmonic functions.* Ibid. 30 (1966), 197–215

Marden, A.

[1] *The bilinear relation on open Riemann surfaces.* Trans. Amer. Math. Soc. 111 (1964), 225–239.

[2] *The weakly reproducing differentials on open Riemann surfaces.* Ann. Acad. Sci. Fenn. Ser. AI No. 359 (1965), 32 pp.

Marden, A.; Richards, I.; Rodin, B.

[1] *Analytic self-mappings of Riemann surfaces.* J. Analyse Math. 18 (1967), 197–225.

Marden, A.; Rodin, B.

[1] *Periods of differentials on open Riemann surfaces.* Duke Math. J. 33 (1966), 103–108.

[2] *Extremal and conjugate extremal distance on open Riemann surfaces with applications to circular-radial split mappings.* Acta Math. 115 (1966), 237–269.

[3] *A complete extremal distance problem on open Riemann surfaces.* Bull. Amer. Math. Soc. 72 (1966), 326–328.

Meehan, H.

[1] *Capacity problems in locally Euclidean spaces.* Doctoral dissertation, Univ. of Calif., Los Angeles, Calif., 1964. 80 pp.

Milgram, A.; Rosenbloom, P.

[1] *Heat conduction on Riemannian manifolds I, II.* Proc. Nat. Acad. Sci. U.S.A. 37 (1951), 180–184, 435–438.

Miranda, C.

[1] *Equazioni alle derivate parziali di tipo ellittico.* Springer-Verlag, Berlin-Göttingen-Heidelberg, 1955. 222 pp.

Mizumoto, H.

[1] *On Riemann surfaces with finite spherical area.* Kōdai Math. Sem. Rep. 7 (1957), 87–96.

[2] *On conformal mapping of a Riemann surface onto a canonical covering surface.* Ibid. 12 (1960), 57–69.

Mori, A.

[1] *On the existence of harmonic functions on a Riemann surface.* J. Fac. Sci. Univ. Tokyo 6 (1951), 247–257.

[2] *On Riemann surfaces on which no bounded harmonic function exists.* J. Math. Soc. Japan 3 (1951), 285–289.

[3] *Conformal representation of a multiply connected domain on a many-sheeted disc.* Ibid. 2 (1951), 198–209.

Myers, S.

[1] *Algebras of differentiable functions.* Proc. Amer. Math. Soc. 5 (1954), 917–922.

Myrberg, P.

[1] *Über die analytische Fortsetzung von beschränkten Funktionen.* Ann. Acad. Sci. Fenn. Ser. AI No. 58 (1949), 7 pp.

NAKAI, M.
[1] *Algebras of some differentiable functions on Riemannian manifolds.* Japan J. Math. 29 (1959), 60–67.
[2] *On Evans potential.* Proc. Japan. Acad. 38 (1962), 624–629.
[3] *Evans' harmonic functions on Riemann surfaces.* Ibid. 39 (1963), 74–78.
[4] *On Φ-bounded harmonic functions.* Ann. Inst. Fourier 16 (1966), 145–157.
[5] *Finite interpolation for analytic functions with finite Dirichlet integral.* Proc. Amer. Math. Soc. 17 (1966), 362–364.
[6] *Harmonic differentials with prescribed singularities* (to appear).
[7] *Potentials of Sario's kernel.* J. Analyse Math. 17 (1966), 225–240.
[8] *Sario's potentials and analytic mappings.* Nagoya Math. J. 29 (1967), 93–101.
[9] *On Evans' kernel.* Pacific J. Math. 22 (1967), 125–137.
[10] *Remarks on Sario potentials* (to appear).
[11] *Principal functions in harmonic spaces* (to appear).
[12] Cf. LARSEN, K.; NAKAI, M.; SARIO, L. [1].

NAKAI, M.; SARIO, L.
[1] *Construction of principal functions by orthogonal projection.* Canad. J. Math. 18 (1966), 887–896.
[2] *Normal operators, linear liftings and the Wiener compactification.* Bull. Amer. Math. Soc. 72 (1966), 947–949.
[3] *Harmonic semifields.* Proc. Nat. Acad. Sci. U.S.A. 56 (1966), 1674–1675.
[4] *Border reduction in existence problems of harmonic forms.* Nagoya Math. J. 29 (1967), 137–143.
[5] *Harmonic fields with given boundary behavior in Riemannian spaces.* J. Analyse Math. 18 (1967), 245–257.
[6] *Completeness and function-theoretic degeneracy of Riemannian spaces.* Proc. Nat. Acad. Sci. U.S.A. 57 (1967), 29–31.
[7] *Classification and deformation of Riemannian spaces.* Math. Scand. 20 (1967), 193–208.
[8] *A parabolic Riemannian ball.* Proceedings of the 1966 Amer. Math. Soc. Summer Institute (to appear).
[9] *Principal fields, semifields and forms on Riemannian spaces.* Ibid. (to appear).
[10] *Point norms in the construction of harmonic forms.* Pacific J. Math. (to appear).
[11] *Classification theory.* D. Van Nostrand Co., Inc., Princeton, N. J. (to appear).
[12] *General principal forms imitating locally square integrable forms* (to appear).

NICKEL, P.
[1] *Canonical mappings of a bordered Riemann surface.* Doctoral dissertation, Univ. of Calif., Los Angeles, Calif., 1959. 81 pp.
[2] *On extremal properties for annular radial and circular slit mappings of bordered Riemann surfaces.* Pacific J. Math. 11 (1961), 1487–1503.

NINOMIYA, N.
[1] *Étude sur la théorie du potentiel pris par rapport au noyau symétrique.* J. Inst. Polytech. Osaka City Univ. 8 (1957), 147–179.

NOSHIRO, K.
[1] *Contributions to the theory of the singularities of analytic functions.* Jap J. Math. 19 (1948), 299–327.
[2] Cf. SARIO, L.; NOSHIRO, K. [1].

OHTSUKA, M.
[1] *Dirichlet problem, extremal length and prime ends.* Lecture notes, Wash. Univ., St. Louis, Mo., 1962. 350 pp.
[2] *Extremal length of families of parallel segments.* J. Sci. Hiroshima Univ. Ser. A-I 28 (1964), 39–51.

OHTSUKA, M.

[3] *On weak and unstable components.* Ibid. 53–58.

[4] *On limits of BLD functions along curves.* Ibid. 67–70.

OIKAWA, K.

[1] *On the stability of boundary components.* Doctoral dissertation, Univ. of Calif., Los Angeles, Calif., 1958. 128 pp.

[2] *A constant related to harmonic functions.* Japan. J. Math. 29 (1959), 111–113.

[3] *On a criterion for the weakness of an ideal boundary component.* Pacific J. Math. 9 (1959), 1233–1238.

[4] *Sario's lemma on harmonic functions.* Proc. Amer. Math. Soc. 11 (1960), 425–428.

[5] *On the stability of boundary components.* Pacific J. Math. 10 (1960), 263–294.

[6] *On the uniqueness and prolongation of an open Riemann surface of finite genus.* Proc. Amer. Math. Soc. 11 (1960), 785–787.

[7] *Remarks to conformal mappings onto radially slit disks.* Sci. Papers Coll. Gen. Ed. Univ. Tokyo 15 (1965), 99–109.

[8] Cf. AKAZA, T.; OIKAWA, K. [1].

OIKAWA, K.; SUITA, N.

[1] *On parallel slit mappings.* Kōdai Math. Sem. Rep. 16 (1964), 249–254.

[2] *Circular slit disk with infinite radius.* Nagoya Math. J. 30 (1967), 57–70.

OW, W.

[1] *Capacity functions in Riemannian spaces.* Doctoral dissertation, Univ. of Calif., Los Angeles, Calif., 1966. 63 pp.

PARREAU, M.

[1] *Sur les moyennes des fonctions harmoniques et analytiques et la classification des surfaces de Riemann.* Ann. Inst. Fourier 3 (1951), 103–197.

PFLUGER, A.

[1] *Theorie der Riemannschen Flächen.* Springer, Berlin-Göttingen-Heidelberg, 1957. 248 pp.

PHELPS, R.

[1] *Extreme positive operators and homomorphisms.* Trans. Amer. Math. Soc. 108 (1963), 265–274.

RADÓ, T.

[1] *Über den Begriff der Riemannschen Fläche.* Acta Szeged 2 (1925), 101–121.

RAO, K.

[1] *Lindelöfian maps and positive harmonic functions.* Doctoral dissertation, Univ. of Calif., Los Angeles, Calif., 1962. 48 pp.

REICH, E.

[1] *On radial slit mappings.* Ann. Acad. Sci. Fenn. Ser. AI No. 296 (1961), 12 pp.

REICH, E.; WARSCHAWSKI, S.

[1] *On canonical conformal maps of regions of arbitrary connectivity.* Pacific J. Math. 10 (1960), 965–985.

[2] *Canonical conformal maps onto a circular slit annulus.* Scripta Math. 25 (1960), 137–146.

DE RHAM, G.

[1] *Sur la théorie des formes différentielles harmoniques.* Ann. Univ. Grenoble 22 (1946), 135–152.

[2] *Variétés différentiables.* Hermann, Paris, 1960. 196 pp.

[3] Cf. BIDAL, P.; DE RHAM, G. [1].

DE RHAM, G.; KODAIRA, K.
[1] *Harmonic integrals*. Mimeographed notes. Institute for Advanced Study, Princeton, N.J., 1950. 114 pp.

RICHARDS, I.
[1] *On the classification of noncompact surfaces*. Trans. Amer. Math. Soc. 106 (1963), 259–269.
[2] Cf. MARDEN, A.; RICHARDS, I.; RODIN, B. [1].

RODIN, B.
[1] *Reproducing kernels and principal functions*. Proc. Amer. Math. Soc. 13 (1962), 982–992.
[2] *Extremal length of weak homology classes on Riemann surfaces*. Ibid. 15 (1964), 369–372.
[3] *The sharpness of Sario's generalized Picard theorem*. Ibid. 15 (1964), 373–374.
[4] *Extremal length and removable boundaries of Riemann surfaces*. Bull. Amer. Math. Soc. 72 (1966), 274–276.
[5] *Extremal length and geometric inequalities*. Proceedings of the 1966 Amer. Math. Soc. Summer Institute (to appear).
[6] *On a paper by M. Watanabe*. J. Math. Kyoto Univ. 6 (1967), 393–395.
[7–9] Cf. MARDEN, A.; RODIN, B. [1–3].
[10] Cf. MARDEN, A.; RICHARDS, I.; RODIN, B. [1].

RODIN, B.; SARIO, L.
[1] *Existence of mappings into noncompact Riemann surfaces*. J. Analyse Math. 17 (1966), 219–223.
[2] *Convergence of normal operators*. Kōdai Math. Sem. Rep. 19 (1967), 165–173.

ROSENBLOOM, P.
[1] Cf. MILGRAM, A.; ROSENBLOOM, P. [1].

ROYDEN, H.
[1] *Some remarks on open Riemann surfaces*. Ann. Acad. Sci. Fenn. Ser. AI No. 85 (1951), 8 pp.
[2] *Harmonic functions on open Riemann surfaces*. Trans. Amer. Math. Soc. 73 (1952), 40–94.
[3] *Some counterexamples in the classification of open Riemann surfaces*. Proc. Amer. Math. Soc. 4 (1953), 363–370.
[4] *A property of quasi-conformal mappings*. Ibid. 5 (1954), 266–269.
[5] *Open Riemann surfaces*. Ann. Acad. Sci. Fenn. Ser. AI No. 249/5 (1958), 13 pp.
[6] *A class of null-bounded Riemann surfaces*. Comment. Math. Helv. 34 (1960), 52–66.
[7] *The Riemann-Roch theorem*. Ibid. 34 (1960), 37–51.
[8] *The boundary values of analytic and harmonic functions*. Math. Z. 78 (1962), 1–24.
[9] *Real analysis*. The Macmillan Co., New York, 1963. 284 pp.
[10] *Function theory on compact Riemann surfaces*. J. Analyse Math. 18 (1967), 295–327.

RYFF, J.
[1] *On the representation of doubly stochastic operators*. Pacific J. Math. 13 (1963), 1379–1386.

SAINOUCHI, Y.
[1] *On the analytic semiexact differentials on an open Riemann surface*. J. Math. Kyoto Univ. 2 (1962/63), 277–293.
[2] *A remark on square integrable analytic semiexact differentials on an open Riemann surface*. Ibid. 4 (1964), 117–121.
[3] Cf. KOBORI, A.; SAINOUCHI, Y. [1]

SAKAI, A.
[1] *On minimal slit domains*. Proc. Japan. Acad. 35 (1959), 128–133.

SARIO, L.

[1] *Existence des fonctions d'allure donnée sur une surface de Riemann arbitraire.* C. R. Acad. Sci. Paris 229 (1949), 1293–1295.

[2] *Quelques propriétés à la frontière se rattachant à la classification des surfaces de Riemann.* Ibid. 230 (1950), 42–44.

[3] *Existence des intégrales abéliennes sur les surfaces de Riemann arbitraires.* Ibid. 230 (1950), 168–170.

[4] *Questions d'existence au voisinage de la frontière d'une surface de Riemann.* Ibid. 230 (1950), 269–271.

[5] *On open Riemann surfaces.* Internat. Congr. Math. Cambridge, 1950. 398–399.

[6] *Linear operators on Riemann surfaces.* Bull. Amer. Math. Soc. 57 (1951), 276.

[7] *Principal functions on Riemann surfaces.* Ibid. 58 (1952), 475–476.

[8] *A linear operator method on arbitrary Riemann surfaces.* Trans. Amer. Math. Soc. 72 (1952), 281–295.

[9] *An extremal method on arbitrary Riemann surfaces.* Ibid. 73 (1952), 459–470.

[10] *Construction of functions with prescribed properties on Riemann surfaces.* Contributions to the theory of Riemann surfaces, Princeton Univ. Press, Prince- N. J., 1953. 63–76.

[11] *Minimizing operators on subregions.* Proc. Amer. Math. Soc. 4 (1953), 350–355.

[12] *Capacity of the boundary and of a boundary component.* Ann. of Math. (2) 59 (1954), 135–144.

[13] *Functionals on Riemann surfaces.* Lectures on functions of a complex variable, Univ. of Mich. Press, Ann Arbor, Mich., 1955. 245–256.

[14] *Extremal problems and harmonic interpolation on open Riemann surfaces.* Trans. Amer. Math. Soc. 79 (1955), 362–377.

[15] *Strong and weak boundary components.* J. Analyse Math. 5 (1956/57), 389–398.

[16] *On univalent functions.* 13 Scand. Congr. Math. Helsinki, 1957. 202–208.

[17] *Stability problems on boundary components.* Seminars on analytic functions II, Institute for Advanced Study, Princeton, N.J., 1958. 55–72.

[18] *Analytic mappings between arbitrary Riemann surfaces.* Bull. Amer. Math. Soc. 68 (1962), 633–637.

[19] *Picard's great theorem on Riemann surfaces.* Amer. Math. Monthly 69 (1962), 598–608.

[20] *Meromorphic functions and conformal metrics on Riemann surfaces.* Pacific J. Math. 12 (1962), 1079–1097.

[21] *Value distribution under analytic mappings of arbitrary Riemann surfaces.* Acta Math. 109 (1963), 1–10.

[22] *Islands and peninsulas on arbitrary Riemann surfaces.* Trans. Amer. Math. Soc. 106 (1963), 521–533.

[23] *Second main theorem without exceptional intervals on arbitrary Riemann surfaces.* Mich. Math. J. 10 (1963), 207–219.

[24] *On locally meromorphic functions with single-valued moduli.* Pacific J. Math. 13 (1963), 709–724.

[25] *Complex analytic mappings.* Bull. Amer. Math. Soc. 69 (1963), 439–445.

[26] *General value distribution theory.* Nagoya Math. J. 23 (1963), 213–229.

[27] *An integral equation and a general existence theorem for harmonic functions.* Comment. Math. Helv. 38 (1964), 284–292.

[28] *Classification of locally Euclidean spaces.* Nagoya Math. J. 25 (1965), 87–111.

[29] *A theorem on mappings into Riemann surfaces of infinite genus.* Trans. Amer. Math. Soc. 117 (1965), 276–284.

[30] Cf. AHLFORS, L.; SARIO, L. [1].

[31] Cf. KLOTZ, T.; SARIO, L. [1].

[32] Cf. LARSEN, K.; NAKAI, M.; SARIO, L. [1].

[33–43] Cf. NAKAI, M.; SARIO, L. [1–11].

[44–45] Cf. RODIN, B.; SARIO, L. [1–2].

SARIO, L.; FUKUDA, N.
[1] *Harmonic functions with given values and minimum norms in Riemannian spaces.* Proc. Nat. Acad. Sci. U.S.A. 53 (1965), 270–273.

SARIO, L.; NOSHIRO, K.
[1] *Value distribution theory.* D. Van Nostrand Co., Inc., Princeton, N.J., 1966. 236 pp.

SARIO, L.; SCHIFFER, M.; GLASNER, M.
[1] *The span and principal functions in Riemannian spaces.* J. Analyse Math. 15 (1965), 115–134.

SARIO, L.; WEILL, G.
[1] *Normal operators and uniformly elliptic self-adjoint partial differential equations.* Trans. Amer. Math. Soc. 120 (1965), 225–235.

SAVAGE, N.
[1] *Weak boundary components of an open Riemann surface.* Doctoral dissertation, Univ. of Calif., Los Angeles, Calif., 1956. 44 pp.
[2] *Weak boundary components of an open Riemann surface.* Duke Math. J. 24 (1957), 79–96.
[3] *Ahlfors' conjecture concerning extreme Sario operators.* Bull. Amer. Math. Soc. 72 (1966), 720–724.

SCHIFFER, M.
[1] *The span of multiply connected domains.* Duke Math. J. 10 (1943), 209–216.
[2] Cf. SARIO, L.; SCHIFFER, M.; GLASNER, M. [1].

SEEWERKER, J.
[1] *The extendability of a Riemann surface.* Doctoral dissertation, Univ. of Calif., Los Angeles, Calif., 1957. 42 pp.

SEGRE, B.
[1] *Forme differenziali e loro integrali.* Vol. I. Docet, Edizioni Universitarie, Roma, 1951. 520 pp.

SELBERG, H.
[1] *Über die ebenen Punktmengen von der Kapizität Null.* Avh. Norske Vid.-Akad. Oslo, No. 10 (1937), 10 pp.

SILVERS, A.
[1] *The use of differential-geometric methods in the study of mappings into abstract Riemann surfaces.* Doctoral dissertation, Univ. of Calif., Los Angeles, Calif., 1963. 64 pp.

SMITH, S.
[1] *Classification of Riemannian spaces.* Doctoral dissertation, Univ. of Calif., Los Angeles, Calif., 1965. 57 pp.

SÖRENSEN, W.
[1] Cf. BADER, R.; SÖRENSEN, W. [1].

SPENCER, D.
[1] *A generalization of a theorem of Hodge.* Proc. Nat. Acad. Sci. U.S.A. 38 (1952), 533–534.
[2] *Heat conduction on arbitrary Riemannian manifolds.* Ibid. 39 (1953), 327–330.
[3] *Real and complex operators on manifolds.* Contributions to the theory of Riemann surfaces, Princeton Univ. Press, Princeton, N.J., 1953. 203–227.
[4] *Dirichlet's principle on manifolds.* Studies in Mathematics and Mechanics presented to Richard von Mises, Academic Press Inc., New York, 1954. 127–134.
[5]–[6] Cf. DUFF, G.; SPENCER, D. [1–2].

SPRINGER, G.

[1] *Introduction to Riemann surfaces.* Addison-Wesley Publishing Company, Inc., Reading, Mass., 1957. 307 pp.

STEIN, K.

[1] Cf. BEHNKE, H.; STEIN, K. [1].

STREBEL, K.

[1] *A remark on the extremal distance of two boundary components.* Proc. Nat. Acad. Sci. U.S.A. 40 (1954), 842–844.

[2] *Die extremale Distanz zweier Enden einer Riemannschen Fläche.* Ann. Acad. Sci. Fenn. Ser. AI No. 179 (1955), 21 pp.

SUITA, N.

[1, 2] Cf. OIKAWA, K.; SUITA, N. [1–2].

TÔKI, Y.

[1] *On the classification of open Riemann surfaces.* Osaka Math. J. 4 (1952), 191–201.

[2] *On the examples in the classification of open Riemann surfaces I.* Ibid. 5 (1953), 267–280.

TSUJI, M.

[1] *On covering surfaces of a closed Riemann surface of genus $p \geq 2$.* Tôhoku Math. J. (2) 5 (1953), 185–188.

[2] *On the capacity of general Cantor sets.* J. Math. Soc. Japan 5 (1953), 235–252.

[3] *On Royden's theorem on a covering surface of a closed Riemann surface.* Ibid. 6 (1954), 32–36.

[4] *A metrical theorem on the singular set of a linear group of the Schottky type.* Ibid. 6 (1954), 115–121.

[5] *Potential theory in modern function theory.* Maruzen Co., Tokyo, 1959. 590 pp.

VIRTANEN, K.

[1] *Über die Existenz von beschränkten harmonischen Funktionen auf offenen Riemannschen Flächen.* Ann. Acad. Sci. Fenn. Ser. AI No. 75 (1950), 8 pp.

WALKER, R.

[1] *Algebraic curves.* Princeton Univ. Press, Princeton, N.J., 1950. 201 pp.

WALSH, B.

[1] *Flux in axiomatic potential theory* (to appear).

[2] Cf. LOEB, P.; WALSH, B. [1].

WARSCHAWSKI, S.

[1–2] Cf. REICH, E.; WARSCHAWSKI, S. [1–2].

WATANABE, M.

[1] *A remark on the Weierstrass points on open Riemann surfaces.* J. Math. Kyoto Univ. 5 (1966), 185–192.

WEILL, G.

[1] *Reproducing kernels and orthogonal kernels for analytic differentials on Riemann surfaces.* Doctoral dissertation, Univ. of Calif., Los Angeles, Calif. 1960. 80 pp.

[2] *Reproducing kernels and orthogonal kernels for analytic differentials on Riemann surfaces.* Pacific J. Math. 12 (1962), 729–767.

[3] *Capacity differentials on open Riemann surfaces.* Ibid. 12 (1962), 769–776.

[4] Cf. SARIO, L.; WEILL, G. [1].

WEYL, H.

[1] *On Hodge's theory of harmonic integrals.* Ann. of Math. (2) 44 (1943), 1–6.

[2] *Die Idee der Riemannschen Fläche.* 3te Aufl., Teubner, Stuttgart, 1958. 162 pp.

WHITNEY, H.

[1] *Geometric integration theory.* Princeton Univ. Press, Princeton, N.J., 1957. 387 pp.

WOLONTIS, V.

[1] *Properties of conformal invariants.* Amer. J. Math. 74 (1952), 587–606.

YOSHIDA, K.

[1] *Functional analysis.* Springer-Verlag, Berlin-Göttingen-Heidelberg, 1965. 458 pp.

AUTHOR INDEX

J stands for the Introduction to a chapter or section

SUBJECT INDEX

J stands for Introduction
Italicized section numbers refer to definitions.